NEURAL NETWORKS FOR VISION

NEURAL NETWORKS FOR VISION AND IMAGE PROCESSING

edited by
Gail A. Carpenter and Stephen Grossberg

A Bradford Book
The MIT Press
Cambridge, Massachusetts
London, England

©1992 Massachusetts Institute of Technology

All rights reserved. No part of this book may be reproduced in any form by any electronic or mechanical means (including photocopying, recording, or information storage and retrieval) without permission in writing from the publisher.

This book was printed and bound in the United States of America.

Library of Congress Cataloging-in-Publication Data

Neural networks for vision and image processing / edited by Gail A. Carpenter and Stephen Grossberg.
     p.   cm.
"A Bradford book."
Includes bibliographical references and index.
ISBN 0-262-53108-9
  1. Neural networks (Computer science) 2. Computer vision. 3. Image processing. I. Carpenter, Gail A. II. Grossberg, Stephen, 1939–   .
QA76.87.N4857 1992
006.3—dc20
                                                                                      92-7243
                                                                                        CIP

DEDICATION

We dedicate this book to our colleagues and students at the Department of Cognitive and Neural Systems

TABLE OF CONTENTS

List of Contributors   ix

Editorial Preface   xiii

Chapter 1   1
Visual Adaptation to a Negative, Brightness-Reversed World:
Some Preliminary Observations
Stuart Anstis

Chapter 2   15
Prevailing Lightness and Hue and Perceived Texture Segregation
Jacob Beck and William Goodwin

Chapter 3   45
Perception: A Biological Perspective
V.S. Ramachandran

Chapter 4   93
The Visual Perception of 3-Dimensional Form
J. Farley Norman and James T. Todd

Chapter 5   111
A New Approach to Shape from Shading
Pierre Breton, Lee A. Iverson, Michael S. Langer and Steven W. Zucker

Chapter 6   133
Dynamic Vision
Alex P. Pentland

Chapter 7   161
Figure-Ground Separation of Connected Scenic Figures:
Boundaries, Filling-In, and Opponent Processing
Stephen Grossberg and Lonce Wyse

Chapter 8   195
Toward a Unified Theory of Spatiotemporal Processing in the Retina
Paolo Gaudiano

Chapter 9   221
Neurodynamics of Real-Time Image Velocity Extraction
David A. Fay and Allen M. Waxman

Chapter 10 247
Why Do Parallel Cortical Systems Exist for the
Perception of Static Form and Moving Form?
Stephen Grossberg

Chapter 11 293
Neural Dynamics of Visual Motion Perception:
Local Detection and Global Grouping
Stephen Grossberg and Ennio Mingolla

Chapter 12 343
Neural Circuits for Visual Attention in the Primate Brain
Robert Desimone

Chapter 13 365
Attentive Supervised Learning and Recognition
by an Adaptive Resonance System
*Gail A. Carpenter, Stephen Grossberg, Natalya Markuzon,
John H. Reynolds and David B. Rosen*

Chapter 14 385
Synchronized Oscillations for Binding Spatially Distributed
Feature Codes into Coherent Spatial Patterns
Stephen Grossberg and David Somers

Chapter 15 407
A Quotient Space Hough Transform for Space-Variant Visual Attention
Alan S. Rojer and Eric L. Schwartz

Chapter 16 437
Optics and Neural Nets
David Casasent

Chapter 17 449
Neural Networks for Image Analysis
Robert Hecht-Nielsen

Subject Index 461

LIST OF CONTRIBUTORS

Stuart Anstis
Department of Psychology
University of California
La Jolla, California 92093

Jacob Beck
Department of Psychology
University of Oregon
Eugene, Oregon 97403

Pierre Breton
Research Center for
 Intelligent Machines
McGill University
3480 University Street
Montreal, Quebec
Canada H3A 2A7

Gail Carpenter
Department of Cognitive and
 Neural Systems
Boston University
111 Cummington Street
Boston, Massachusetts 02215

David Casasent
Center for Excellence in Optical
 Data Processing
Department of Electrical and
 Computer Engineering
Carnegie Mellon University
Pittsburgh, Pennsylvania 15213

Robert Desimone
Laboratory of Neuropsychology
National Institute of Mental Health
Building 9, Room 1E104
Bethesda, Maryland 20892

David A. Fay
Machine Intelligence Group
MIT Lincoln Laboratory
Lexington, Massachusetts 02173

Paolo Gaudiano
Department of Cognitive and
 Neural Systems
Boston University
111 Cummington Street
Boston, Massachusetts 02215

William Goodwin
Department of Psychology
University of Oregon
Eugene, Oregon 97403

Stephen Grossberg
Department of Cognitive and
 Neural Systems
Boston University
111 Cummington Street
Boston, Massachusetts 02215

Robert Hecht-Nielsen
HNC, Inc.
5501 Oberlin Drive
San Diego, California 92121

Lee Iverson
Research Center for
 Intelligent Machines
McGill University
3480 University Street
Montreal, Quebec
Canada H3A 2A7

Michael Langer
Research Center for
  Intelligent Machines
McGill University
3480 University Street
Montreal, Quebec
Canada H3A 2A7

Natalya Markuzon
Department of Cognitive and
  Neural Systems
Boston University
111 Cummington Street
Boston, Massachusetts 02215

Ennio Mingolla
Department of Cognitive and
  Neural Systems
Boston University
111 Cummington Street
Boston, Massachusetts 02215

J. Farley Norman
Department of Psychology
Brandeis University
Waltham, Massachusetts 02154

Alex Pentland
The Media Laboratory
Massachusetts Institute of Technology
20 Ames Street
Room E15-387
Cambridge, Massachusetts 02139

V.S. Ramachandran
Department of Psychology
University of California
9500 Gilman Drive
La Jolla, California 92093

John H. Reynolds
Department of Cognitive and
  Neural Systems
Boston University
111 Cummington Street
Boston, Massachusetts 02215

Alan S. Rojer
Computational Neuroscience Laboratory
Department of Psychiatry
New York University Medical Center
550 First Avenue
New York, New York 10016

David B. Rosen
Department of Cognitive and
  Neural Systems
Boston University
111 Cummington Street
Boston, Massachusetts 02215

Eric L. Schwartz
Computational Neuroscience Laboratory
Department of Psychiatry
New York University Medical Center
550 First Avenue
New York, New York 10016

David Somers
Department of Cognitive and
  Neural Systems
Boston University
111 Cummington Street
Boston, Massachusetts 02215

George Sperling
Department of Psychology
New York University
6 Washington Place, Room 980
New York, New York 10016

James T. Todd
Department of Psychology
Brandeis University
Waltham, Massachusetts 02154

Allen M. Waxman
Machine Intelligence Group
MIT Lincoln Laboratory
Lexington, Massachusetts 02173

Lonce Wyse
Department of Cognitive and
 Neural Systems
Boston University
111 Cummington Street
Boston, Massachusetts 02215

Steven Zucker
Research Center for
 Intelligent Machines
McGill University
3480 University Street
Montreal, Quebec
Canada H3A 2A7

EDITORIAL PREFACE

This book provides a resource for teaching and research about the vitally important areas of biological vision and image processing technology. Vision is one of the most important sources of information for supporting intelligent human behavior, as well as a key competence for designing new types of intelligent computers and machines. The book brings together recent research contributions from leading experimentalists and modelers, who presented their results at a conference in May, 1991 at the Wang Institute of Boston University. The interdisciplinary nature of the conference is reflected in its range of topics, from visual neurobiology and psychophysics through neural and computational modelling to technological applications. Such a program format acknowledges the important role that biological data and models have had on the development of technological applications, and the role that computational models have had on guiding the progress of experimental vision research.

The book's chapters mirror this interdisciplinary perspective. They are grouped according to the phenomena and problems that they address, rather than the methods that are used to analyse them. The first eleven chapters concern visual processes that are often described as "preattentive". Such processes tend to occur at earlier stages of brain processing, albeit stages that can include visual cortex, and can proceed automatically without the intervention of attention, learning, and object recognition. The remaining six chapters consider "attentive" processes, that tend to occur at later stages of brain processing, and that critically involve mechanisms of attention, learning, and object recognition. The distinction between preattentive and attentive processes is one of emphasis, at least in psychophysical studies of human vision, because every human response to a sensory input engages all the neural stages that occur between input and output. The distinction is nonetheless a useful one, and it has become increasingly well articulated with every advance in correlating visual percepts with the neural processing stages that generate them.

The "preattentive" chapters are themselves broken into two groupings. The first six chapters primarily consider processes underlying the perception of static images. Brightness, color, texture, shading, stereo, and form are subjects of inquiry here. The next five chapters consider processes underlying the perception of moving images. Some of these chapters also analyse why static and moving images need to be processed by different, but parallel, mechanisms.

All of the chapters address the fact that visual properties are not processed independently. For example, Stuart Anstis describes some of the remarkable effects on perceptual recognition of reversing image brightness and color. Jacob Beck and William Goodwin document how texture segregation depends on the interchange of light and dark values, or red and green hues in an image. Farley Norman and James Todd discuss psychophysical experiments that shed light on a number of the perceptual constraints and processes that are relevant in human perception of three-dimensional form, both of static and moving images. V.S. Ramachandran explores the perception of form through experiments that include contrast reversals, illusory contours, filling-in, and apparent motion.

Alex Pentland provides an analysis of models for analysing vision as a dynamic system, rather than a static one, and discusses a model for surface representation that includes concepts from Kalman filtering and orthogonal wavelets, as well as their possible neural interpretation in terms of receptive fields and cortical networks. Another new approach to understanding surface representation, especially in the context of the shape-from-shading problem, is presented by Pierre Breton, Lee Iverson, Michael Langer, and Steven Zucker, who combine properties of shading flow fields, a concept related to networks of oriented receptive fields, with relaxation labelling, a form of cooperative computation, to cope concurrently with surface, material, and light source compatibility constraints. The final chapter in the static vision cluster is by Stephen Grossberg and Lonce Wyse, who describe a biologically motivated model for parallel separation of multiple scenic figures from one another and from their background onto different surface representations, while suppressing scenic noise, by using a combination of opponent processing, boundary segmentation, and filling-in mechanisms.

The first chapter concerning the perception of moving objects is by Paolo Gaudiano, who also uses mechanisms of opponent processing to provide a unified explanation of retinal data about X and Y cells, particularly differences in their sustained and transient responses. The fact that both static figure-ground separation and the retinal processing of transient information both use similar opponent processing mechanisms illustrates the need for interdisciplinary vision research that clarifies how similar neural mechanisms can be used for different perceptual purposes. David Fay and Allen Waxman consider the neurodynamics of image velocity extraction using a multi-level model that also begins with retina-like processing, and implement a model for video rate image velocity extraction on a PIPE computer.

The next chapter, by Stephen Grossberg, analyses why parallel cortical systems are needed to process information about static and moving objects, and outlines a unified model for these parallel streams in terms of a global symmetry principle that has enabled explanations to be offered of many previously intractable perceptual and neural data, including data about illusory contours, filling-in, apparent motion, and perceptual aftereffects. The chapter by Stephen Grossberg and Ennio Mingolla further develops a model of the cortical motion processing stream to suggest a solution of the global aperture problem, which clarifies how cooperative processes can contextually eliminate local ambiguities in the computation of motion direction.

The chapter by Robert Desimone begins the cluster concerned with attentive vision and pattern recognition by summarizing recent neurobiological experiments and concepts concerning attentive processing by the inferior temporal (IT) cortex, area V4 of the prestriate visual cortex, and the pulvinar. Gail Carpenter, Stephen Grossberg, Natalya Markuzon, John Reynolds, and David Rosen then describe how adaptive resonance theory (ART) networks can learn to categorize and focus attention upon predictive combinations of visual features, and can use predictive feedback to drive a memory search leading to new attentional foci that conjointly minimize predictive error and maximize predictive generalization. The chapter by Stephen Grossberg and David Somers describes how cortical models of preattentive boundary segmentation and attentive object recognition can rapidly bind spatially distributed feature detectors into synchronized oscillations within a single processing cycle, and thereby prevent these features from becoming attached to representations of the wrong visual objects. Alan Rojer and Eric Schwartz describe a computational model of visual attention capable of rapidly selecting motor fixation points for a space-variant sensor which uses Bayesian and Hough transform concepts in a hardware system for reading license plates of moving vehicles in real time. David Casasent describes neural networks for learning and recognition that are amenable to implementation in optical hardware. Robert Hecht-Nielsen describes a neural network for image analysis that has been applied to problems of sorting, target recognition, and medical image analysis.

We wish to thank the authors for their careful selection and preparation of the material in their chapters, Cynthia Bradford and Diana Meyers at the Center for Adaptive Systems for their help in preparing the text and index, and the Life Sciences Directorate of AFOSR for its financial support of the conference.

Gail A. Carpenter
Stephen Grossberg
Boston, Massachusetts
January, 1992

NEURAL NETWORKS FOR VISION AND IMAGE PROCESSING

VISUAL ADAPTATION TO A NEGATIVE, BRIGHTNESS-REVERSED WORLD: SOME PRELIMINARY OBSERVATIONS

Stuart Anstis, Dept of Psychology, UC San Diego, La Jolla CA 92093+[+]

Abstract

The author adapted for three days and nights to a visual world in which brightness and colour were reversed by a video technique which turned black into white and vice versa. Shape from shading was disrupted, and so was the ability to recognise faces and to read their expressions. Reflectance edges caused by highlights and shadows were misperceived. In the negative picture, highlights were black and shadows were white, so highlights looked like black bugs, fuzzy shadows looked like reflections, and sharp-edged white shadows looked like real objects attached to the objects that cast them. The observer gradually learned to read facial expressions, extract shape from shading, and perceive sharp-edged shadows, but the ability to identify faces and perceive fuzzy shadows and highlights showed little improvement. The improvements felt subjectively like inferences rather than direct perceptions.

Introduction

We use luminance, more than anything else, to find our way around in the world. Other cues such as color, texture, or motion can specify objects but luminance is more important. We have no trouble identifying objects in a black and white picture, but perception is greatly impoverished in an equiluminous colored world and there is a considerable loss of visual depth, motion and form perception (Livingstone and Hubel 1987). In particular, luminance information is crucial to such visual tasks as extracting shape from shading (Pentland 1989) and recognising faces. If a picture of bumps and hollows that are defined by luminance shading is

---

[+] Supported by the Defence and Civil Institute of Environmental Medicine through Contract W7711-9-7080/01-XSE from Supply & Services Canada, and by Grant A0260 from NSERC. I particularly thank Patricia Hutahajan and Larry Franch for their unfailing assistance in running the experiments, and Patrick Cavanagh and Alan Ho for helpful comments and suggestions. Thanks also to Harvey Smallman, Doug Willen, Trang Thien Ngo and Agatha Sims. Some of this work was briefly presented at the ARVO conference, Sarasota, FL, May 1991.

turned upside down, or reversed in brightness, the perceived depth of the bumps and hollows reverses (Ramachandran 1988). Extraction of depth by shape from shading seems to be an early process which precedes perceptual grouping (Ramachandran 1988) and pop-out in visual search tasks (Enns 1990: Enns and Rensinck 1990).

Face recognition has been intensively studied in recent years. For reviews, see Penry 1971: Harmon 1973: Davies, Ellis & Shepherd 1981: Ekman 1982: Bruyer 1986: Bruce 1988. Face recognition becomes harder if the visual stimulus is upside down (Yin 1969: Valentine 1988: Kemp, McManus & Pigott 1990: Rock 1966) or spatially filtered (Harmon 1973: Hayes, Morrone & Burr 1986: Hayes 1988). The importance of shape from shading has been investigated, in particular, it has been reported that recognising photographic *negatives* of famous faces is very difficult (Galper 1970: Phillips 1972: Hayes, Morrone & Burr 1986: Hayes 1988: Kemp, McManus & Pigott 1990). The Fourier power spectra are identical to those of positives, and the phase spectra nearly so. The difficulty arises only with 3-D lith (black/white) or half tone (gray scale) portraits, not with outline drawings. Probably, luminance reversal disrupts shape from shading which is a crucial step in recognising the 3-D shape of faces. Cavanagh and Leclerc (1989) have shown that faces can be interpreted using shape from shading information, provided only that the shadowed regions are made darker than the non-shadowed regions. They presented a photograph of a face that had been thresholded so that it contained only two levels of lightness. If the uniform grey areas that corresponded to the lighter areas were replaced by random dots, or moving random dots, or even moving random dots in a separate depth plane, then the percept of a face was maintained, provided only that the lightness information was retained by controlling the intensity of the random dots. If the portions of the face that corresponded to the darker area were also replaced by dots, even dots moving in the opposite direction from those in the lighter area, the perception of a face continued. However, as soon as the lightness information was removed, the percept of a face vanishes. This was true despite the fact that any of the other cues such as motion and depth are perfectly capable of eliciting the outlines of the various facial features. They are just not automatically combined to form a face by our visual system. The implication is that lightness information is uniquely important in the recognition of real-world objects. (Quoted from Savoy 1987).

Is shape from shading affected by perceptual experience? Granrud, Yonas and Opland (1985) found that babies would reach for convex shapes defined by shading at the age of 7 months, but not at 5 months. This suggests that the ability to perceive depth from pictorial cues first develops between 5 and 7 months, but it does not tell us whether this is a consequence of maturation or of perceptual experience. Hess (1950, 1961) and Hershberger (1970) reared chicks in an environment illuminated from below. Hess then presented his chicks with photographs of grains, some with shadows extending upwards (as if lit from below), others with shadows extending downwards (as if lit from above). The chicks gradually came to prefer pecking at the photographs in which the grains extended upwards. However, Hershberger points out that the chicks may not have been using shadows as a depth cue, but simply acquired a preference for shadows extending upwards. Hershberger also reared chicks in cages lit from below, but then trained them to discriminate between physically convex and concave dents, pecking at one but not the other, in a training setup with diffuse shadowless lighting. The test condition was a pair of photographs of the same bump (or hollow), one erect and the other inverted, which to humans at least gave the characteristic illusion of depth reversal. Result: All birds saw photographs shaded on top as concave and ones shaded on the bottom as convex. So the experimental chicks responded to the shading just as normal chicks do, namely as if there were an overhead source of illumination. <u>Hershberger concluded that the chicks' extraction of depth from shading is innate.</u>

Adaptation to inversion

Real world

Stratton 1897: upside down inversion

Anstis 1991: brightness inversion

Figure 1

Even though chicks showed no perceptual learning, we wondered whether humans would. There have been many studies of visual adaptation to spatial rearrangements, starting with Stratton's (1896, 1897a,b) classic studies on adaptation to an upside-down world. This literature has been reviewed by Rock (1966) and Howard (1982). In order to learn whether humans could perform luminance-based visual tasks when normal brightness relations were disturbed, we have studied visual adaptation to a world that was reversed not in position but in brightness (Fig. 1), using video techniques. We wondered whether the observer would learn to reinterpret shading cues to depth as a result of his exposure to a negative world, in which shading with pale at the top corresponded to a concave, not a convex surface. Human performance in such situations is acquiring practical importance with the advent of computer graphic devices such as satellite photographs, night vision goggles, medical imaging and the like, especially when they arbitrarily assign "pseudo colours" and luminance levels to encode information.

Method

We examined the effects of a three-day exposure to a visual world in which brightnesses were *reversed* by means of a closed circuit video link. During passive adaptation the observer watched TV in negative. During active adaptation he walked about, or sat and viewed his own hands, conversed and interacted with others, while viewing a negative monitor fed from a TV camera. Perceptual phenomena studied before, during and after this adaptation included the perception of highlights and shadows and perception of 3-D shape from shading in convex and concave surfaces. The observer was also asked to recognise facial emotions such as anger, surprise or happiness, and to identify celebrities from their negative faces. The observations are given in the Diary section.

The main device used was a Hitachi VM-5350A VHS color video camera. This camera has a switch that converts the camera output, and also the view through the viewfinder, from positive to negative. To view the world in negative while walking around, the observer simply held the camera up to his eye and looked through the viewfinder all the time with the other eye patched. The weight of the camera (2.55 kg) was supported on a TriStar Camporter shoulder brace. This direct viewing through the viewfinder was versatile, since any desired scene could be inspected, and the observer's movements gave good motion parallax cues. A disadvantage was that the view through the finder subtended only 17° wide x 10° high, and like all TV cameras gave only a monochrome (black/ grey/ white) view. The camera zoom was pre-set to give a life sized image, and the camera's focus, iris and white-balance controls were set on automatic.

A more comfortable arrangement was used when the observer was seated. Images of the visual world, including the observer's own hands, were inverted in brightness and color by the camera and displayed on a TV screen. Thus a light pink face in positive was converted to a dark cyan face in negative. The negative visual image was optically superimposed on the actual position of the object by means of a double sided mirror (Fig. 2, after Howard 1982, Fig 12.13, page 519). The observer could reach under the mirror and watch his own hands as he manipulated various objects, and he practised visuomotor coordination by assembling jigsaw puzzles, sculpting modelling clay, playing solitaire with marbles and so on. He also conversed with friends, seeing their faces in negative, and played two-player hand games with them such as shaking hands and thumb wrestling. The TV screen was viewed binocularly. For non-interactive visual experience, the subject wore a view-restricting hood and watched TV in negative.

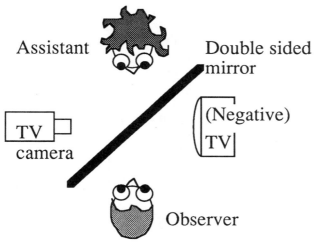

Figure 2

A diary of informal observations was kept by speaking into a portable tape recorder. The tapes were transcribed later, and provide the basis for what follows.

Diary

When I first sat in the apparatus I felt claustrophobic. The narrow space between the mirror and TV screen hem one in, reducing one's freedom and mobility. I am obliged to sit in a small space on a not too comfortable chair, seeing the world only on a TV screen a couple of feet from my nose. At first the negative picture is confusing and unpleasant. The pictures look like uninterpretable nonsense, and it feels like landing in a foreign country where one speaks not a word of the language. After a little, however, one begins to recognise that one is seeing familiar objects such as chairs, pairs of legs walking about and so on, albeit in disguise, and it feels as though it will become possible to make sense of this new world.

I hold my hands up in front of me and wiggle them. They look familiar, except that they are dark blue, and they definitely feel a part of me. The only odd feature is the black dots on my finger nails, which seem to float a bit loose as I wave my hands about, like flecks of soot which are not securely anchored in place. I suppose they must be highlights reflecting off my nails, although they certainly do not look as if they are. I hold my hands near my chest, and notice that the underside of my hands seem to give off a glow which illuminates my chest. Each hand is like a weak lantern that can illuminate nearby objects, as though my palms were painted with luminous paint. It takes a while to realise that this glow is really the shadow of my hand produced by the diffuse overhead lighting.

As Patricia walks across the room toward me, I notice that her feet are reflected in the polished marble floor. Then I remember that the floor of the laboratory is carpeted, and I realise that these bright "reflections" are really shadows of her feet, seen in negative.

Patricia comes and sits opposite me and I watch her face on the negative TV as we talk. She has dark blue skin and white hair. When she smiles she has "fine black teeth" like Queen Elizabeth I; this is disconcerting, and so is that fact that when she opens her mouth it looks white inside as though she has a lamp inside her mouth. She has white irises in her eyes as though she is wearing silvered contact lenses. When she smiles and frowns I can read her expression quite easily; yet it is hard to recognise that she is Patricia, not Donna or Louise. It seems that in this new negative world emotions are easier to recognise than personal identity.

Patricia goes off to lunch and I play solitaire. The marbles are placed in an array of saucer-shaped depressions on a board and I must jump one marble over another, removing from the board each jumped-over marble. My aim is to clear the board leaving only one solitary marble. The idea is that playing with spheres and hollows will provide good visuomotor training that will rapidly recalibrate the contribution of the luminance input into 3-D shape from shading. I am a little clumsy in manipulating the pieces; perhaps the TV image is inadvertently shifted slightly sideways from where it should be, and slightly larger than life size. I lose at solitaire, ending up with an irreducible five marbles on the board.

The marbles are made of shiny white plastic. On the TV screen each one seems to have a dark nipple painted on it, which remains in the same position even when I rotate the marble with a fingertip. Or perhaps it looks more like an eyeball on which the dark iris remains facing up even when the marble is rotated. The iris is actually a highlight from the overhead lights.

Patricia brings me a sandwich for lunch, but she will not tell me what it is. I am not allowed to look directly at the (positive) sandwich for fear of losing my adaptation, so I have to eat it with my eyes shut. It takes me some time to recognise the taste.

I'm out of doors taking a walk, and I can see a black smudge which I assume is a lamp, and rays of darkness are coming out of it as though it is spraying black paint everywhere. The whole scene looks like a snowstorm-- the dark patches on the pavement look like ice or melting snow, although I know that they are really patches of light cast on the ground by the street lamps. Now when I first come out I think it is still daylight, and I see intensely black spots which I think were reflections of the sun that might burn the camera tube. I do not realise it is dark outside (it is about 7 pm) and that the black spots are street lamps. So the scene looks like a daylit snowscape to me. As for this smudge -- as Frank leans over I see his hair, which is made dark by the smudge source (in positive: illuminated by the light source). I see bright shadows stream away from his feet, rather like rays of light emanating from him. I'm looking into the tree and walking back and forth; there seems to be white stuff among the branches of the tree, like clouds of interstellar gas, although I suppose they are really shadows. Motion parallax gives me very nice 3-D as I move back and forth. I am walking up to the staircase and then suddenly think: Who is that in front of me, getting in my way? I take another look and see that it was my own shadow falling on the balustrade. This is very sharp-edged, and colored white not black, and of course moves in a very lifelike way (it moves in just the same way as I do). It looks exactly like a person. When Frank walks from into the shadow of a building into the sunlight, he is moving in the negative picture from a light into a dark zone, so at first I wrongly perceive him as moving from light into shadow. But as he steps into the sunlight his shadow suddenly appears attached to his feet. It's not obviously a shadow because it is sharp and white, in fact it looks like a separate object. The fact that it moves exactly in step with him seems to make it easier for me to start perceiving it as a shadow rather than a separate object with a life of its own. As I achieve the perception of a shadow I also correctly perceive the sunlit area as really sunlit -- although it still looks weird! I wonder why shadows and highlights are so fascinating in negative. Instead of identifying them as illumination edges, I mistake them for reflectance edges.

While watching TV I am struck by how beautifully crisp all the moving shapes are, and somehow the movements seem to enhance that. I am watching a man dancing on the screen, and I can see every movement he makes-- his gait is very clear and all his clothes swing around him, and I can see that he is tap dancing to woo a woman. All this is very clear -- he is wearing a straw hat and striped blazer. What is not clear to me is the man's identity. Someone tells me that it is Bob Hope, whose face I know very well. I completely failed to recognise him in negative. So gait and bodily movements seem to read very well in negative, facial identity not well at all. As I watch an actress near the end of the experiment, I feel that I can read her expressions very well-- smiling, looking away, downcast eyes and so on -- but half an hour later, when I come out from the experimenal setup and see her in positive, I can see many more graduations of color in her face, which gave good 3-D moulding, than in negative. In positive I can see the shadows on her face, and the little red patches of rouge on her cheeks which I could not see in negative. I can read much more subtle changes in emotional expression than was possible in negative. I can recognise her identity in negative -- it is Kate Jackson in the program "Scarecrow and Mrs. King", and I recognise the face of the principal male actor in the same series, although I never knew his name. Again, I immediately recognise Cassius Clay in negative, admittedly in a commercial for Sports Illustrated which gives a strong context effect. So some identities do shine forth in negative. Colors seem much more washed out and desaturated in negative. I don't know if this is a true visual effect, perhaps based on the fact that we are very good at discriminating flesh tones but are perhaps not so good at the opposite regions of the color circle, or whether there is something uninteresting about the TV set so that it does not show up the colors too well. I think the latter, because I have some differently colored felt pens, which are brilliantly colored in real life, pretty highly colored in positive on TV, but washed out and desaturated in negative, even when we turn up the color controls.

I am wearing the blindfold at night. Being blind is not fun. From now on I am going to be very nice to blind people. Have you ever taken a shower in complete darkness? And walking into the bathroom and hanging the towels tidily on the rail, which is trivial for a sighted person, is quite an undertaking when you are blind. How does a blind person deal with a plate of meat and potatoes when he cannot see what he is cutting up? He does not know whether he has gristle or what, on the end of his fork. I soon found myself eating only foods that were easy to handle. Even cereal seems to fall off the spoon more easily when you cannot see it, so I soon found myself falling back on cookies, fish sticks and ice cream bars -- unhealthy junk food, but easy to eat because I can put it straight in my mouth without having to use a knife and fork.

When I watch a face in negative on TV I tend to take the visual message at its "face value" -- I tend to perceive the face as wearing dark blue make up, lit from below. So I am not really seeing it as a negative, but as an unfamiliar positive. As for eyes, I notice that the pupil is a white disk (dark in positive) and the bright highlight or corneal reflex is a tiny black dot in negative, which I often misperceive as if it were a small, beady, malevolent pupil. So I am misinterpreting the whole face, and I have to make a conscious effort to perceive this black spot as a highlight and to recognise the white disk as the pupil that it is. When I do try to see things in negative, I am translating deliberately from darks into lights, rather as when first learning French one mentally translates phrases from French into English. What I hope to achieve later is to start to think in the French language, so that I will perceive things correctly from the negative picture without having to translate. Around lunch time -- I am looking at a Mooney picture, which Harvey presented to me upside down. I immediately knew that it is upside down, but I am reading the white parts as though they really are white, -- I am dealing with the negative as though it is a positive.

Saturday afternoon -- Harvey takes me out for a walk to look at the traffic. The cars look like Dinky toys -- they lack reality. Each car travels along on a little white platform [which is really its shadow] but somehow being in negative they look like paper cutouts which are harmless. However, they are accompanied by a real man-sized roar of noise which seems much too real to go with the visuals. The sound track is much more real and compelling than the visuals. For example I would not be frightened to cross the road visually, but I would be frightened by the sound, which is just as well because it will keep me safely on the pavement. I then ask Harvey to talk to me, and I found something similar -- his voice is too real to go with his negative picture. This is odd because when I watch negative

TV all day, with voices attached to negative faces, I do not notice this phenomenon. On TV it seems OK because, I think, the faces are manipulated electronically but so are the voices, so they both suffer a concomitant loss in presence and reality. So the voices sound just as artificial as the pictures look, and everything matches.

I visit the laundry room and I am really struck by the smell and feel of the hot, scented blast of air that comes out as I open the door. Again, the realities do not seem to match. The negative TV pictures of the machines that I see through the viewfinder seem rather detached from reality and the smells seem too real to belong. So this reality mismatch applies between smell and vision, as well as between hearing and vision.

When I watch people swimming in the pool I notice that when swimmers kick up foam, the water appears to turn black as though ink is momentarily mixed into it, but disappear when the disturbance subsides and the water falls back into the pool. In real life, of course, foam looks white when bubbles of air mix into the water, and I suppose we must have learned that this change is a temporary physical change that can be ignored, not an addition of white pigment to the water. In negative I lost this perceptual constancy and did not make this perceptual connection. An analogy comes into my mind; we generally ignore echoes and reverberations, but when we hear a tape played backwards the echoes precede the sounds that cause them and are extremely intrusive.

When I walk along I find the smallest slope disturbing -- I like to know about them ahead of time and not just encounter them with my feet as I go along. When viewing in these restricted conditions I hate surprises. When a branch brushes across my face I freeze and find it rather upsetting. It is even worse when my hip bumps into a wall. So I do not like when somebody said go on, march forward, as though they were making fun of my cowardice, because only I know what the hazards are from my point of view. I find that the loss of peripheral vision is even more of a handicap than the video inversion of brightness. Seeing in negative is not a big problem if you are merely trying to avoid bumping into obstacles. You can still see the obstacles perfectly well in negative, although identifying the details within the obstacle is another story. The real problem lies with the tube -- maybe this is what life is like for patients with retinitis pigmentosa feels like [in which peripheral vision is lost but foveal vision is spared]. There is also some loss of perspective because of the restricted view, and also because of fish-eye distortion from the wide angle TV lens that I am using in an attempt to compensate for the restricted field of view.

Monday afternoon: Coming out from the adaptation: I am looking at a negative TV image of Patricia. Now I flip the picture to positive. Ah, there she is! Her eyes seem to flip across as her pupils change suddenly from light to dark. She looks much far brilliantly colored and saturated. Her hair looks normal [from white in the negative picture to its real-life black]. Everything looks more solid and real, more meaty, and there is no time lag in perceiving things normally again on leaving the negative setup. A toy plush toucan that was first presented to me during the experiment has changed into colors that I don't like so much -- its green body in the negative has changed to a dull pink. The toucan looks less real than in negative, but everything else looks more real. [I had hitherto seen the toucan only in negative, never before in positive]. Patricia's face looks much more real, more molded in 3-D and highly rounded, and I can read her facial expressions much more clearly now. My own hand looks much pinker. I can make sense without any trouble of reflections on the glass crystal of my watch as I tilt it back and forth, it has lost the sense of paradox that it had in negative when the reflections looked black. Shadows, both cast and attached, on the wooden building blocks give good relief and I see the blocks standing up off the table with a nice 3-D look. Objects look nicer, good enough to eat, candy colored and more plastic; it's a pleasure to see them back. Everything looks impossibly real, like seeing a technicolor film after watching black and white films for a long time, although that's not an exact description of what happened. Now I am looking at positive TV to record my impressions. Not a cartoon, I want to see real people please. It is often hard to tell whether a cartoon is in negative or not, whereas with real people you know immediately. Now I am looking at real faces and they look ordinary but more so. Facial expressions look much more vivid than before, I can recognise people easily. Candy bright, saturated colors, ordinary faces instead of puzzles. I don't have to reason any more, they are just people -- there is no need to study their faces and figure them out, I can identify people effortlessly in positive. Here is somebody who looks like Jimmy Carter but I know at once that it is not Carter himself; I immediately recognise that I've never seen that person before! This ability, which I completely lost in negative, must be analogous to our well-known ability to recognise that *carinel* is not a word but a nonsense string of letters that we have never seen before. This surely tells us something facinating about the way in which we store and retrieve both words and faces.

I have taken the view-restricting hood off. I feel a little dizzy and the world moves slightly when I turn my head, but I don't know whether that is a result of the experiment or just from lack of sleep. I can't tell you how solid and real the world looks. The room looks messy but it feels so good to be out of my self-imposed prison.

Conclusions

In summary, here are some of the perceptual effects that I observed:

At first the negative picture is confusing and unpleasant like a foreign land, but gradually one begins to recognise familiar objects and make sense of this new world.

Gait and bodily movements read very well in negative. Facial expressions of emotions become clear after some practice, but faces cannot be identified well at all. Performance in negative never comes close to that in positive. In positive, but not in negative, the moment I first see a face I can read its facial expressions and identify the face as familiar or as belonging to a stranger.

Shadows and highlights are strangely fascinating in negative. A diffuse shadow looks like a luminous glow, while a sharp-edged shadow looks like a second white object attached to the primary object. However, the way that the shadow moved in step with its object rapidly establishes it as a shadow.

The sensory "qualia" never change. Black never comes to look white.

I sense a "reality mismatch" between the diminished reality of negative visual images and the strong, clear sounds such as voices or motor sounds that they produce.

References

Bruce, V. (1988) *Recognising faces.* Hillsdale, N. J.: Lawrence Erlbaum Associates.

Bruyer, R. (Ed.) (1986) *The Neuropsychology of Face Perception and Facial Expression.* Hillsdale, N. J.: Lawrence Erlbaum Associates.

Cavanagh, P. and Leclerc, Y. (1989) Shape from shadows. *Journal of Experimental Psychology: Human Perception & Performance* 15 3-27

Davies, G. M., Ellis, H. D. and Shepherd, J. W. (Eds) (1981): *Perceiving and remembering faces.* London: Academic Press.

Ekman, P. (Ed) (1982): *Emotion in the human face* (Second Edition). Cambridge University Press.

Enns, J. T. (1990) Three-dimensional features that pop out in visual search. In D. Brogan (Ed.), *Visual search*, pp. 37-45. London: Taylor and Francis.

Enns, J. T. and Rensinck, R. (1990). Scene-based features influence visual search. *Science*, 247, 721-723.

Galper, R. E. (1970) Recognition of faces in photographic negative. *Psychonomic Science* 19 207-208

Granrud, C. E., Yonas, A. and Opland, E. C. (1985) Infants' sensitivity to the depth cue of shading. *Perception and Psychophysics* 37 415-419.

Hanley, J. R., Pearson, N. A. and Howard, L. A. (1990) The effects of different types of encoding task on memory for famous faces and names. *Quarterly Journal of Experimental Psychology,* 42A, 4, 741-762.

Harmon, L. D. (1973) The recognition of faces. *Scientific American* (November), 71-82. Reprinted in: *Image, Object and Illusion: Readings from the Scientific American,* pp. 101-112. San Francisco: W. H. Freeman.

Hayes, A. (1988) Identification of two-tone images; some implications for high- and low-spatial-frequency processes in human vision. *Perception* 17 429-436.

Hayes, T. Morrone M. C. & Burr, D. C. (1986) Recognition of positive and negative bandpass-filtered images. *Perception* 15, 595-602.

Herschberger, W. A. (1970) Attached shadow orientation perceived as depth by chickens reared in an environment illuminated from below. *Journal of Comparative & Physiological Psychology* 73, 407-411.

Hess, E. H. (1950) Development of the chick's responses to light and shade cues of depth. *Journal of Comparative & Physiological Psychology* 43, 112-122.

Hess, E. H. (1961) Shadows and depth perception. *Scientific American* 204, 138-148.

Howard, I. P. (1982) *Human visual orientation.* New York: John Wiley.
Kemp, R., McManus, C. & Pigott, T. (1990) Sensitivity to the displacement of facial features in negative and inverted images. *Perception* **19**, 531-543.
Livingstone, M. and Hubel, D. (1987) Psychophysical evidence for separate channels for color, motion and depth. *Journal of Neuroscience* **7**, 3416-3468
Penry, J. (1971) *Looking at faces and remembering them.* London: Elek Books.
Pentland, A. (1989) Shape information from shading: A theory about human perception. *Spatial Vision* **4** 166-182.
Phillips, R. J. (1972) Why are faces hard to recognise in photographic negative? *Perception and Psychophysics,* 12, 425-426.
Ramachandran, V. S. (1988) Perception of shape from shading. *Nature* **331** 163-166.
Rock, I. (1966) *The nature of perceptual adaptation.*
Savoy, R. (1987) Contingent aftereffects and isoluminance: Psychophysical evidence for separation of color, orientation, and motion. *Computer vision, graphics, and image processing* **37**, 3-19.
Stratton, G. (1896) Some preliminary experiments on vision. *Psychological Review* **3**, 611.
Stratton, G. (1897a) Vision without inversion of the retinal image. *Psychological Review* **4**, 341.
Stratton, G. (1897b) Vision without inversion of the retinal image. *Psychological Review* **4**, 463.
Valentine, T. (1988) Upside-down faces: a review of the effect of inversion upon face recognition. *British Journal of Psychology* **79**, 471-491.
Yin, R. K. (1969) Looking at upside-down faces. *Journal of Experimental Psychology* **81** 141-145.

PREVAILING LIGHTNESS AND HUE AND PERCEIVED TEXTURE SEGREGATION

Jacob Beck and William Goodwin
Department of Psychology
University of Oregon
Eugene, OR 97403

Abstract

We investigated the perceived texture segregation of patterns composed of elements of the same sizes and shapes. What distinguished the two types of elements was that the light and dark lightnesses (or the red and green hues) of the two areas composing the elements in a texture pattern were interchanged. Perceived segregation was in accordance with the salience of the lightness and hue differences of the elements. The outputs of DOG and Gabor filters failed to account for the perceived segregation.

Introduction

Texture segregation has been explained by spatial-frequency analyzers operating on image intensities and by higher-order preattentive grouping processes operating on elementary properties such as edges and lightness values. Numerous investigators have shown that differences in two-dimensional spatial-frequency content account for how well different texture regions perceptually segregate (Beck, Sutter, & Ivry, 1987; Bergen & Landy, 1990; Chubb & Sperling, 1988; Daugman, 1987, 1988; Fogel & Sagi, 1989; Graham, Beck, & Sutter, 1989, in press; Malik & Perona, 1989; Nothdurft, 1990, 1991; Sutter, Beck, & Graham, 1989; Turner, 1986). Perceived segregation also occurs as a result of preattentive grouping processes. Perceived segregation resulting from the alignment of element edges into a line-like pattern (Beck, Rosenfeld & Ivry, 1989) and the segregation of randomly interspersed populations of light and dark squares into subpopulations (Beck, Graham, & Sutter, 1991) is not explainable by the differential stimulation of quasi-linear spatial-frequency analyzers such as simple cells.

Spatial-frequency channels and preattentive grouping processes may be distinguished by how they are affected by stimulus variables. First, there is a tradeoff between area × contrast when texture segregation is a direct consequence of the way in which intensities in a texture pattern stimulate quasi-linear spatial-frequency analyzers. Figure 1 shows an example of the element-arrangement patterns studied originally by Beck, Prazdny, and Rosenfeld (1983) and investigated further by Beck, Sutter, & Ivry (1987) and by Sutter, Beck, & Graham (1989). (Figures showing texture patterns do not accurately reproduce the lightness values. They are shown to illustrate the arrangements of the elements.) The two types of elements, large and small squares, are arranged in stripes in the top and bottom regions and in a checkerboard in the center region. Sutter, Beck and Graham (1989) found that area and contrast are not independent but can cancel each other. As the contrast of the small square increased from

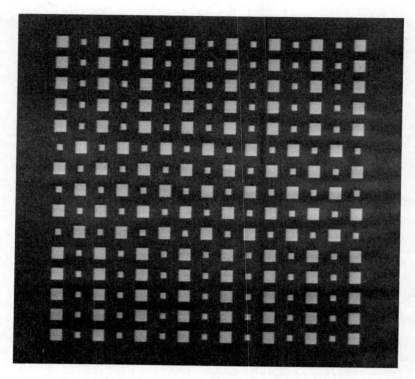

Figure 1. An example of an element-arrangement pattern. The squares are of equal contrast and their area ratio is 4:1. The squares are arranged in stripes in the top and bottom regions and in a checkerboard in the center region.

a value equal to the contrast of the large square, perceived segregation first decreased and then increased. The minimum perceived segregation occurred when the product of area × contrast was approximately equal. Beck, Rosenfeld, and Ivry (1989), in contrast, failed to find an area × contrast tradeoff when segregation was due to the preattentive grouping of aligned edges.

Second, the receptive fields showing strikingly different outputs to the different arrangements of the squares in the striped and checkerboard texture regions are the receptive fields that are sensitive to the fundamental frequency and orientation of the striped and checkerboard textures (Sutter, Beck & Graham, 1989). These are the large receptive fields which when one column of squares falls in the excitatory center, the neighboring column of squares falls in the inhibitory region (and vice versa). These receptive fields are not sensitive to edge alignment and we have found that the perceived segregation of a pattern is not diminished by the misalignment of edges (Poulson, 1988). Figure 2 illustrates the aligned (top) and misaligned (bottom) element-arrangement patterns. The abscissa plots the different kinds of patterns presented in the experiment[1]. Figure 3 shows that the mean ratings of perceived segregation were highly similar for the aligned and misaligned patterns. Beck, Rosenfeld, & Ivry (1989), in comparison, found that edge alignment, edge length, and principal axis orientation affected perceived line segregation.

Third, Beck, Graham, & Sutter (1991) compared the perceived segregation of striped and checkerboard arrangements of light and dark squares (Figure 4) into regions with the segregation of two randomly interspersed populations of light and dark squares into sub-

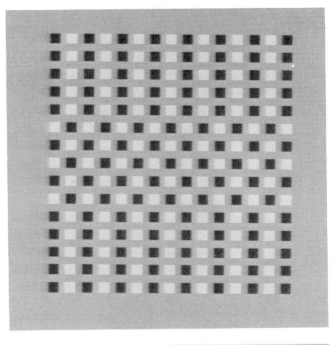

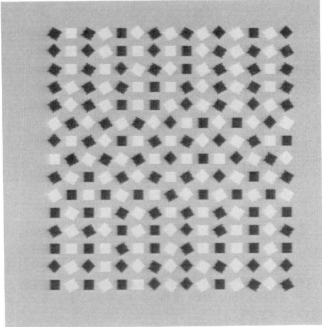

Figure 2. Aligned (top) and misaligned (bottom) displays of an element-arrangement pattern. The misalignment of the squares changed randomly on each trial (Poulson, 1988).

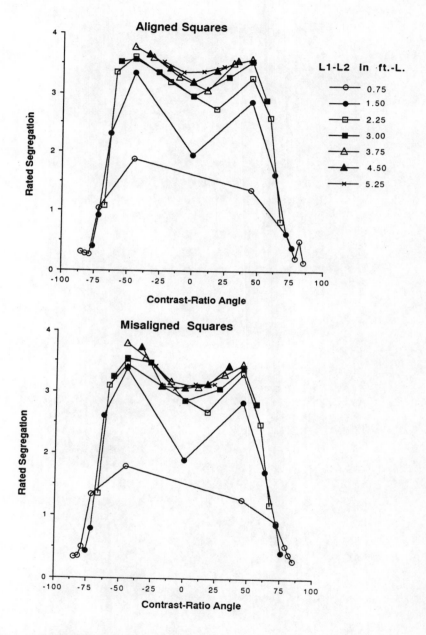

Figure 3. Mean segregation ratings of the aligned and misaligned patterns (Poulson, 1988). The legend shows the luminance difference of the two types of elements in a texture pattern.

populations (Figure 5). Subjects both rated the segregation of a pattern and matched the lightnesses of the two squares composing a pattern. Perceived lightnesses were the same for a given set of squares whether they were in texture regions or in intermixed populations. However, the perceived segregation of the population and region patterns differed. Perceived population segregation tended to be a simple monotonic function of the lightness differences of the squares but perceived region segregation was not[2]. Perceived region segregation was largely determined, but not completely, by the ratio of the contrasts of the squares, except when the background luminance was between the luminances of the squares[3]. The reason for the difference between population segregation and region segregation may again be that region segregation is mediated by the large oriented receptive fields that are sensitive to the fundamental frequency and orientation of the texture region (Beck, Sutter, & Graham, 1991). These receptive fields cannot be responsible for population segregation because the light and dark squares are distributed randomly throughout these patterns and therefore do not define a consistent arrangement of any particular spatial frequency and orientation. The light and dark squares in the population patterns fall equally on excitatory and inhibitory regions of these large receptive fields. Perceived segregation appears to be the result of a preattentive grouping of the light and dark squares based on their lightness difference.

Spatial-Frequency, Lightness and Hue

We report experiments further investigating the role of spatial-frequency content and of lightness and hue differences on perceived segregation.

I-Beam, Pedestal, Grating, Center-Surround Elements

The stimuli were computer generated patterns composed of four quadrants and were displayed for 1000 msec on a RGB monitor. The elements in one quadrant differed from those in the other quadrants. The discrepant quadrant appeared equally often in each of the four corners. The two types of elements composing a pattern were equal in size (24 pixels on a side—1 pixel=1.08 minutes) and were composed of the same two luminances. What distinguished the elements in a discrepant quadrant from those in the non-discrepant quadrants of a pattern was that the light and dark areas of the elements were interchanged. The two areas composing an element were equal. A subject rated the immediate perceived segregation of the discrepant quadrant on a scale from 0 to 4. A rating of 4 indicated that the discrepant quadrant segregated strongly and a rating of 0 indicated that the discrepant quadrant failed to segregate. Ten subjects served in each experiment and mean ratings were based on 8 judgments per subject. The procedure and instructions were similar to Sutter, Beck, & Graham (1989). The results reported differ at the .05 level of significance or greater.

The top row in Figure 8 illustrates the elements in the discrepant quadrant, and the bottom row the elements in the non-discrepant quadrants. There were 9 element pairs. Two elements were composed of 12 × 24 pixel bars. In the Bars-Same pattern, the bars were the same through out the display, the top bar was dark (d) and the bottom bar was light (l). In the Bars-Reversed pattern, the top bar was dark and the bottom bar light in the discrepant quadrant; in the non-discrepant quadrants the top bar was light and the bottom bar dark. Three elements were composed of pedestals on a 3 × 24 pixel base. In the Low-Pedestal pattern, the pedestal was 2 × 12 pixels; in the High-Pedestal pattern 8 × 12 pixels; and in

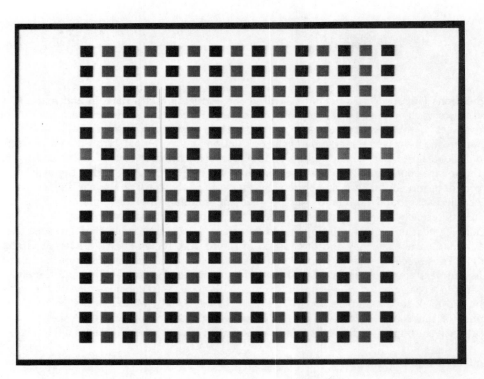

Figure 4. An example of the element-arrangement patterns investigated by Beck, Graham, and Sutter (1991).

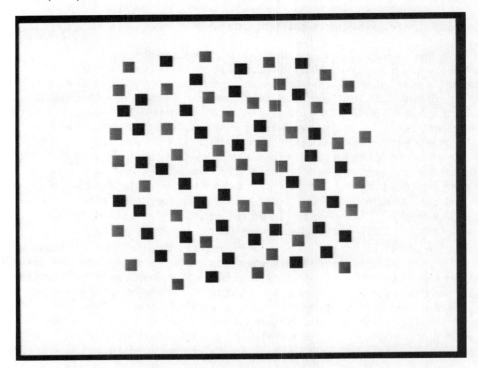

Figure 5. An example of the population patterns investigated by Beck, Graham, and Sutter (1991).

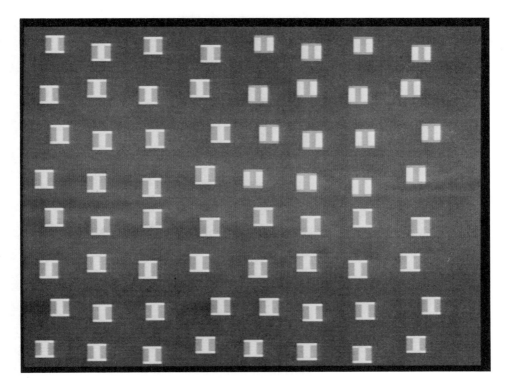

Figure 6. The I-Beam quadrant pattern in Experiment 1.

the Pedestal-to-Top pattern 21 × 10 pixels. The I-Beam was composed of a center beam 24 × 8 pixels and crossbeams 3 × 24 pixels. The Grating element was composed of a center area 24 × 12 pixels and flanking areas of 24 × 6 pixels. The Center-3 pattern was an 18 × 16 pixel rectangle surrounded on three sides. The Center-4 element was a 17 × 17 pixel square surrounded on four sides. The arrangement of the elements was the same in all patterns.

Sparse and Dense Patterns

In Experiment I, the patterns contained 16 elements in each quadrant (sparse pattern). Figure 6 illustrates the I-Beam and Figure 7 the Center-4 patterns with the discrepant quadrant in the upper right. The luminance of the background was 9.8 ft-L., the lower intensity area in an element 14.3 ft.-L. (a contrast of .35), and the higher intensity area in an element 16.3 ft.-L. (a contrast of .54). Figure 8 shows the mean of subjects' segregation ratings. The perceived segregation of the Center-4 and the I-Beam patterns are strikingly different. The Center-4 pattern segregated strongly with a mean rating of 3.3, while the I-Beam segregated weakly with a mean rating of 1.6.

We examined whether the outputs from DOG and Gabor filters predicted the segregation ratings. The excitatory sigma of the DOG filters increased in steps of powers of the square-root of 2 from 1 to 16 giving a total of 9 channels. The inhibitory sigma was 1.25 the excitatory sigma. The on-center even-symmetric oriented receptive fields were modelled by two-dimensional Gabor functions after Daugman (1985) and Watson (1983). The filters' center spatial-frequencies increased in steps of powers of the square-root of 2 from .25 to 16 cycles/degree for three different orientations—vertical, 45 degrees, and horizontal giving a

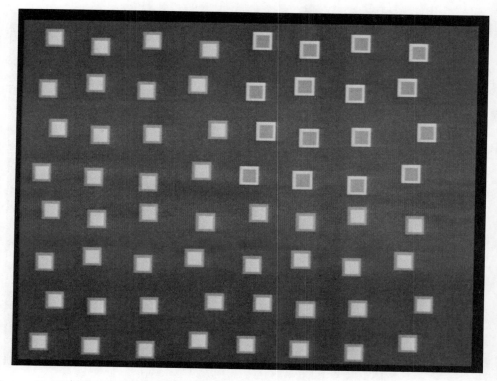

Figure 7. The Center-4 quadrant pattern in Experiment 1.

total of 39 channels. The spatial-frequency and orientation half-amplitude full-bandwidth of each filter was one octave and 38 degrees, respectively.

The modulation of outputs in each channel was assessed by computing the standard deviation of the outputs for the different spatial positions of the DOG and Gabor filters. The difference between each channels standard deviation to the discrepant and non-discrepant quadrants yielded a within channel difference for each stimulus. The within-channel differences were weighted by the contrast sensitivity function. The contrast-sensitivity function, and the Gabor weighting functions are described fully in Sutter, Beck and Graham (1989). The vertical axis of the DOG and Gabor graphs plot the square-root of the sum of the squares of the weighted within channel differences:

$$\sqrt{\sum_i^{N_c}[\text{Diff}_i \bullet S_{obs}(f_i)]^2}$$

where Diff_i is the within channel difference for a given frequency and orientation, $S_{obs}(f_i)$ is the sensitivity of the observer to a given frequency and orientation, and N_c is the number of channels (frequencies x orientations). Perceived segregation should be monotonically related to the filter outputs.

Figure 9 shows the outputs of the DOG filters to the discrepant and non-discrepant quadrants. Plotted are the statistics computed for various functions of the output. "Normal" refers to all the outputs. "Positive" to the outputs greater than zero. "Negative" to the outputs less than zero. "Half-Wave Rectification (on cell)" to setting the negative outputs to zero. "Half-Wave Rectification (off cell)" to taking the absolute value of the negative

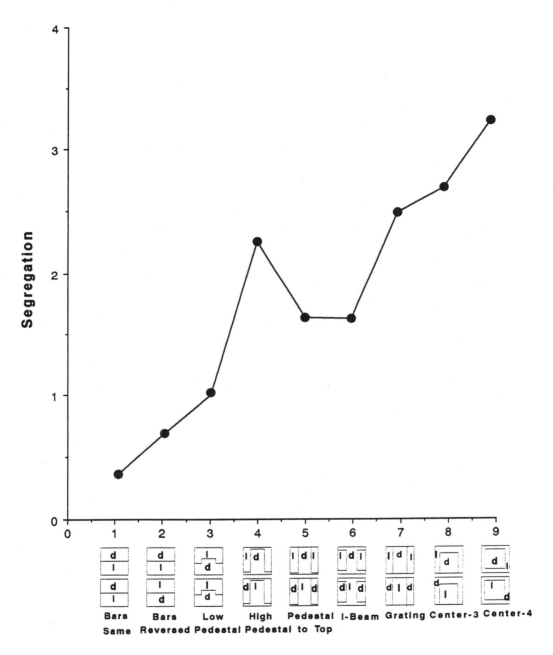

Figure 8. Mean segregation ratings in Experiment 1.

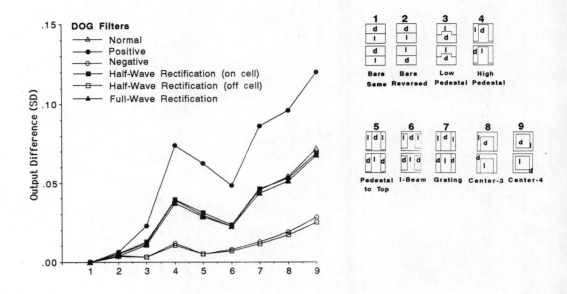

Figure 9. The responses of the DOG filters to the patterns in Experiment 1.

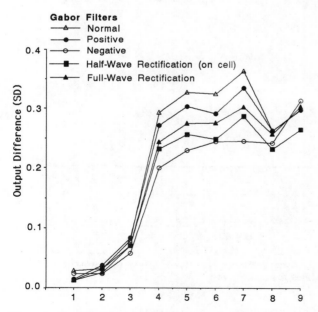

Figure 10. The responses of the Gabor filters to the patterns in Experiment 1. The off cell statistics were not computed.

outputs and setting the positive outputs to zero. "Full-Wave Rectification" to taking the absolute value of the outputs. The various measures of output modulation are in accord with the perceived segregation ratings. Figure 10 shows the outputs of the Gabor filters. As can be seen, they fail to predict the perceived segregation ratings.

Experiment 2 was conducted to determine the effects of pattern density on perceived segregation and the outputs of the DOG filters. Sparse and dense patterns were presented. The sparse patterns, as in Experiment 1, contained 16 elements in each quadrant. The dense patterns contained 32 elements in each quadrant. Figure 11 shows the perceived segregation ratings for the sparse and dense patterns. The shapes of the overall curves are very similar. For both the sparse and dense patterns, the Center-4 pattern (mean ratings 3.3 and 3.7, respectively) segregated more strongly than the I-Beam (mean ratings 1.6 and 1.9, respectively). The segregation ratings for the dense patterns tend to be greater than for the sparse patterns for the High-Pedestal through Center-4 elements. Figure 12 shows the predicted segregation when the dense patterns were convolved with the DOG filters. The overall pattern of the outputs are similar to those of the sparse pattern (Figure 9). Consistent with subjects judgments of perceived segregation, the outputs for the dense pattern tend to be greater than for the sparse pattern from the High-Pedestal through Center-4 elements.

Isoluminant Patterns

Experiment 3 investigated whether the outputs of concentric receptive fields predict perceived segregation when the elements were composed of red and green areas of equal luminance. The elements composing the patterns were the same as in Experiment 1 and are shown below the horizontal axis in Figure 13. The arrangement of the elements was as in Experiment 1. The red and green areas of the elements in the non-discrepant quadrants were the reverse of those in the discrepant quadrant. The patterns were presented on white and black backgrounds. The perceived segregation was similar to that with achromatic elements. The Center-4 pattern segregated strongly (mean ratings 3.8 and 3.7) while the I-beam segregated weakly (mean ratings 1.5 and 1.2) with the black and white backgrounds, respectively. The results do not depend on isoluminance. The overall shapes of the functions remained the same when small luminance differences were added to the hue difference.

We convolved the patterns with concentric filters modeled after the Type I center-surround cells described by Ts'o and Gilbert (1988). These cells have an opponent color center-surround organization—e.g., red is excitatory in the center and green is inhibitory in the surround. However, red in the surround and green in the center have no effect. The excitatory sigma of the Type I receptive fields increased in powers of 2 from 1 to 16 giving a total of 5 channels. The inhibitory sigma was 1.25 the excitatory sigma. The within channel differences were weighted by the contrast sensitivity function taken from Mullen (1985, Figure 6). Figure 14 shows that the outputs of the Type I filters are in accord with the mean perceived segregation ratings.

DOG and Gabor Filters

Though the DOG and Type I filters predict the perceived segregation ratings, the discrepancy between the DOG and Gabor filters suggests that the correspondence between perceived segregation and the outputs of the concentric filters is accidental. Several observations supported this suggestion.

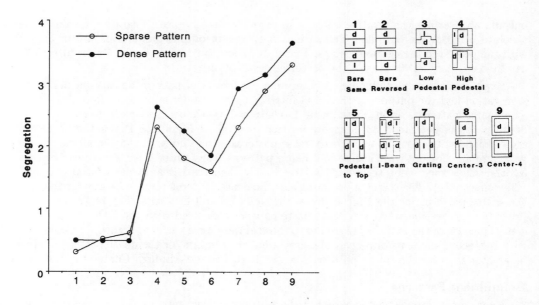

Figure 11. Mean segregation ratings of the sparse and dense patterns in Experiment 2.

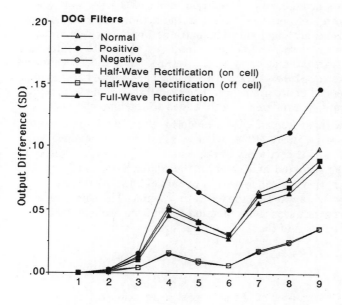

Figure 12. The responses of the DOG filters to the dense patterns in Experiment 2.

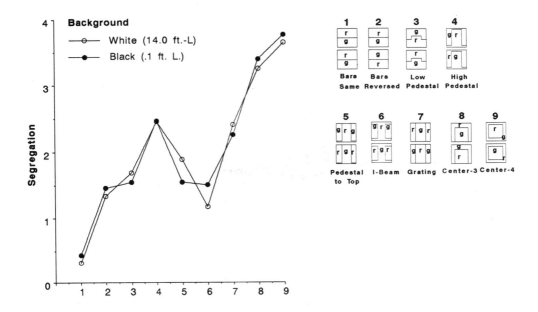

Figure 13. Mean segregation ratings of the isoluminant patterns in Experiment 3.

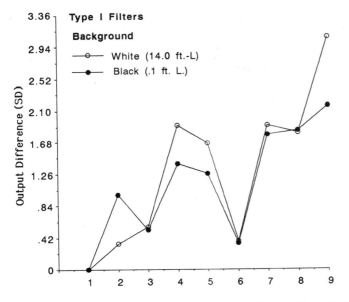

Figure 14. The responses of the Type I filters (Ts'o & Gilbert, 1988) to the isoluminant patterns.

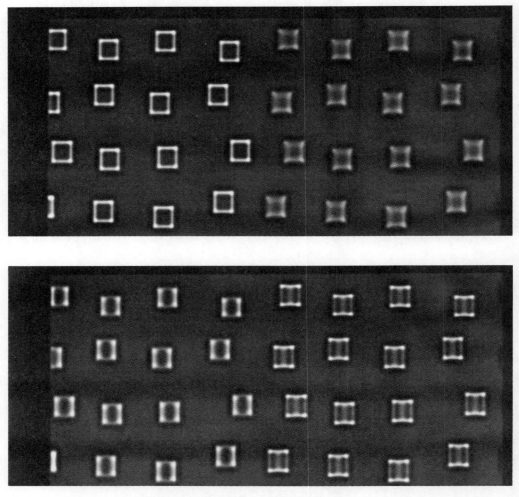

Figure 15. The outputs of the 23 pixel DOG filter to the discrepant (left) and non-discrepant (right) Center-4 (top) and I-Beam (bottom) elements in Experiment 1.

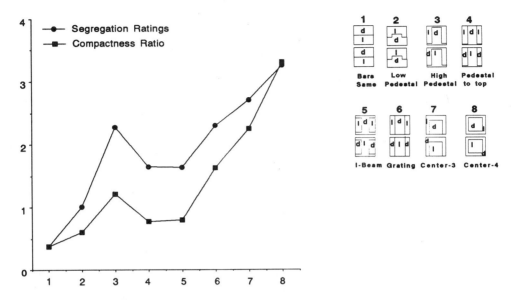

Figure 16. The mean segregation ratings in Experiment 1 and the ratio of the compactness values of the two areas composing an element.

First, the outputs of the DOG filters predicting the perceived segregation do not correspond with the perceptual experience. The strong segregation of the Center-4 patterns appears to be the result of the salience of the dark squares in the discrepant quadrant and of the light squares in the non-discrepant quadrants. The output differences predicting the greater perceived segregation of the Center-4 pattern are, however, not the result of the filters responding to the center square in the Center-4 element. They arise from the responses of the 11 and 23 pixel diameter DOG filters to the edges and corners of the Center-4 element. Figure 15 shows the outputs of the 23 pixel filter (Normal outputs) to the discrepant (left) and non-discrepant (right) Center-4 (top) and I-Beam (bottom) elements. As can be seen, the output is greater to the edges and corners of the Center-4 discrepant element than to the Center-4 non-discrepant element. The outputs of the 23 pixel filter to the discrepant and non-discrepant elements in the I-Beam pattern are more similar.

Second, there is a correspondence between perceived segregation and the compactness of the areas composing the texture elements. Compactness is given by the square of the perimeter divided by the area of a region. Perceived segregation occurs strongly when there is a large difference in the compactness of the two areas composing a texture element, and occurs weakly when the compactness of the two areas composing a texture element are approximately equal. Figure 16 plots the ratio of the compactness values of the two regions composing a texture element. (The smaller compactness value was placed in the denominator.) The compactness ratios were linearly transformed so that the lowest compactness ratio, 1.0, for the Bars-Same element was made equal to the segregation rating for Bars-Same pattern and the highest compactness ratio, 2.4, for the Center-4 element was made equal to the

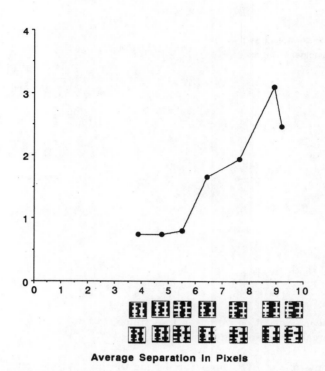

Figure 17. Mean segregation ratings of the irregular figures in Experiment 4.

segregation rating of the Center-4 pattern in Experiment 1. For comparison, the perceived segregation ratings in Experiment 1 are also plotted. The overall similarity to the perceived segregation ratings in Experiment 1 suggests that perceived segregation is affected by the difference in the compactness of the two regions making up a texture element.

Third, though each of the texture elements are composed of equal areas of the same two lightnesses or hues, the elements differ in their impression of prevailing color. For example, the Center-4 element in the discrepant quadrant looks prevailingly dark (or red in the isoluminant patterns), and in the non-discrepant quadrants prevailingly light (or green). In contrast, the I-beam elements look light and dark (or red and green) in both the discrepant and non-discrepant quadrants. The absence of a prevailing color in the I-Beam appears to be a consequence of the equal widths of the center beam and the flanking rectangles in the I-Beam element. Perceived segregation would appear to be a function of this difference in the prevailing lightnesses and hues of the discrepant and non-discrepant elements in a texture pattern. The conspicuousness of a texture depends on the degree to which it contrasts with adjacent textures. The difference in the compactness of the Center-4 square and annulus gives rise to a difference in prevailing color while the similarity in the compactness of the I-Beam and flanking rectangles does not.

Prevailing Lightness

We have investigated the hypothesis that it is the difference in lightness that yields strong perceived segregation.

Irregular Figures

The elements in Experiment 4 were irregular figures and are shown below the abscissa in Figure 17. The arrangement of the elements was as in Experiment 1. The top row shows the light and dark areas composing an element in the discrepant quadrant and the bottom row shows the light and dark areas composing an element in the non-discrepant quadrants. The areas of the light and dark areas of an element were equal. Figure 17 plots perceived segregation in an experiment in which the luminance of the background was midway between the luminances of the light and dark areas composing each element. The background luminance was 10 ft.-L.; the high and low luminances 20 and 0 ft.-L (contrasts of +1.0 and −1.0). The horizontal axis plots the average distance between the vertical bars in each element. In general, the greater the separation between the vertical bars the stronger is the perceived segregation. The exception is Irregular-Figure 6 in which the horizontal bars were inadvertently aligned yielding strong subjective contours. Figure 18 shows the predicted segregation ratings when the patterns are convolved with the DOG filters. Unlike the I-Beam set of patterns, none of the output measures correctly predict the order relations in the data. Figure 19 shows that the outputs of the Gabor filters also fail to predict perceived segregation. The results suggest that perceived segregation occurs when there is a distinctive wide area which is seen as light or dark. In contrast, segregation fails to occur when the vertical bars making up a pattern are approximately of equal width.

Element–Arrangement Patterns

Experiment 5 also supports the hypothesis that the correspondences between the DOG and Type I filters and perceived segregation are accidental. The two types of elements were arranged in stripes and checks. The Center-4 and I-Beam patterns are shown in Figures 20 and 21. Patterns composed of solid dark and light squares were also presented. The Center-4 (Figure 20) and I-Beam elements (Figure 21) were arranged in stripes and checks. The stripes and checkerboard textures appeared equally often in the top and bottom regions. The background luminance was set at 14 ft.-L. Four contrasts were used. The high and low luminances were 14.8 and 13.2 ft.-L (contrasts of +.06 and −.06), 15.8 and 12.2 ft.-L (contrasts of +.13 and −.13), and 17.5 and 10.5 ft.-L. (contrasts of +.25 and −.25), 21.3 and 6.7 ft.-L. (contrasts of +.52 and −.52). Figure 22 shows that the subjects' mean ratings of perceived segregation are similar to the quadrant patterns. As mentioned above, the element-arrangement patterns respond primarily to the fundamental frequency of the pattern. Since the luminances and areas of the two elements are the same and both elements are present in the striped and checkerboard textures, the DOG and Gabor filters should respond approximately the same to the two textures. Figures 23 and 24 show that the outputs of the DOG and Gabor filters fail to correspond to the perceived segregation.

Experiment 6 investigated the effects of setting the background luminance between the high and low luminance areas of the elements with a quadrant pattern. The .06, .13, and .25 contrasts in Experiment 5 were used. The arrangement of the elements was as in Experiment 1. Figure 25 shows the mean perceived segregation ratings. As in the previous experiments, the I-Beam patterns segregated less strongly than the Center-4 patterns. Figure 26 shows the outputs of the DOG filters. As would be expected, the Normal and Full-wave rectification outputs fail to predict perceived segregation when the background luminance was between the luminances composing the elements. However, unlike the element–arrangement patterns, the Positive, Negative, and Half-wave rectification statistics predicted the overall order of the perceived segregation ratings.

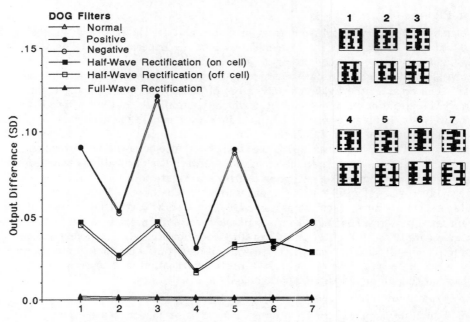

Figure 18. The responses of the DOG filters to the irregular figures in Experiment 4.

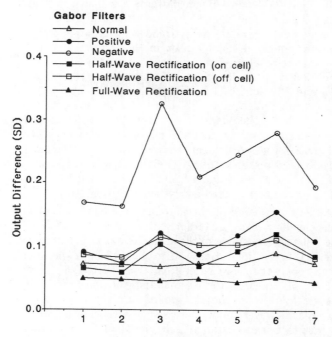

Figure 19. The responses of the Gabor filters to the irregular figures in Experiment 4.

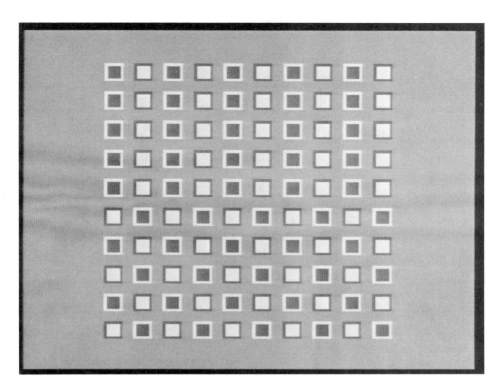

Figure 20. The element-arrangement pattern with the Center–4 element.

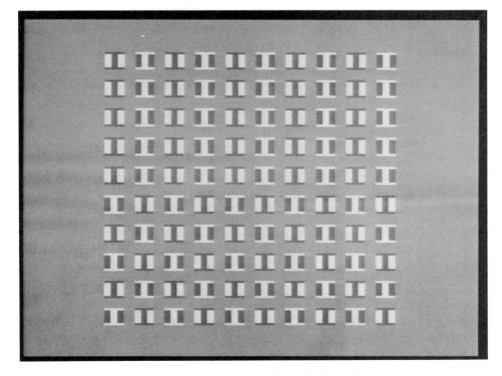

Figure 21. The element-arrangement pattern with the I-Beam element.

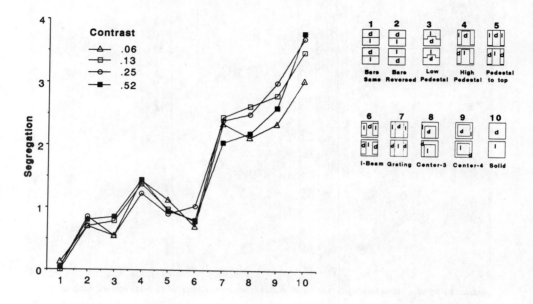

Figure 22. Mean segregation ratings of the element-arrangement patterns in Experiment 5.

I-Beam and Center-4 Elements

Experiment 7 varied the luminances of the rectangles in the I-Beam element and of the Center-4 annulus. The background luminance was 28 ft.-L., the I-Beam and Center-4 figures 2 ft.-L., and the luminances of the rectangles on the sides of the I-Beam and of the Center-4 annulus were varied from 6 to 15 ft.-L. The elements were presented in a quadrant pattern. In addition to the segregation ratings, subjects matched the lightnesses of the I-Beam and Center-4 figures in the discrepant and the non-discrepant quadrants to a gray scale on the RGB monitor. The gray scale and the texture pattern were viewed alternately by depressing a mouse button. The procedure was similar to that in Beck, Graham, & Sutter (1991).

Figure 27 shows the outputs of the Gabor Filters (Normal outputs) were similar for the I-Beam and Center-4 patterns. Figure 28 plots perceived segregation as a function of the difference between subjects' lightness matches of the I-Beam and Center-4 elements in the discrepant and non-discrepant quadrants. Subjects also matched the gray scale on the RGB monitor to a Munsell Value scale. The abscissa–values in Figure 28 are the Munsell Value differences computed from subjects' matches of the monitor gray scale to the Munsell scale. The perceived segregation of the I-Beam increased with the lightness difference. In contrast, the perceived segregation of the Center-4 element was strong except for the smallest lightness difference. What is suggested is that the segregation of the Center-4 pattern is due to the salience of the lightness difference and that a large uniform area (the center square) increases the salience of this lightness difference.

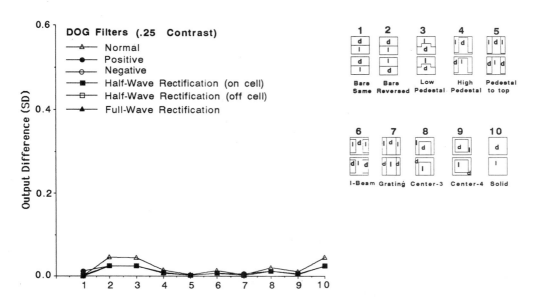

Figure 23. The responses of the DOG filters to the .25 contrast element-arrangement patterns in Experiment 5.

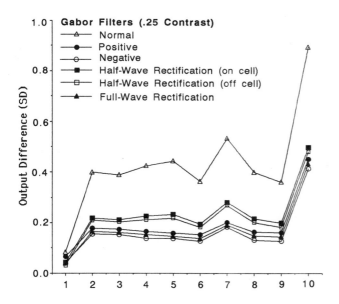

Figure 24. The responses of the Gabor filters to the .25 contrast element-arrangement patterns in Experiment 5.

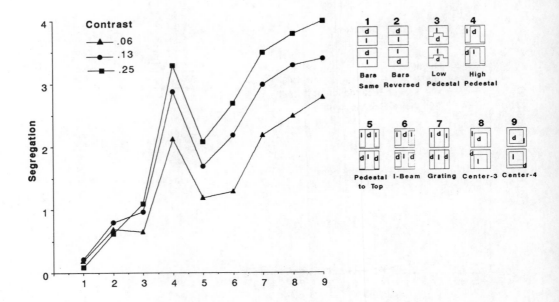

Figure 25. Mean segregation ratings of the quadrant patterns in Experiment 6.

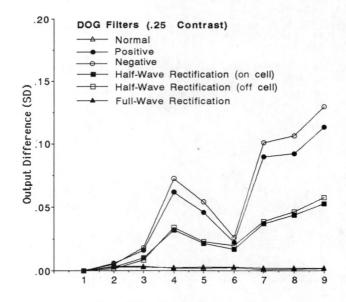

Figure 26. The responses of the DOG filters to the .25 contrast quadrant patterns in Experiment 6.

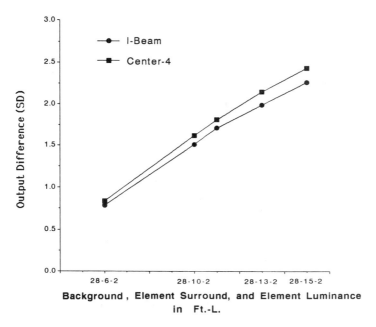

Figure 27. The responses of the Gabor filters to the I-Beam and Center-4 patterns in Experiment 7.

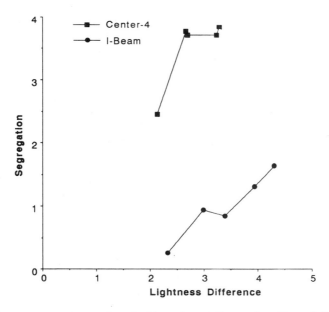

Figure 28. Mean segregation ratings in Experiment 7 as a function of the lightness differences between the discrepant and non-discrepant Center-4 and I-Beam elements.

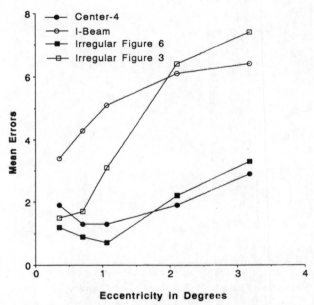

Figure 29. Peripheral discriminability as a function of eccentricity in Experiment 8.

Peripheral Discriminability

Texture segregation produced by figural differences have been shown to correspond to their discriminability when presented peripherally in a multielement display in which a subject is uncertain about the position of the target and can not focus attention (Beck, 1982). There is a greater sensitivity, for example, to differences in line orientation than to differences in line arrangement in peripheral vision (Beck, 1972; Beck & Ambler, 1972, 1973; Saarinen, 1987). Experiment 8 examined the peripheral discriminability of the I-Beam and Center-4 elements, and of the Irregular-Figures 3 and 6 (Figure 18). Four figures were presented in the corners of an imaginary square. A target consisted of either all figures the same (the figure in the non-discrepant quadrants) or of three figures the same (the figure in the non-discrepant quadrants) and a disparate figure (the figure in the discrepant quadrant). The disparate figure could appear in any of the four corners. The angular separations between a fixation dot in the center of the imaginary square and the center of the figures in the corners of the imaginary square were .35, .70, 1.05, 2.11, and 3.18 degrees. Eight trials were presented with a disparate figure and eight trials without a disparate figure. The displays were flashed for 200 msec.

Figure 29 shows the means of subjects' errors. Peripheral discriminability under uncertainty was greater for the Center-4 and Irregular-Figure 6 which yielded strong texture segregation than for the I-Beam and the Irregular-Figure 3 which yielded weak texture segregation. Differences in lightness or hue of the Center-4 and Irregular- Figure 6 are salient and, therefore, give strong preattentive texture segregation. Similar lightness changes in the I-Beam and Irregular-Figure 3 are not salient and fail to yield strong texture segregation.

As with changes in line arrangement, the peripheral visual system is not sensitive to changes in figure-ground *per se*.

Discussion

Our experiments indicate that texture segregation of the lightness and hue reversal elements is not explainable as a direct consequence of the differential stimulation of quasi-linear spatial-frequency analyzers. The application of spatial-frequency analysis to texture segregation can not be explained in terms of solely linear operations, and involves nonlinearities. Two early nonlinearities have been shown to occur. One nonlinearity is an intensity dependent nonlinearity which can be accounted for by sensory adaptation occurring before the channels or by a compressive intracortical interaction among the channels (Grossberg & Mingolla, 1985). The second nonlinearity is a rectification-like nonlinearity that is like that presumed to occur in complex cells. Nonlinear operations on the outputs of linear spatial-frequency channels have been suggested by the work of many investigators (Bergen & Adelson 1988; Chubb & Sperling, 1988; Grossberg & Mingolla, 1985; Fogel & Sagi, 1988; and Malik & Perona, 1990). A scheme followed by many texture segregation models involves three processing stages: an initial filtering by localized linear filters, a nonlinear transformation, and a second refiltering by localized linear filters. Graham, Beck & Sutter (in press) present a three-stage model which they call a complex-channels model. There are many parameters in such a model and we have not been able to definitely reject this more complex model though we have not been able to see how to explain the experimental results using this kind of model.

Texture segregation has been shown to occur as a result of differences in the outputs of simple-cell like filters that operate on intensity values and as a result of property differences such as lightness differences and alignment differences (Beck, 1992). Spatial frequencies compute a spatial average and do not differentiate constituent elements from each other or from the background. Perceived segregation based on property differences requires the explicit representation of individual features and properties such as edges and lightness values. Wertheimer (1923) proposed that grouping is an associative process based upon property similarities. Grouping, however, may also be described as a segregative process based upon property differences. Beck (1972, 1982) hypothesized that preattentive grouping occurs in terms of stimulus differences rather than in terms of stimulus similarity. Nothdurft (in press) provides further evidence that preattentive processing involves stimulus differences rather than similarities.

Two unresolved issues need to be noted. One issue is whether perceived segregation depends on spatially local computations. Sagi & Julesz (1987) argue that preattentive processing depends on local differences, but there is evidence that preattentive processing can occur also in terms of spatially separated region differences (Bacon & Egeth, 1991). The second issue is whether preattentive segregation is affected by regularities such as the collinearity and parallelism of lines. Beck, Rosenfeld, & Ivry (1989) hypothesize that preattentive segregation takes place with respect to both elementary features and "emergent" features. The length of a long line, for example, due to the preattentive grouping of short lines is an emergent feature segregating it from the surrounding short lines. Perceived segregation depends not only on the degree to which elementary features contrast with the adjacent region but also on the contrast of emergent features. Emergent features in turn depend on the kinds of regularity that can be processed preattentively (e.g., Grossberg & Mingolla, 1985; Sha'ashua & Ullman, 1988).

Previous research with 2D perceived shapes found that differences in the spatial relations between features such as the arrangement of lines in a shape that leave the slopes of the component lines the same do not generally yield strong texture segregation (Beck, 1982). Enns (1990) showed that this generalization does not hold for visual search when the shapes appear three-dimensional. Parallel visual search was possible for targets and distractors equated for 2D features that differed in their perceived 3D orientation. Texture segregation may also be affected by the stimulus representation. Ramachandran (1990) found that convexity and concavity conveyed by gradients of shading yielded perceptual segregation. Similar lightness changes that did not look three-dimensional did not perceptually segregate. Changes in the orientation of a stimulus which keeps the slopes of the component features constant also yields stronger texture segregation when the figures were seen as three-dimensional than when the figures were seen as two-dimensional (Beck, 1992).

Texture segregation occurs for stimulus properties that are processed in parallel or if serially very quickly (Beck & Ambler, 1972, 1973). Texture segregation does not occur if discrimination requires focussed attention. Perceived segregation based on the perceived 3D orientations of 2D shapes depends on shape information and appears, at least in some circumstances, to require attentional processing. For example, the discrimination of convex and concave based on shading gradients take several seconds and phenomenally appears to require attention. How does this occur? One possibility is that attention acts as a trigger. Attention is required to see a 2D shape as three-dimensional. Once a 3D interpretation is achieved, this interpretation is applied in parallel globally to all the figures in a display. A model for the rapid recovery of 3D properties has been presented by Enns & Rensink (in press). Another possibility is that attention allows limited parallel processing. Pashler (1987), for example, showed that observers could attend to up to eight items and process them in parallel. Attention more broadly spread augments the processing of information over a larger area but requires more salient stimulus differences than when focussed more narrowly (Eriksen, 1990). According to this hypothesis, attention facilitates the rapid parallel processing necessary for perceptual segregation. The above is speculative and we are not in a position to decide between these alternatives or other alternatives. The discussion is intended only to point out the issue. Future experimental work should be able to clarify the matter.

NOTES

The research was supported by AFOSR Grant AFOSR-88-0323. We are indebted to Norma Graham for valuable discussions.

1. There were three types of patterns: patterns with one-element-only, patterns with elements having opposite-sign-of-contrast, and patterns with elements having same-sign-of-contrast. The Contrast–Ratio Angle was computed by the equation

$$\text{Contrast–Ratio Angle} = 135° - arctan \Delta L_1 / \Delta L_2$$

where ΔL_1 and ΔL_2 represent the differences in the luminances of the squares from the background luminance. All patterns with one-element lighter than the background are plotted at the horizontal coordinate +45 degrees. Patterns with one-element darker than the background are plotted at the horizontal coordinate −45 degrees. For patterns in which the contrasts are opposite but equal the horizontal coordinate is zero. For patterns in which the contrasts are the same sign, the horizontal coordinates are between ±45 and ±90 degrees. As the luminances in the same-sign-of-contrast patterns get further from the background, the horizontal coordinate moves further away from ±45 degrees until it reaches ±90 degrees for patterns in which the luminances are infinite. See Graham, Beck, and Sutter (in press).

2. In our experiments the perceived lightness depended heavily, though not completely, on the ratio of the luminance of a square to the background luminance. Contrast was defined as the difference between the luminance of a square and the luminance of the background divided by the background luminance. The perceived lightness depended on 1 plus the contrast of the square:

$$L_{sq}/L_{bkg} = (L_{sq} - L_{bkg})/L_{bkg} + 1 = C_{sq} + 1$$

One interpretation of this equation is that lightness is computed by a more central processes in which the luminance of a square is evaluated relative to the background luminance. The value of 1 represents the background luminance in terms of the background luminance as the unit of measurement.

3. When the squares' luminances are relatively far above or below the background, perceived segregation depends on the ratio of the contrasts of the two squares rather than the difference in contrasts. Our explanation is that it is the result of an intensity-dependent nonlinearity See Graham, Beck, & Sutter, (in press) for more details of this dependence. An intriguing result that has not yet been completely explained is the strong segregation of opposite-sign of contrast squares in an element-arrangement pattern (Beck, Sutter, & Ivry, 1987; Graham, Beck, & Sutter, in press).

REFERENCES

Bacon, W.F. & Egeth, H.E. (1991) Local processing in Preattentive Feature Detection. *Journal of Experimental Psychology: Human Perception and Performance*, **17**, 77-90.

Beck, J. (1972). Similarity grouping and peripheral discriminability under uncertainty. *American Journal of Psychology*, **85**, 1-19.

Beck, J. (1982). Textural segmentation. In **Organization and Representation in Perception**, J. Beck (Ed.) Erlbaum, Hillsdale, N.J. (pp. 285-317).

Beck J., (in press) Visual processing in texture segregation. In **Visual Search II**, D. Brogan (Ed.) London: Taylor & Francis.

Beck, J. & Ambler, B. (1972). Discriminability of differences in line slope and in line arrangement as a function of mask delay. *Perception and Psychophysics*, **12**, 33-38.

Beck, J. & Ambler, B. (1973). The effects of concentrated and distributed attention on peripheral acuity. *Perception and Psychophysics*, **14**, 225-230.

Beck, J., Prazdny, K., & Rosenfeld, A. (1983). A theory of textural segmentation. In **Human and Machine Vision**, J. Beck, B. Hope, & A.Rosenfeld, (Eds.). New York: Academic Press (pp.1-38).

Beck, J., Rosenfeld, A. & Ivry, R. (1989). Line segregation. *Spatial Vision*, **4**, 75-101.

Beck, J., Sutter, A. & Ivry, R. (1987). Spatial frequency channels and perceptual grouping in texture segregation. *Computer Vision, Graphics, and Image Processing*, **37**, 299-325.

Beck, J., Graham, N. & Sutter, A. (1990). The effects of lightness differences on the perceived segregation of regions and populations. *Perception and Psychophysics*, 49, 257-269.

Bergen, J.R. & Adelson, E.H. (1988) Early vision and texture perception. *Nature*, **333**, 363-364.

Bergen, J. R. & Landy, M.S. (1991). Computational modeling of visual texture segregation. In **Computational Models of Visual Processing**, M.S. Landy & J. A. Movshon, (Eds.). Cambridge, MA: MIT Press (pp. 253-271).

Chubb, C. & Sperling, G. (1988). Processing stages in non-Fourier motion perception. *Supplement to Investigative Ophthalmology and Visual Science*, **29**, 266.

Daugman, J.G. (1985). Uncertainty relation for resolution in space, spatial, frequency, and orientation, optimized by two dimensional visual cortical filters. *Journal of the Optical Society of America A*, **2**, 1160-1169.

Daugman, J.G. (1987). Image analysis and compact coding by oriented 2D Gabor Primitives. **S.P.I.E. Proceedings, 758**, 19-30.

Daugman, J.G. (1988). Complete discrete 2-F Gabor transforms by neural networks for image analysis and compression. *IEEE Transactions on Acoustics, Speech, and Signal Processing*, **36**, 1169-1179.

Enns, J.T. (1990). Three dimensional features that pop out in visual search. In **Visual Search**, D. Brogan (Ed.) London: Taylor & Francis, (pp. 37-45).

Enns, J.T. & Rensink, R.A. (in press) A model for the rapid interpretation of line drawings in early vision. In **Visual Search II**, D. Brogan (Ed.) London: Taylor & Francis.

Eriksen, C.W. (1990) Attentional search of the visual field. In **Visual Search II**, D. Brogan (Ed.) London: Taylor & Francis (pp. 3-19).

Fogel, I. & Sagi, D. (1989). Gabor Filters as texture discriminator. *Biological Cybernetics*, **61**, 103-113.

Graham, N., Beck, J. & Sutter, A. (1989). Two nonlinearities in texture segregation. *Supplement to Investigative Ophthalmology and Visual Science*, 30, 361.

Graham, N., Beck, J., & Sutter, A. (in press) Contrast and spatial variables in texture segregation: testing a complex spatial frequency model. *Vision Research.*

Grossberg, S. & Mingolla, E. (1985). Neural dynamics of perceptual grouping: textures, boundaries, and emergent features. *Perception & Psychophysics*, 38, 141-171.

Malik, J. & Perona, P. (1990). Preattentive texture discrimination with early vision mechanism. *Journal of the Optical Society of America A*, 2, 923-932.

Mullen, K. (1985). The contrast sensitivity of human color vision to red-green and blue-yellow chromatic gratings. *Journal of Physiology*, 359, 381-400.

Nothdurft, H.C. (1990). Texton segregation by associated differences in global and local luminance distribution. **Proceedings of the Royal Society, London B**, 239, 295-320.

Nothdurft, H.C. (1991) Different effects from spatial frequency masking in texture segregation segregation and texton detection tasks. *Vision Research*, 31, 299-320.

Nothdurft, (in press) Feature analysis and the role of similarity in preattentive vision. *Perception & Psychophysics.*

Pashler, R. (1987). Detecting conjunctions of color and form: Reassessing the serial search hypothesis. *Perception and Psychophysics*, 41, 191-201.

Poulsen, D. (1988) A study of preattentive texture segregation of tripartite patterns. Undergraduate honors thesis. University of Oregon, Eugene, Oregon.

Ramachandran, V. S. (1990). Perception of depth from shading. *Scientific American*, 269, 76-83.

Saarinen, J. (1987). Perception of positional relationships between line segments in eccentric vision. *Perception*, 16, 583-591.

Sagi, D. & Julesz, B. (1987) Short-range limitation on detection of feature differences. *Spatial Vision*, 2, 39-49.

Sha'ashua, A. & Ullman, S. (1988). Structural saliency: the detection of globally salient structures using a locally connected network. **IEEE Second International Conference on Computer Vision**, 321-327.

Sutter, A., Beck, J., & Graham, N. (1989). Contrast and spatial variables in texture segregation: testing a simple spatial frequency channels model. *Perception & Psychophysics*, 46, 312-332.

Ts'o D.Y. & Gilbert, C.D. (1988). The organization of chromatic and spatial interactions in the primate striate cortex. *The Journal of Neuroscience*, 8, 1712-1727.

Turner, M.R. (1986). Texture discrimination by Gabor functions. *Biological Cybernetics*, 55, 71-82.

Watson, A.B. (1983). Detection and recognition of simple spatial forms. In **Physiological and Biological Preprocessing of Images**, O.J. Braddick & A.C. Sleigh, (Eds.). Springer-Verlag, New York, N.Y. (pp. 110-114).

Wertheimer, M. (1923). Untersuchungen zur Lehre von der Gestalt, II. *Psychologische Forschung*, 4, 301-350.

Perception: A Biological Perspective

by

V.S. Ramachandran
Psychology Department 0109
University of California at San Diego
9500 Gilman Drive
La Jolla, CA 92093-0109
(619) 534-6240
FAX (619) 534-7190

INTRODUCTION

What goes on inside your head when you perceive the world? Ask a man on the street this question and he will probably give you an account somewhat along the following lines. The optical elements in the eye produce an image of the world on the retina - an image that is picked up by photosensitive receptors in the retina and transmitted to the brain. A replica of the image is then displayed on a neural screen called the visual cortex and the whole process results in seeing. So, for example, if an object were to move in the visual field there would be a corresponding motion of the image in the inner "neural screen" and this will result in a vivid *perception* of movement that corresponds in every way to the motion of the original object.

A moment's reflection will reveal the logical fallacy in this account. If an image were to be displayed in the visual cortex you would obviously need another "inner eye" to perceive that image and yet another inner eye to scan the image in *that* eye and so on and so forth - *ad infinitum*. As Gregory (1977) points out, the outcome would be an endless regress of eyes and images and that, of course, does not really solve the problem of perception. Or, to put it differently, there is no use in recreating images on a neural screen in the brain since there is no little man ("homonculus") inside the head to perceive these images.

So the first thing to do, in order to understand perception, is to get rid of the idea of images in the head and to think in terms of symbolic representations or *transforms*. A good example of a symbolic description is a written paragraph like the one you are reading now. If you were to write a letter to a friend describing your apartment, you could do so quite successfully even though the actual words and paragraphs in your letter bear no *physical* resemblance to your apartment. The paragraph conveys a symbolic description of your apartment. It seems likely that the visual system does the same thing when conveying information to the brain - i.e. it conveys symbols rather than images. The purpose of studying perception is to understand the nature of these symbols and to "crack" the code used by the brain. In this sense the role of the psychologist studying perception is similar to that of a cryptographer or archaeologist trying to decipher the script of an alien language.

One approach to the problem of perception that has become very popular in the last two decades is the computational approach which was pioneered by David Marr at MIT. I think it is fair to say that the main strength of his approach, the computational approach, is that it allows a much more rigorous formulation of perceptual problems, at least in some cases, than what one could achieve by just doing some psychophysics or physiology alone. And this has been a real contribution, but there are also several major pitfalls - several important differences between machine and human vision. I have spelled out these differences in some detail elsewhere (Ramachandran, 1990; 1991), and so I will confine myself in this chapter to just one or two major points that are worth repeating.

My first point concerns the role of *image segmentation* in perception. According to David Marr's (1982) principle of modular design, certain elementary visual functions such as stereopsis, motion, color etc. are mediated relatively early in visual processing by specialized modules, whereas segmentation of the visual scene into separate objects is assumed to be a more complex process that can actually use the output of these early vision modules. Since the modules perform their functions prior to image segmentation, the argument goes, one can successfully model them and study them experimentally without worrying about segmentation. Contrary to this view, our evidence suggests that image segmentation can profoundly influence a number of early visual processes such as stereopsis (Ramachandran, 1986). Structure from motion (Ramachandran, Cobb & Rogers-Ramachandran, 1988), motion correspondence (Ramachandran, 1985; Ramachandran, Rao and Vidyasagar (1973) and shape-from-shading (Ramachandran, 1988a,b). The implication is that the early vision modules are not autonomous - they interact significantly with each other and with segmentation (See also Grossberg and Mingolla, 1988; Nakayama and Shimojo, 1992). Any program of research on perception must take these facts into account.

Second, let us consider the relevance of the neural hardware that must be used to implement the "algorithms" of perception. Many of you will remember Marr's point that any complex information processing system such as the human visual system can be understood at three different levels: (a) the very abstract level of the computational problems the system is trying to solve, (b) the level of algorithm, a sequence of steps that you can use to solve the problem and finally and least important, (c) the level of hardware implementation. His point was you can understand each of these levels without necessarily paying any attention to the other levels. I would like to submit this may be true for some simple machines, but when we are talking about biological systems, in fact, it is important to deliberately get confused between these levels - deliberately make what orthodox philosophers might call "category mistakes". In fact, I would like to show you several examples to illustrate the idea that our perceptual experience of the world is powerfully constrained by the actual neural machinery, the hardware that mediates perception. And, in general, I think no important discovery in science has ever been made by respecting the distinctions between different levels. In fact, if you think about Mendelian inheritance, you can't think of two more different levels than the behavior of pea plants on the one hand and the structure of the molecules on the other hand. But it is by bridging these two levels that the science of biology was transformed completely. I think the same sort of thing is beginning to happen or ought to happen in psychology and in physiology.

In this essay, I will describe several examples of attempts that we have made, in our laboratory, to develop such links between psychophysics and physiology. Many of these "links" are, admittedly, still somewhat tenuous but if they serve to stimulate new lines of enquiry they will have served their purpose adequately. I will describe 5 sets of experiments which are only loosely related to each other.

1) Phantom contours; a new class of stimuli that selectively activates the magnocellular pathway in man.
2) A new visual aftereffect induced by two dimensional visual "noise."
3) "Filling in" of scotomas and blind spots in the visual field and possible physiological correlates.
4) Temporal constraints on the perception of illusory contours.
5) Spatial interactions in the perception of shape from shading.
6) Interactions between attention and perception.

I hasten to add that although the first three sets of experiments can be directly linked to the physiology, the physiological mechanisms underlying (4), (5) and (6) are still quite obscure. I mention them only because they were recently discovered in our laboratory and should be of interest to any audience of computer scientists and visual neurophysiologists.

I. PHANTOM CONTOURS
Separate Visual Pathways for Contours and Surfaces

If you glance at an object in front of you, you will notice that it has two distinct attributes: a) a surface of a specific color and texture, and b) a boundary or contour that delineates the object from its background. It has been argued on grounds of computational feasibility (Grossberg & Migolla, 1987) and physiological grounds (Livingston & Hubel, 1987; Ramachandran, 1986) that these two attributes might be extracted and processed by separate mechanisms. Indeed, it is sometimes assumed that the visual system begins by using the contour extracting system (b) to create a sort of cartoon "sketch" of the visual scene and then "fills in" the surface attributes such as color and texture.

In this section I will present some new psychophysical evidence for the existence of such a distinction. Specifically, we find that there are two separate systems in human vision: a) a fast contour extracting system that is insensitive to sign, i.e. it is indifferent to the surface colors that define the contour; and b) a slow system that is sensitive to signs and can therefore signal surface colors (including black and white). It seems plausible to assume that these two systems correspond closely to the magnocellular and parvocellular (M and P) pathways of physiologists, but, as we shall see later, there are problems with this simple interpretation.

The stimulus we used is shown in Figure 1, a texture border composed of conspicuous black spots on the left and white spots on the right displayed on a homogeneous grey background of intermediate luminance. (Each spot subtended 20" of arc.) We then replaced all the black spots with white, and white spots with black and repeated the sequence in a continuous cycle at about 15 Hz so that you have two sets of spots flickering out of phase. If you compare any two spots in the

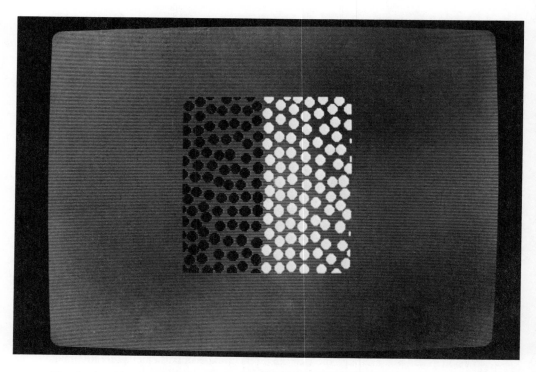

Figure 1: A single frame of a sequence used to generate phantom contours (Rogers-Ramachandran and Ramachandran, 1991). In the second frame, all the black spots are replaced with white spots (without changing their locations) and the black spots are replaced with white ones. If the two frames are presented in a continuous cycle at about 15 Hz, the spots on the two sides of the border become indistinguishable but the border itself remains clearly visible. We have dubbed this border a *phantom contour* or *magno contour*.

The entire stimulus subtended 13.4° wide by 15.4° tall and was viewed from a distance of 0.4m. The spots themselves subtended 34 min of arc and their luminances were 0.22 cd/m^2 (black spots) and 3.66 cd/m^2 (white spots). The luminance of the background grey was 2.36 cd/m^2. The flicker rate was 15 Hz. All stimuli were presented on a CRT using a Commodore Amiga microcomputer.

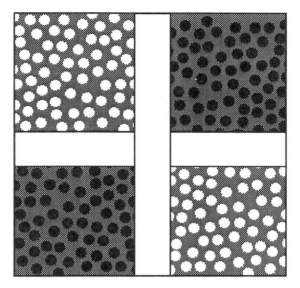

Figure 2: The four quadrant configuration used to test whether the individual elements are discriminable (see text). The spots subtended 38', the same as those used to construct the phantom contours. Flicker rate was 15Hz.

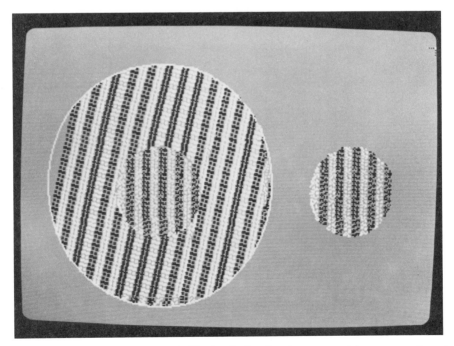

Figure 4: An example of the "tilt illusion" stimulus that we used in our experiment. The two small circular grating patches are both vertical and parallel but the one on the left looks tilted. The illusion remained unaffected even at high speeds of alternation (>15 Hz). Both inducing and "test" fields in the left-hand panel were flickering and subjects adjusted the orientation of the flickering right-hand panel to match the one on the left.

display they look identical - it is impossible to see whether they are flickering in phase or out of phase and they acquire a peculiar lustrous quality. Yet, surprisingly, the border defined by the two sets of spots is clearly visible! We call this a "phantom border" since the two regions that define it cannot be discriminated. (Rogers-Ramachandran and Ramachandran, 1991)

A more formal experiment was conducted on eight naive subjects who were run on twenty trials each. On each trial a vertical or a horizontal phantom contour was presented for 1 second (see Figure 1 for details). The subject's reported the orientation of the contour correctly on 99.4% of trials. But can we really be sure that this discrimination is based on the phantom contour itself and not on the elements that define it? To explore this we constructed the square matrix shown in Figure 2 in which the phantom contours were occluded by a "cross" configuration. On any given trial either a) dots in the top two squares were out of phase with dots in the bottom two squares; b) the left two squares were out of phase with the right two squares; or c) two diagonally opposite "corner" squares were out of phase with the other two squares. The subject's task was to report which of these he was seeing. The stimulus duration was 1 second. The same eight subjects were run on ten trials each and we found that their performance on this task was at chance level (25 i.e. 32% of 80 trials correct). To explain these results, we postulate that there are two distinct systems in human vision: a contour system that is insensitive to the *sign* of the border (whether white is to the left or right of the edge) and a surface system that can potentially report edge polarity but only at slow speeds (<8Hz). Since the two sets of spots differ only in *temporal phase* and since the flicker rate (15 Hz) is too high for the surface system to follow, the border ought to be visible only to the "fast" contour system (although it cannot inform you at any given instant which side is black and which side is white). The surface system, on the other hand, can tell you what the surface color attributes are but the flicker rate is too high for it to follow. What you end up seeing, therefore, is a contour that is defined by two surfaces that look completely identical!

Loss of Phantom Contours at Equiluminance

Next, we repeated the experiments using red and green spots. (Their luminances were 1.03 cd/m^2 and 1.19 cd/m^2 respectively and the background was about 2 cd/m^2.) The spots were made equiluminous to each other (but not to the background) using the "Minimally Distinct" borders technique . (A vari-color or "color flow" filter was used.) Eight subjects were used and 20 trials of vertical and horizontal texture borders defined by red and green spots were presented in a random sequence (160 trials total) with each stimulus lasting for 1 sec. Performance was correct on only 52.5% (S.D. = 10.35) of trials - a finding which suggests that the contour system responds mainly to luminous defined edges. We also tried gradually reducing the speed of alternation and found that the color-defined borders could be seen only when the speed was slow enough (<7 Hz) that the spots themselves were clearly discriminable. (At this speed the border was seen to rapidly reverse color.) A phantom contour - i.e. a contour defined by indiscriminable areas - was not seen at any speed.

Is the inability to discriminate orientation conveyed by phase-alternating colored spots simply due to a reduction in contrast? To find out we simply lowered the contrast of our black-white spots. Even at contrast levels that were *lower* (8%) than the effective contrast of the colored stimuli - so that the spots became difficult to resolve - we found that the phantom contours could be clearly discriminated on over 95% of trials. We also tried varying dot density and dot size and found that this did not critically influence the illusion.

Temporal Phase Discrimination

Using an additional six subjects we measured the speed of alternation at which they could perform accurately on the four quadrant phase-discrimination task (Fig. 2). Subjects adjusted flicker rate (five ascending and five descending trials) and found that for colored dots the mean speed was 5.06 Hz and for black/white dots it was 6.9 Hz. The phantom contour orientation, on the other hand, could be discriminated at arbitrarily high speeds (e.g. 25 Hz) for the black-white spots but only at 5.0 Hz for the colored spots. Thus the two thresholds - temporal phase discrimination vs seeing contours - are nearly identical for colored stimuli but grossly dissimilar for black/white spots. These findings provide additional evidence for our view that phantom contours selectively stimulate a "fast" system that is concerned with extracting borders but is insensitive to the *sign* of the borders.

Effects of optical blur and peripheral viewing
The visibility of phantom contours was also enhanced considerably by either by viewing the display peripherally or by blurring it using a sheet of velum. The enhancement obtained in peripheral vision might have something to do with the fact that the density of M-cells is higher in the periphery of the retina than near the macula.

Phantom contours also become less vivid if the density of dots is reduced and this enabled us to measure the effects of optical blur and eccentric viewing more systematically (Fig. 3). Six naive subjects were used and they viewed the stimuli binocularly under three different conditions: sharp, blurred, and sharp-peripheral. Each condition contained 100 trials (25 trials of each of the four dot densities). A trial consisted of the presentation of 250 msec of a mask, followed by 250 msec of the stimulus, followed by another 250 msec mask, and concluded by two seconds of a blank homogeneous grey screen. In the sharp condition, subjects viewed the stimulus sequence by simply fixating approximately in the center of the CRT. Under the blurred condition, a sheet of velum was placed over the CRT. Lastly, in the sharp-peripheral condition, subjects were told to fixate on a blue dot placed 5.5 cm to the right of the stimuli, thus viewing the stimuli using their peripheral vision. The order of the conditions that each subject viewed the movie under was counter-balanced to ensure that performance was not biased by a learning affect. The subject's task was to simply report whether the phantom contour was vertical or horizontal (Fig.3). The percentage of correct identifications of a horizontal or vertical line served as the dependent measure. Notice that the subjects show increased overall accuracy of performance on the orientation discrimination task in the blurred and sharp-peripheral conditions.

Taken collectively, these findings suggest that the "fast sign invariant" system we have identified may indeed correspond to the M-pathway of physiologists. We hasten to add, however, that there are several problems with this simple interpretation. First, although P-cells are more sluggish on the whole the M-cells, some P-cells can follow very high rates of flicker. Second, the presence or absence of sensitivity to sign has not been studied very carefully at various points along the two pathways and there have been some conflicting reports in the literature. One way out of this difficulty would be to suggest that the psychophysical stimulus we have devised reflects the properties of an *M-recipient* zone and a *P-recipient* zone rather than the M- and P-pathways themselves. Perhaps these M- and P-recipient regions are much more clearly segregated in their temporal and spatial properties than the M- and P-pathways. A P-recipient region concerned with extracting surface colors, for instance, might have a much lower temporal cut-off than P-cells in the LGN.

The possibility of using phase-reversing edges to selectively stimulate the magno system was first suggested by Livingstone and Hubel (1987) based on their pioneering physiological studies. The main difference between their stimulus and the present one is that we have used a texture border composed of spots instead of a real luminance edge and this has the advantage of getting rid of potential edge artifacts (e.g. Mach Bands) that could arise from spatial non-linearities. Such artifacts would be impossible to avoid if "real" edges are used. (Also, since the spots that define the texture are themselves indiscriminable, our stimulus may be the only known example of a "texture border" defined by elements that look completely identical.) And finally, by using texture edges composed of dots, we have been able to explore the effects of dot density, contrast, blur, etc. By examining the effects of these variables and by comparing temporal phase discrimination with orientation discrimination, we were able to convince ourselves that these contours do indeed selectively stimulate a separate "fast" pathway in the brain - a pathway that signals the presence of borders but not the surface attributes that define the border.

Finally, it is worth emphasizing that the importance of our psychophysical results does not depend exclusively on its correlation with the physiological segregation of M- and P-pathways. Indeed, even if there were no correlation at all with the physiology, our psychophysical findings

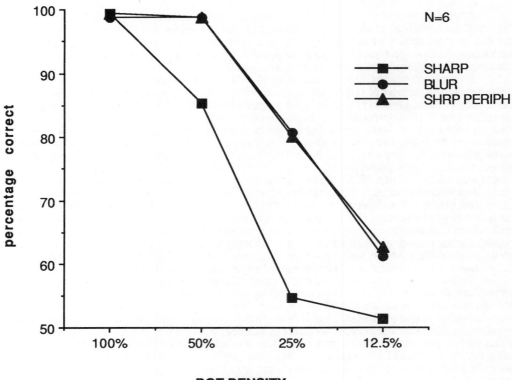

Figure 3. We constructed an array of phantom contour stimuli with decreasing dot densities. Orientation discrimination (horizontal vs vertical) was measured for six subjects on 25 trials at each of four dot densities. At 25% dot density, mean performance fell to chance level (55%). We repeated the experiment using two different viewing conditions 1) blurring the targets and 2) peripheral presentation. Performance was improved considerably in both instances.

would stand on their own since they imply that there must be two very distinct *systems* - one concerned with contours and the other with surfaces.

Can phantom contours "drive" classical visual illusions?

To determine whether phantom contours selectively excite the M-pathway (or some M recipient zone) in the brain one would need to do two kinds of experiments: a) present these stimuli to single cells at various points along the visual pathways, and b) study the performance of lesioned monkeys on the phantom contours orientation discrimination task. If it turns out that the phantom contours do indeed stimulate the M-pathway exclusively as we have suggested, then they could be used for making a reversible parvo "lesion" in an intact human subject. One could therefore, repeat, in a single (albeit very long) afternoon, all traditional monkey lesion experiments without ever having to actually train monkeys or make lesions. We found, for example, that one could make a "movie" consisting exclusively of phantom contours by using a sequence of frames in which the contours were shifted progressively in successive frames but the spots defining the contours were themselves completely uncorrelated. The result was a striking impression of smoothly drifting phantom contours whose constituents were flickering and indiscriminable. The effect was especially pronounced when we blurred the stimulus using a sheet of velum. One wonders whether motion detecting cells in the middle temporal area (MT) can also signal the motion of phantom contours.

Next, we found that the "tilt illusion" (Blakemore, 1973) could also be conveyed using phantom contours. (Figure 4) Subjects were asked to rotate a circular "test" grating patch until its orientation matched the one that looked tilted. Mean induced tilt (for six subject's who made ten settings in each condition) was 5.1°. We conclude that the inhibition between orientation selective cells that underlies this effect (Blakemore, 1973) must occur either within the M-pathway itself (e.g. in the "broad stripes" in 18) or higher up-stream in the magno recipient "form" areas in the temporal lobes. It is unlikely to occur in the parvo-recipient zones of 17 and 18 - a conclusion that is also consistent with the observation that the illusion deteriorates at a equiluminance (Livingstone, M.; personal communication). We also found that the "size contrast" illusion could also be conveyed by using counterphase flickering dots instead of dots that were continuous visible. Indeed, the flicker seemed to actually produce a strong *enhancement* of the effect; an observation that suggests that simultaneous activation of the P-pathway (in the case of non-flickering patterns) actually inhibits or at least "dilutes" the contribution of the m-pathway to perceived size. Obviously, further experiments would be needed to confirm this idea more directly. One wonders whether other visual illusions such as the tilt after effect, the "barber pole" illusion, the McCollough effect, and stereopsis (Julesz, 1971) can also be conveyed by using phantom contours.

Finally, we tried to determine whether binocular rivalry can be driven by phantom contours. Using 5 naive subjects we presented vertical phantom contours to the left eye and horizontal ones to the right eye and found, to our surprise, that a plaid pattern was seen. (For this experiment we used "real" square wave gratings rather than ones composed of dots.) Rivalry was restored only when we reduced the frequency of alternation to 6 Hz, suggesting that the process is driven mainly by contours that excite the P-pathway.

Taken collectively, our findings suggest that phantom contours can be used as a psychophysical "scalpel" to isolate and study the functions of the contour system; i.e. to determine psychophysically whether a given perceptual function (e.g. motion, orientation etc.) can be driven by the contour system alone. Also they might be useful as a probe to determine whether single cells in any given visual area (e.g. MT or V4) receive a M-cell contribution or not. And finally, it has not escaped our notice that they might provide a convenient non-invasive clinical test for revealing the early loss of magnocellular function that is thought to occur in glaucoma (Quigly, et. al., 1982) and dyslexia (Livingstone and Galaburda, 1991). They could, in fact, be regarded as equivalent to Ishihara's pseudoisochromatic test plates except that they test the integrity of the magno rather than chromatic pathways.

II. A NEW AFTEREFFECT INDUCED BY 2-D VISUAL NOISE

I will now describe a striking new visual aftereffect that we discovered while studying the "filling in" of artificially induced scotomas (see next section).

My interest in scotomas dates back to the time I spent in neurology clinics over a decade ago. Scotomas are "gaps" or holes in the visual field which are produced by small, discrete lesions in the primary visual cortex - usually the result of a stroke or a tumor. Curiously enough, the patient himself is often blissfully unaware of the gaping hole in his visual field. If he gazes at a red wall, for instance, he doesn't see a black hole corresponding to the scotoma; the visual field looks uniformly red. Indeed, even if the patient gazes at a pattern of wallpaper no gap or hole is seen - the wallpaper seems to somehow mysteriously "fill in" from the surround. Yet if he gazes at a companion seen against a background of wallpaper, the companion's head may vanish (if the scotoma is positioned correctly) and get "replaced" by the surrounding wallpaper.

What is the neural basis of this "filling in" process? Indeed, do we even need to postulate a specific filling in mechanism? It is true that we don't ordinarily notice blind spots and scotomas, but is this any more mysterious than the fact that we don't notice the gap behind our heads? If you stand in your bathroom you don't see a black hole behind your head, but the reason for this surely is that the visual areas of your brain simply don't represent events behind your head. You certainly wouldn't want to conclude that the wallpaper in the bathroom had somehow "filled in" behind your head. This distinction is not merely semantic. For events behind our head what we have is a *propositional* representation similar to a logical inference. For the region corresponding to the blind spot, on the other hand, there may be an actual *perceptual* representation. How can we distinguish between these two possibilities without becoming inextricably tangled in a philosophical conundrum?

The clinical literature on the "filling in" phenomenon is characteristically vague and so we decided to take a second look at the phenomenon. Instead of using neurological patients we developed a novel technique for creating an *artificial* scotoma in an intact human subject (Ramachandran & Gregory, 1991). The display we used is shown in Figure 5- a small grey square displayed on a background of twinkly 2-D visual noise. When we stared steadily at a fixation spot on the center of the screen for about 5 seconds the square vanished completely and the space corresponding to it was "filled in" with the twinkly noise from the surround. This effect is different from classical "Troxler fading" because the dark/light edges at the borders of the square were being continuously refreshed and this eliminates any possibility of retinal receptor fatigue. The "fading" of the grey square was very similar to the fading of kinetic edges that was originally observed by us in 1987 (Ramachandran & Anstis, 1987; Ramachandran, 1989a).

We noticed that the fading of the square was especially pronounced with eccentric viewing. This may simply reflect the progressive increase in receptive field size with retinal eccentricity. If the fading occurs as a result of "fatigue" of neurons that extract the border of the grey square, then even a tiny eye movement will "restore" the square by stimulating a new set of neurons. Since the receptive fields are smaller near the fovea, a smaller eye movement will be sufficient to restore the square near the center of gaze, and this might explain why the square doesn't fade as easily in central vision. We tested this hypothesis directly by waiting until the square disappeared and then moving it to see how large a displacement would be needed to make it reappear (Fig. 6). We found, as expected, that much greater displacements were required in peripheral vision than near the center of gaze.

Physiological Basis of "Filling In"

As soon as the square disappeared, the region corresponding to it was "filled in" with the surrounding twinkle. What are the physiological mechanisms underlying this "filling in" process? Our recent experiments suggest that filling in occurs fairly early in visual processing and must involve creating an actual neural representation of the surround in the region of the scotoma (Ramachandran & Gregory, 1991; see below). This view is consistent with a remarkable series of physiological experiments performed recently by Gillbert and Wiesel (1991). They destroyed a small patch of retina and recorded from the area of visual cortex to which this patch would normally project. These cells were initially silent, of course, as one would expect, but the

Figure 5: Stimulus used to produce an artificial perceptual scotoma. The background consisted of twinkling spots of eight different grey levels. The square subtended 1.5° X 1.5° and it had the same mean luminance (50 cd-m^{-2}) as the twinkling texture. The fixation spot was about 6° away from the border of the square. On steady fixation the square vanished in about 5 s and was filled in by the twinkling noise in the surround. A similar fading and filling in of texture was originally described by Ramachandran and Anstis (1987) and Ramachandran (1989) but in their stimulus the square was a "window" filled with horizontally moving dots rather than a homogeneous grey. Also, we found that if the square was very small (<0.2°) it could be seen to vanish even if it was very bright or dark, that is non-equiluminous with the surround. (The effect could then be seen even if the fixation spot was only 2° from the square.) The fading occurred even more quickly (<2 s) if the square was in a different stereoscopic plane (nearer or further) than the twinkling texture. This was confirmed by all four subjects.

researchers found that within a few minutes the same cells now had receptive fields that were outside the zone of lesion; they could be excited by visual stimuli that lay outside the scotoma! More recently they tried repeating these experiments using an artificial scotoma instead of a real one. Instead of destroying retinal receptors they used a display similar to Figure 5. (The display they used had little line segments moving around randomly outside the area occupied by the grey square.) To their surprise they found that cells which initially had receptive fields within the grey square now had receptive fields that were much larger and included regions outside the square. These rapid changes in receptive field organization may provide an explanation for some of the perceptual "filing in" effects that we have described in this article. Since these cells were originally responding to stimuli inside the scotoma, perhaps higher brain centers are "fooled" into thinking that stimuli immediately outside the scotoma are now actually inside it and this would correspond roughly to what we call "filling in". Furthermore, these results imply that receptive fields are not fixed anatomical entities formed by retinal receptors funnelling information on to single cells in the cortex. They suggest, instead, that the classical receptive field may be just the tip of an iceberg. Each cell may have thousands of silent synapses which can be reactivated or inhibited - perhaps in seconds - in response to ongoing visual stimulation and this can manifest as dynamic changes in receptive field organization. Studying these dynamic changes may give us novel insights into the neural mechanisms underlying perception.

A New Visual Aftereffect
The perceptual "filling in" of twinkle from the surround is interesting in itself, but what we observed next came as a complete surprise. After gazing at the fixation spot for 10 seconds we suddenly switched off the entire display and replaced it with a homogeneous grey field which had the same mean luminance as the twinkle. To our astonishment we now found that there was a square shaped patch of twinkling dots in the region that was originally occupied by the grey square! What was truly remarkable about this percept was its dynamic nature - that it was actually seen to twinkle and persist for about 10 seconds even though there was no stimulus on the retina. One wonders whether the numerous reciprocal cross-connections between different extrastriate visual areas are somehow included in maintaining this dynamic percept.

The perceived twinkle appeared to have the same spatial and temporal characteristics as the original inducing noise in the surround. When we lowered the temporal frequency of the surrounding noise the perceived temporal frequency of the twinkling square patch was also reduced correspondingly. Also, we tried using twinkling vertical lines instead of 2-D noise and found that many of our subjects could see a persisting patch of twinkling lines or "streaks" instead of spots.

We suggest that the dynamic noise somehow induces a peculiar state of adaptation in the surround which subsequently induces a percept of twinkling noise in the region corresponding to the "scotoma" (subjects occasionally report seeing stationary spots inside the scotoma instead of twinkling ones). Whatever the nature of the neural representation corresponding to this persistent patch of noise it is unlikely to be very different from the process causes the "filling in" in the first place. It is quite possible, for example, that the neural changes observed by Gilbert & Wiesel might persist for several seconds after the inducing stimuli have been removed but this idea has not been tested directly.

III. FILLING IN OF ARTIFICIAL SCOTOMAS

The stimulus described in the previous section enabled us to study a wide range of "filling in" effects in perception. We have described these observations in detail elsewhere (Ramachandran & Gregory, 1991) and I will therefore confine myself, here, to describing just one or two new experiments. One can ask, for example, whether filling in is a unitary process or whether it can occur separately for color and texture. To find out, we modified Figure 5 so that the surround consisted of sparse black dots twinkling on a homogeneous pink background, and the square consisted of black spots moving horizontally against a grey background that was equiluminous with the pink. This square, defined by a difference in motion as well as color, also faded completely in a few seconds, but we found that the filling-in occurred in two distinct stages. First,

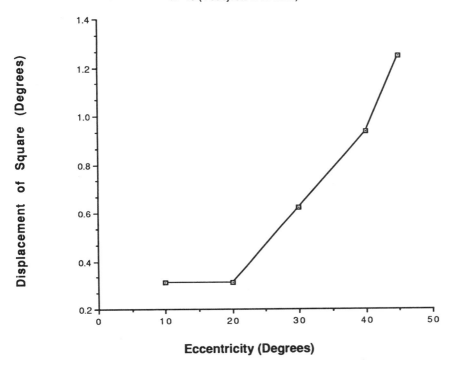

Figure 6: Perceptual fading occurs more readily in peripheral vision. At each eccentricity (x axis) we waited till the square faded. We then displaced the square to determine the amount of displacement that was required to restore the square (y axis).

the homogeneous grey region in the square vanished and was filled in by the pink from the surround, so that one now had the experience of seeing the black dots moving against the filled-in pink that did not actually exist on the retina. Once this had occurred, the moving spots also faded and were replaced a few seconds later by the twinkling spots in the surround. Thus, there may be separate fill mechanism for color and texture corresponding, perhaps, to the different extrastriate visual areas that are thought to be specialized for color or motion (twinkle). Perhaps the color border fades first, and the surround color is then assigned to this region (in the "color areas" such as V4). Because the moving spots have not yet faded, the visual system "assumes" that the spots must be moving against the filled-in pink.

Next, we wondered whether there was something special about twinkling dots or whether such filling in effects also be observed in stationary non-flickering patterns? We tried to answer this question by using patterns such as Figure 7A and Figure 7B. In Figure 7A the surround consists of a stationary horizontal square wave grating, and in Figure 7B it was composed of ordinary type-written English, Latin, or "nonsense" text. The filling in of text was especially striking and was reported by several subjects, although, needless to say, none of them could actually read the text within the filled-in region. Again, this effect is different from classical Troxler fading since it results in a *selective* fading of the square - the text itself does not disappear.

Equiluminous chromatic borders and luminance edges also tend to fade during optical image stabilization (Yarbus, 1967) or steady fixation (Livingstone and Hubel, 1987). We found that this was true even if an achromatic (grey) disc was displayed on an equiluminous pink background. The disc (subtending 2°) faded in about 3-4 s and was replaced by the pink from the background (Figure 8). The presence of the pink within the filled-in region cannot be explained in terms of conventional color contrast or adaptation effects. Surprisingly, if a thin black ring (0.8" diameter) was introduced in the center of the disc, the pink filled the *interior* of the ring as well, its spread was not "blocked" by the ring. This observation is especially interesting since it implies that the phrase "filling in" is merely a metaphor. If there had been an actual neural process that even remotely resembled "filling in" then one would have expected its progress to be blocked by the black ring but no such effect occurred. Therefore, we would be better off saying that the visual system "assigns" the same color as the surround to the faded region instead of implying that some neural process actually begins from outside and invades or "fills in" the interior of the disc.

What would happen if the grey disc were to straddle the boundary between two different colors (Fig. 9a)? Would the red "fill in" from the left and the yellow from the right so that one ends up with a yellow/red border between the two filled in colors? Curiously, we found that this never occurred. The two colors started filling in from the sides as one might expect but they never actually formed a border. Subjects reported, instead, that they appeared to "fight" or that there was a diffuse nebulous "haze" in this region. For some reason, the visual system will not ordinarily accept a border between two "filled in" colors!

Our last experiment in this series demonstrates an intriguing interaction between illusory contours and the filling in mechanism (Figure 9b). As in the previous demonstration, we began with a circular grey disc that straddled the vertical border between two equiluminous colors (pink and yellow). A vertical illusory contour was then introduced that coincided with the chromatic border and also continued across the center of the grey disk. On steady eccentric fixation the colors filled in the disk from the two sides but unlike the previous experiment they formed a crisp illusory color border corresponding to the illusory contour. We may conclude, therefore, that the filling in process is strongly influenced by early image segmentation produced by the illusory contours. This is somewhat surprising since most vision researchers have assumed in the past that the fading and filling in of colors is an early or "primitive" visual process that depends only on such factors as the presence or absence of luminance defined edges.

Filling in the Blind Spot
We have until now discussed "artificial scotomas" but can similar filling in effects be demonstrated for the *natural* blind spots that we all have in our visual fields? If you shut one eye and look at a red wall the entire wall looks homogeneous and the region corresponding to the blind

Figure 7A: A grey square (or disc) displayed against a background of nonflickering horizontal lines. Also fades and gets "filled in" by lines. Again this effect is different from classical "Troxler fading" in that it is *selective* and does not involve the whole image; the borders of the square fade but the lines do not. On more prolonged viewing the whole image fades, as one would expect.

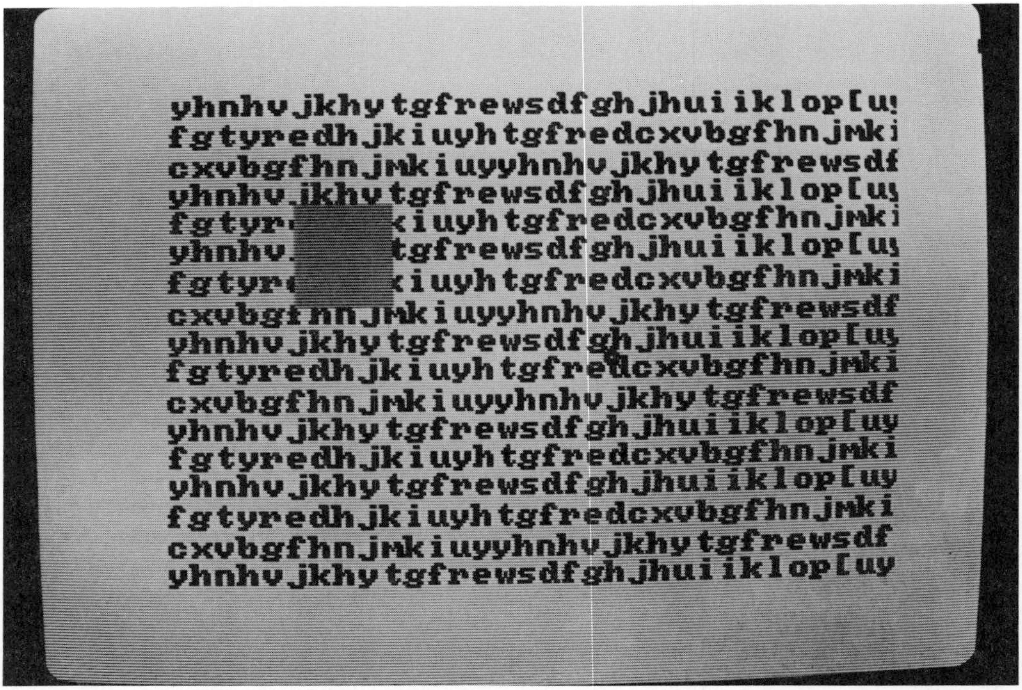

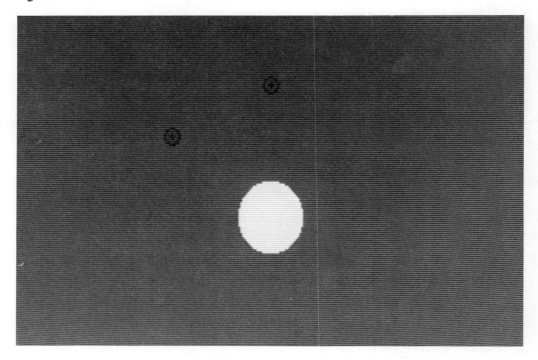

Figure 7B: Fading and filling in can also be observed if the square is displayed on English, Latin or "nonsense" text. Needless to say, subjects could not actually read the text in the "filled in" region.

Figure 8: A grey disc (depicted light grey here) was displayed on an equiluminous pink background (depicted dark grey). The grey disk faded in 5 seconds and was filled in by the pink from the background. If a thin black ring was displayed on the disc it did not block the "filling in" - the pink was seen both inside and outside the ring.

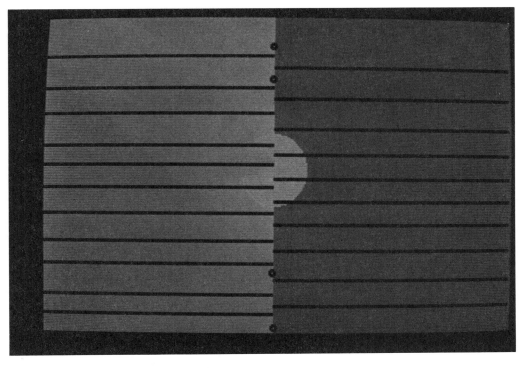

Figure 9: Stimulus configuration used to demonstrate the effect of illusory contours on the filling-in process. (a) The background was pink on the left (depicted light grey here) and green (depicted dark grey here) on the right and the circular disk was an equiluminous grey. Filling in occurred from both sides and formed a "border" between the two filled-in colors that coincided with the illusory contour. In the absence of the central vertical illusory contour (b) the border appeared indistinct and unstable.

spot has the same color (red) as the background. This was first noticed in 1862 by the English physicist Sir David Brewster who remarks "we would expect whether we use one or both eyes, to see a black or dark spot upon every landscape within 15° of the point which most particularly attracts our notice. The Divine Artificer, however, has not left his works thus imperfect...the spot, in place of being black has always the same color as the ground." Curiously, the blind spot remains invisible even when you look at a repetitive visual texture such as wallpaper and this has led many authors to suppose that the pattern from the surround somehow mysteriously fills in the gap corresponding to the blind spot.

I was initially very skeptical of the idea that our visual system actually "fills in" the blind spot. I can hold my finger up against any textured background and make the finger tip disappear by placing it in my blindspot. But if you were to ask me whether I can actually *see* the texture near the missing finger-tip my answer would be that I don't know. In fact, what I experience here is exactly the same feeling of ignorance that I do about events behind my head - a vague inkling feeling that the texture might be there - but certainly no clear perception of a texture.

Now we can raise the same question about the natural blind spot as the ones we raised for artificial scotomas. Do we really "fill in" the blind spot or do we simply fail to notice *absence* of neural signals from this region of the visual field? When I informally polled my colleagues, I discovered that most of them believed the "failure to notice" hypothesis rather than the "filling in" hypothesis. (This was true of both physiologists and psychologists whom I talked to.) Indeed, the distinguished American philosopher Daniel Dennett has recently argued (Dennett, 1991) that "filling in" is an inappropriate phrase since it necessarily assumes that there is an audience inside a Cartesian theatre.

I will now describe several experiments which prove decisively that we do, in some sense, actually "fill in" the blind spot - contrary to the views of my colleagues. The correct question to ask, in this regard, is how rich is the neural representation corresponding to the blind spot? At what stage in visual processing is this representation created and what does it have in common with other types of perceptual completion or surface interpolation? Can we explore the spatial and temporal limits of this "filling in" process - if it really does exist? My purpose in this section will be to try to answer these questions. First, we will begin with a simple demonstration that suggests that at least discontinuities in straight lines do get "filled in" (Fig. 10). Hold this illustration in front of you and move it to and fro while fixating the red X (with the right eye shut). You will find that the black disc on the left disappears as soon as it falls in the blind spot, but there is no obvious gap or discontinuity in the line. Yet if you place the upper end of a line inside the blind spot (Fig. 11) the line does indeed get chopped off - you don't see the line continuing into the blind spot. So your visual system can bridge a discontinuity caused by the blind spot but cannot extrapolate a line that is truncated or "chopped off" by the blind spot.

The completion of lines across gaps seems to depend only on the contours themselves - the color of the line does not seem to matter. We demonstrate this in figure 12 - in which the upper line is black, the lower line is white. There is no question that the line itself appears continuous even though, paradoxically, one cannot actually see the border between the black and white segments (curiously, many subjects describe this region as appearing "lustrous" or "metallic"). The paradox arises presumably because although somewhere in the visual pathways there are nerves signaling that the line is continuous (or, at least, not discontinuous), there are no nerves signaling the horizontal black/white border that falls on the blind spot.

How sophisticated is this "filling in" process? To explore this I devised the series of stimulus shown in Figure 13. Notice that even the center of the cross appears normal and uninterrupted when positioned on the blind spot. Indeed, if you position the blind spot correctly the pattern hardly looks different from a normal complete pattern; you may even see the lines converging to an imaginary "point" in the center. (Remember to keep the right eye closed.) We may conclude, therefore, that the visual system is able to "complete" the gaps even in these relatively complex types of patterns. Yet we found that there are also clear limits to what can be achieved. For example, when we superimposed the blind spot on the arc of a circle, it did not appear complete. It was clearly "chopped off" by the blind spot (Ramachandran 1992 a&b). The

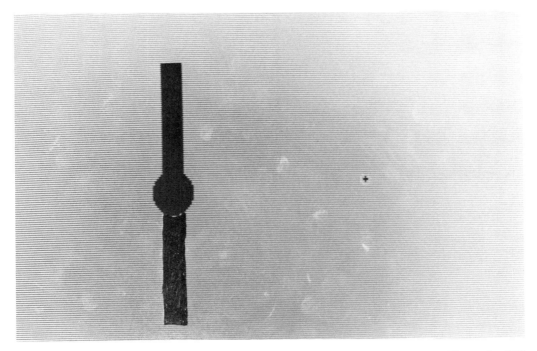

Figure 10: A vertical line passing through the blind spot looks complete. Fixate the "X" on the right and "aim" your left eye's blind spot on the disc (with the right eye shut). The line will look complete and continuous even though a portion of it ought to be invisible.

Figure 11: If one end of a line falls inside the blind spot then the tip does indeed get "chopped off" so that it now has a virtual terminator.

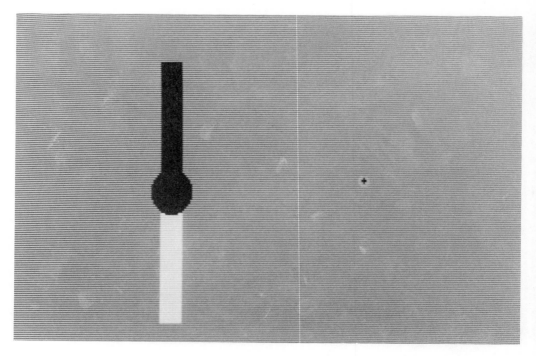

Figure 12: Different colors (or luminances) used for the upper and lower line segments do not prevent the completion process. The line looks complete, yet, paradoxically, you cannot actually see a border between the black and white segments. Many observers report that this region looks "lustrous" or "metallic."

Figure 13: A more complex pattern such as a cross is also completed (A). Interestingly if the two lines are of opposite contrast (i.e. black vs white) then they tend to "compete" for completion across the blind spot (B). The line that one is attending to tends to complete across the blind spot and is seen to subjectively "occlude" the other line. If the two lines of the cross are of unequal lengths, the longer arc is usually seen to complete the gap and occlude the shorter one.

implication of these observations is that the filling in of the blind spot is a primitive process that occurs at a relatively early stage in visual processing. Consequently, you can bridge discontinuities in contours and surface textures across a blind spot but you cannot complete objects. This effect contrasts sharply with observations we made recently on patients with cortical scotomas. If the arc of a circle or corner of a square falls on a cortical scotoma, the figure looks 'bitten off' initially, but after 6 or 7 seconds the visual system 'completes' the figure. (Ramachandran, 1992b).

Our next experiment takes advantage, once again, of "illusory contours" - edges that are usually invoked by the visual system to account for otherwise inexplicable gaps or discontinuities in the image. A vertical black line, we have seen, is readily completed across the blind spot (see Figure 10). But what if one were to place a vertical illusory contour on the blind spot? The reader can answer this question for himself by placing the black disc inside the blind spot so that it vanishes (see Figure 14). The question is, do you then complete the "real" horizontal line that defines the illusory contour or do you complete the vertical illusory contour? Most observers report that they tend to see the illusory contour as complete - not the real contour! But if all the horizontal lines, except three, are removed then subjects do see the horizontal line as complete. (Figure 15) Information from the illusory contour must, therefore, somehow "veto" the completion of the single horizontal black line that runs through the blind spot.

Our last experiment on the blind spot takes advantage of a phenomenon that visual psychologists call "pop-out". (Julesz, 1971; Treisman, 1988) It has been known for some time now that certain elementary features in the visual environment will "pop-out" perceptually, i.e. appear very conspicuous when displayed against a background of a large number of distractors that differ from the target along some specified dimension such as color, orientation or motion. For example, if you have a display in which a diagonal line is embedded in a matrix of vertical lines, then the single diagonal line 'pops-out' i.e. you will have no difficulty spotting it. The same is true for the single red spot displayed against a background of green spots. On the other hand, it is rather difficult to spot the single T against a background of L's. It has been suggested, therefore, that popout can occur only for features that are extracted relatively early in visual processing, such as orientation or color, but not for more complex features such as corners (i.e. L's) or T junctions.

We can now ask, does the filling in of the blind spot occur before or after "pop-out"? To explore this we created the display shown in Figure 16 in which one of the rings alone is placed on the blind spot and the others are not. (The ring was larger than the blind spot so that its inner margins just overlapped the edges of the blind spot.) Of course, this ring looks like a disc but the interesting thing is that it also pops out even though on the retina it is actually identical to all the other rings! The implication, then, is that the "filling in" process must occur quite early in visual processing since it actually precedes visual search and pop-out. Also, this demonstration effectively demolishes the view that you don't actually fill in the blind spot, and that you simply "ignore" what is going on there. How can something you ignore actually *pop-out* at you, i.e. how can it look *more* conspicuous than everything else in the visual field?

So far, we have considered the filling in of the natural blind spot and of artificial scotomas. Obviously one would like to know whether similar principles would also hold for scotomas caused by retinal and cortical damage. We have not studied these in detail yet but our preliminary work suggests certain interesting differences between different types of filling in. Patients with retinal scotomas (and recent cortical scotomas), for example, often report partial filling in of homogeneous colors, stationary low contrast stripes and non-flickering random-dot patterns but almost no filling in when they look at television "snow" or even a homogeneous flickering field. Indeed when looking at "snow" they can even draw their own scotoma on the screen. Further experiments along these lines - comparing different types of scotomas - may give us new diagnostic tools as well as new insights into the "filling in" process.

Functional Significance of "Filling In"

Finally, one can ask why study the blind spot or the "filling in" of artificial scotomas? What is the biological purpose of these processes? It seems unlikely that the visual system has evolved a mechanism for the sole purpose of filling in blind spots and scotomas. A much more

Figure 14: Illusory contours are seen in this display. Instead of seeing the lines terminating abruptly the visual system interprets this image as arising from an opaque yellow vertical strip occluding the horizontal lines in the background. Interestingly when the number of lines is large, the illusory strip is seen to "complete" across the blind spot rather than the horizontal lines that define the illusory contour.

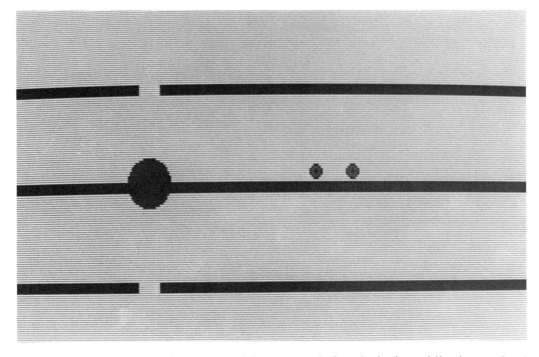

Figure 15: If a smaller number of lines are used, than the horizontal line is completed instead of the illusory contour.

Figure 16: A yellow ring looks like a homogeneous "filled" disc when its inner border is made to just overlap the outer margins of the blind spot. Even if this disc is surrounded by rings it "pops out" conspicuously even though on the retina it is actually identical to the other rings.

plausible interpretation is that the process is simply a manifestation of a more general perceptual mechanism - one that the AI chaps call "surface interpolation". When we look at a chair or table, for instance, the visual system extracts information mainly about its contours or edges and creates a representation that resembles a sort of cartoon sketch. The color and texture of the chair is perhaps then "filled in" by a process, akin to the filling in of scotomas and blind spots. this would allow the visual system to avoid the computational burden of having to create a detailed representation of surface textures and colors - a process that may be too time-consuming for an organism that is trying to jump on a surface or avoid an obstacle.

IV. TEMPORAL DYNAMICS OF ILLUSORY CONTOURS

Physiologists and AI researchers usually take it for granted that perception occurs in stages (e.g. see Marr, 1982) but there are really very few psychophysical experiments that provide direct evidence for this view. In this section I will describe a new approach to this problem that takes advantage of illusory contours.

A neglected aspect of illusory contours concerns the time course of their development. To investigate this, we created the two displays shown in Figure 17 and 18. Figure 17 shows that when the illusory figure is superimposed on the checkerboard pattern there is a striking enhancement of the subjective contours (Ramachandran 1986, 1990) - an effect that occurs only if the edges of the check are collinear (in phase) with the cut edges of the sectored disks. On the other hand, if the checks are deliberately misaligned (Fig. 18), the illusory contours disappear altogether - since the stimulus configuration is no longer compatible with a "cognitive" occlusion hypothesis.

We were intrigued by these interactions between the illusory figure and the background texture (checkerboard) and wondered what would happen if one were to switch back and forth between these two figures (Figs. 17 & 18). Notice that the sectored discs do not change location but that the checkerboard is alternately misaligned and lined up in the two frames. How fast would the frames need to be alternated in order to preserve the continuous appearance of illusory contours?

The two movie frames were generated on an RGB monitor using an Amiga 2500 microcomputer. We began by viewing frame 1 (in phase) for several seconds and then suddenly introduced Frame 2 (misaligned). Not surprisingly, perhaps, the illusory contours and depth stratification vanished almost instantly. The experiment was then repeated on 6 naive subjects who were asked to press a key to initiate each trial and to press a second key as soon as they lost the stratification and illusory contours. They were instructed to look at a small red fixation spot off to one side of the illusory square; the spot being continuously visible throughout the procedure. The mean reaction time for these subjects was about 400 msec.

The surprise came when we reversed the movie sequence so that the "misaligned" conditioned was presented first and then followed by the "in phase" condition. As in the previous experiment we fixated on the red spot and viewed frame 1 for several seconds before switching on frame 2. This time we found that the illusory square took several seconds to appear. Again we repeated the experiment on the same six subjects and found that on average the square took about 1.3 seconds to emerge. Surprisingly, on some individual trials the square did not appear for 4 or 5 seconds and the subjects themselves noted the long delay. This long latency was somewhat reminiscent of the latency that occurs when viewing Julesz' random-dot stereograms for the first time (Julesz, 1971; Ramachandran, 1976). The difference is that much of the latency observed in stereopsis is probably due to vergence eye movements whereas the latency experienced with our illusory figure is unlikely to involve eye movements.

If the transition from misaligned to "in phase" conditions really takes as long as 1.5 seconds, then it must follow that if the two frames are alternated at greater than about 0.7 Hz, the illusory contours should always remain invisible because the frames would change before the illusory contours had time to "form". Using the same six subjects, we were able to confirm that this was indeed the case.

Figure 17: An illusory (Kanizsa) square is displayed on a background of checks. When the check's edges are lined up with the illusory square there is striking enhancement of perceived depth and of the illusory contours (Ramachandran, 1986). This enhancement effect disappears if equiluminous red/green checks are used instead of grey/white checks (Ramachandran, 1990).

Figure 18: If the checks are misaligned with the edges of the "Pacmen" the illusory contours disappear completely.

Some interesting effects can be obtained by varying the degree of misalignment and the spatial frequency of the checks. If the misalignment is only slight the region corresponding to the illusory square is seen to "bulge" forward as the visual system tries to arrive at a compromise solution. If very high spatial frequency checks are used, the inhibition of illusory contours is reduced considerably.

The experiments we have discussed so far have involved alternating between two frames - one of which conveyed an illusory figure and one of which didn't. What would happen, instead, if one were to present just a single illusory figure? How much time would it take to simply perceive an illusory figure starting from a blank frame? To answer this question we simply flashed Figure 17 on the screen for varying intervals and asked our subjects to report whether they saw the illusory figure or not. We found that the square was seen on 90% of the trials even when the duration of the stimulus was as brief as 200 msc. We may conclude, therefore, that seeing the illusory stratification itself can happen extremely fast - in less than 200 msc. What seems to take time is the transition from one perceptual "hypothesis" to another. Furthermore, there appears to be an interesting asymmetry in that the "vetoing" or inhibiting an illusory figure appears to take considerably less time than "rebuilding" contours that have been inhibited or destroyed.

In summary we find that a) when illusory figures are superimposed on a checkerboard there is a striking enhancement of the illusory figure and the contours associated with it ("stratification"). This stratifications can occur extremely rapidly since it can be seen in 200 msc flashes. b) When the check's edges are misaligned with those of the illusory figure, the stratification (and associated illusory contours) disappear completely. c) If after inspecting the misaligned pattern for a short while (e.g. 10 seconds) we switched to the aligned version, the illusory square took several seconds to emerge. d) When we switched to the aligned version to the misaligned one the illusory contour disappeared instantly.

These results complement the recent findings of Reynolds (1981), who superimposed an illusory triangle on a pattern resembling a brick wall. As in Figure 18 the brick wall rendered the illusory triangle invisible, but only if the stimulus was flashed on for durations longer than 200 msc. If the stimulus was flashed for less than 200 msc, on the other hand, subjects could see the illusory contours quite clearly. Reynolds (1981) interprets these results to mean that the illusory contours as signalled initially at an earlier stage in the system and then subsequently "vetoed" by the bricks at a later stage.

Our results, taken together with those of Reynolds, suggest that the perception of illusory surfaces occurs in stages and that these stages can be revealed by using sequentially presented stimuli. They also flatly contradict the Gibsonian view that perception does not involve active information processing - a view that has a surprisingly large following in North America. An especially surprising result is that "vetoing" an illusory can take place very quickly (<200 msec), whereas, once this vetoing has occurred, seeing the illusory contours again can sometimes take several seconds. Whatever the interpretation of this asymmetry, our results suggest an interesting physiological experiment on single cells in area 18 which are known to signal illusory contours (Von der Heydt, 1990). One could confront such a cell with Figure 17 and then switch to Figure 18. Would the cell fire instantly or would it take several seconds to respond? This might tell us where "vetoing" and subsequent reactivation of the cells' response occurs in area 18 itself or at a later stage where different depth cues are combined in the visual system to create a cue invariant 3-D representation of the world.

V. SPATIAL INTERACTIONS IN THE PERCEPTION OF SHAPE FROM SHADING

In this section I would like to discuss shape from shading, a problem that has been very much neglected by both physiologists and psychologists. In fact the only person to have thought about it seriously was the Renaissance scholar, Leonardo da Vinci, ages ago, and since then there have been only two or three psychophysical experiments on shape from shading (e.g. see Todd & Mingolla, 1983). This is quite surprising. I think this is because most physiologists and psychologists think of shading as a high level "cognitive" process and not something that's handled by what we regard as the early visual system. But I would like to argue contrary to this that shading is extracted very early in the visual system.

Of the numerous mechanisms used by the visual system to recover the third dimension, the ability to use shading is probably the most primitive. One reason for believing this is that in the natural world animals have often evolved the principle of countershading to conceal their shapes from predators; they have pale bellies that serve to neutralize the effects of the sun shining from above. The widespread prevalence of countershading in a variety of animal species (including many fishes and caterpillars) suggests that under usual circumstances shading must potentially be a very important source of information about three dimensional shapes.

The starting point of our own investigations in this field was to create a set of simple computer-generated displays (Fig. 22). The impression of depth perceived in these displays is based exclusively on subtle variations in shading and we made sure that they were devoid of any complex objects and patterns. The purpose of using such displays, of course, is that they allow us to isolate the brain mechanisms that process shading information from higher-level mechanisms that may also contribute to depth perception in real-life visual processing. So, in a sense, the displays serve the same role in the study of shape from shading that Julesz Stereogram's do in the study of stereopsis.

Now what we decided to do is to start from first principles, start from scratch, and we created a very simple display, just a bunch of eggs (Fig. 19). Most readers should see this display as a set of eggs illuminated from the left. But actually of course this display is ambiguous, because it could equally depict a set of holes or cavities, which some of you may be able to see with some effort of will, illuminated from the right. Now the interesting thing is: either they are all seen as eggs or all seen as cavities. You can't simultaneously see one as an egg and one as a cavity. Well why not? Well, if you were to ask my colleagues in the psychology department, they would say there is a Law of Common Fate, that all things tend to do the same thing.[1] But is it really the Law of Common Fate, or is it because the visual system has a built-in assumption that the light usually shines only from one direction: there is only one light source illuminating the whole visual image? To you find out, we did the simple experiment shown in Figure 20. Here we have two rows which are mirror images of each other. And when you look at them initially you see them all as eggs. But as you gaze at it for awhile, you will find that when you see one row as eggs you tend to see the other row as cavities. Hence the effect seen in Figure 19 is not the result of the a common depth assumption but the result of an assumption there is only one light source illuminating the entire visual image, or at least in that portion. And we call this the single light source constraint (Ramachandran, 1988). Obviously it's a reasonable thing to assume because we live on a planet which has only one sun.

Notice that in Figure 20 either row can be seen as convex or concave if the other row is excluded. When both rows are viewed simultaneously, however, seeing one row as convex actually forces the other row to be perceived as concave. Some powerful inhibitory mechanisms must be involved in generating these effects.

The single light source assumption is, of course, implicit in many AI models of Shape from Shading but Figure 20, as far as we know, is the first clear-cut demonstration that such a rule actually exists in human vision. In addition to the single-light-source constraint there also appears to be a built-in assumption that the light is shining from above - a principle that is well-known to artists. This would explain why, in Figure 21, objects in the top panel are always seen as convex whereas those in the bottom panel are often perceived as "holes" or "cavities". The sign of depth can be readily reversed by simply turning the page upside down. The effect is a weak one, however, since either panel can in fact be seen as convex or concave with some effort of will (especially if the other panel is excluded from view to eliminate the single-light source constraint). On the other hand, if a mixture of such objects is presented it is almost impossible to perceptually reverse any of them because of the combined effect of two constraints - the single light source constraint and the "top" light source constraint (Fig. 22).

These observations suggest that the visual system assumes that the sun is shining from above. But how does the visual system know 'above' from 'below'? Is it the object's orientation in relation to the retina that matters, or its orientation with respect to gravity? This question was first raised by Yonas, Kuskowski, & Sternfels (1979) who tested both 3 year old infants and 7 year old children using ambiguous stimuli. (They used photographs of real objects rather than computer generated images). They found that the response of the 3 year olds depended almost

---

[1] As Peter Medawar might have phrased it, explanations of this kind are "mere analgesics; they dull the ache of incomprehension without removing the cause.

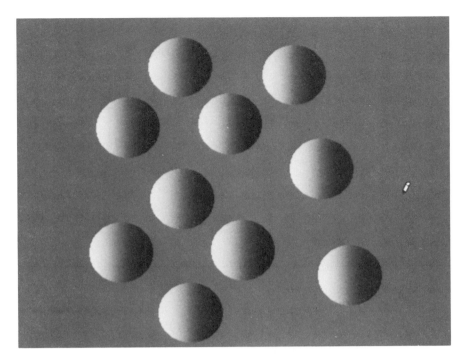

Figure 19: This display depicts a set of objects which conveys a strong impression of depth. The sign of perceived depth, however, is ambiguous since the visual system has no way of "knowing" where the light source is. Consequently the display can be perceived as consisting of either convex objects illuminated from the left or concave objects lit from the right ("eggs" or "egg-crate"). The reader can mentally shift the light source and generate a depth inversion. Interestingly, when a depth inversion occurs it tends to occur simultaneously for all objects in the display.

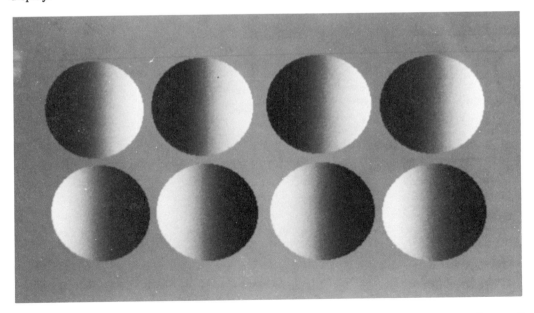

Figure 20: This figure demonstrates the single light source assumption using a mixture of shaded objects which are mirror images of each other. Objects in one row can be seen as either convex or concave if the other row is excluded. But when both rows are viewed simultaneously, seeing one row as convex forces the other row to be perceived as concave.

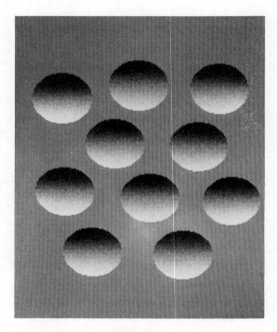

Figure 21: This computer generated photograph demonstrates that the visual system has a built-in "assumption" that the light source is shining from above. Notice that the depth in these displays is conveyed exclusively through shading with no other depth cues present. The shaded objects in the top panel are unusually seen as convex while those in the bottom are usually seen as concave. Notice, however, that the illusion (i.e. the difference between convex and concave) is not as pronounced as in Figure 22 in which the objects are intermixed.

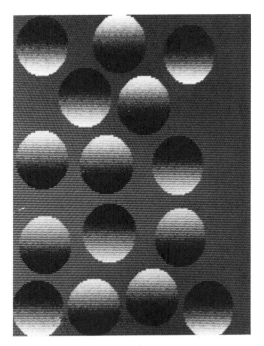

Figure 22: A random mixture of eggs and cavities. Notice that the depth effects are enhanced and also that you can mentally group and segregate all the eggs from all the cavities. The effect is reduced considerably if you rotate the page by 90°. If you turn the page upside down, on the other hand, all the eggs become cavities and the cavities become eggs.

Surprisingly, the grouping effect can be obtained even if different eggs and cavities are in different (random) stereoscopic planes suggesting that it is based directly on 3-D shape from shading rather than on the depth that is conveyed by the shading.

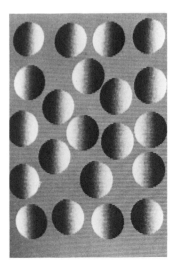

Figure 23: Is similar to 22 except that we have left-right differences in shading instead of top-bottom differences. Grouping is difficult to obtain suggesting, again, that the grouping observed in 22 cannot be due to differences in luminance polarity alone.

exclusively on retinal orientation whereas seven year olds showed roughly equal dependance on both retinal and gravitational frames of reference. They suggested that as children grow older they progressively shift their responses towards more abstract frames of reference. Curiously, Yonas, et al., did not test adults but their results imply that if the same trend continues then adults should show an even higher dependence on gravitational (rather than retinal) 'upright' than 7 year olds.

To test this prediction I tried presenting stimuli such as Figure 22 and 23 to adult subjects and asking them to rotate their heads by 90°. Instantly, all the objects which were 'top' lit <u>in relation to the retina</u> were seen as Convex and the others were seen as Cavities. The effect was a striking one; even a head tilt of as little as 15° to 20° was sufficient to generate the unambiguous percept of eggs and cavities (Ramachandran, 1988). When the head was tilted by 15° in the opposite direction the eggs and cavities reversed depth instantly. I recently showed a slide of this display to a lay audience of several thousand spectators. (Ramachandran, 1989b) and all of them, without exception, reported the perceptual switch. We may conclude from this that the interpretation of shape-from-shading depends primarily, if not exclusively, on retinal rather than gravitational cues. The implication is that shape from shading is probably extracted fairly early in visual processing since it must occur prior to vestibular correction for head-tilt.

To measure these effects more carefully we created a display in which a set of eggs was arranged in the form of an "O" displayed against a background of cavities (Fig. 24). The subject's task was to discriminate this from an incomplete "O" which had a "bite" taken out . Not surprisingly, the discrimination was very difficult if the shading was left-to-right instead of top-to-bottom . We measured the ability of 6 naive subjects to perform this discrimination while sitting upright and then repeated the experiment with the subjects lying on their sides (Figure 25). Notice that in each case discrimination is easiest when the shading is vertical in relation to the *retina* rather than in relation to gravitational coordinates (Kleffner & Ramachandran, 1992).

Now to me this is one of the exciting things about studying perception; that you can begin with a set of almost "axiomatic" first principles and arrive at interesting generalizations. I will now show you several new displays which demonstrate some striking spatial interactions in the extraction of shape from shading by the visual system. These interactions bear a tantalizing resemblance to the classic 'center surround' effects that have been reported in other stimulus domains such as luminance, color (Land, 1983), and motion (Nakayama & Loomis, 1974). They reveal, once again, the importance of the single-light-source constraint.

First, we wondered what would happen to the interpretation of shape-from-shading; if one were to give the visual system conflicting information about light source location. To explore this we created the display shown in Figure 26. The display was shown to an audience of 48 students and they were asked to examine the 2 panels (A & B) carefully and to compare the two central discs. Their task was to judge which of the two central figures (A & B) appeared more convex and to jot down their answers on a piece of paper. The experiment was then repeated with the position of the two panels (A & B) interchanged.

The results were clear-cut; the central disc in panel B almost always appeared more convex than in panel A. (72 out of 96 trials). In fact many subjects spontaneously reported that the disc in panel A almost appeared flat.

These results suggest that the magnitude of depth perceived from shading is enhanced considerably if objects in the surround have the opposite polarity; a spatial contrast effect that is vaguely reminiscent of the center-surround effects that have been reported for other stimulus dimensions such as motion and color. Another way of saying this would be that the perception of shape from shading is enhanced considerably if the information in the scene is compatible with a single light source. When the information from the majority of objects (e.g., panel A) suggests that the light source is on the left (or right) the shading on the central object is perceived as a variation in reflectance rather than depth.

Another example can be seen in Figure 27. In this display subjects usually saw the central disc in panel A as being more convex than in panel B (44 out of 48 trials). The reason for this, again, may be an interplay between 3 constraints or 'rules' - the rule that the light is probably shining from above, the rule that there is probably a single light source and, finally, a tendency to

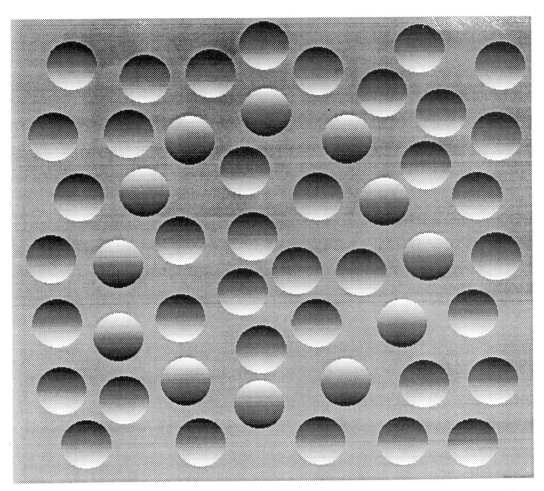

Figure 24: Sample stimulus used to investigate figure and ground segregation. A circle was constructed from the target items ("eggs") and presented against a background of distractors ("cavities"). On half of the trials the circle was incomplete (three consecutive eggs missing) and the subjects' task was to determine whether the circle was complete or incomplete within a fixed presentation time.

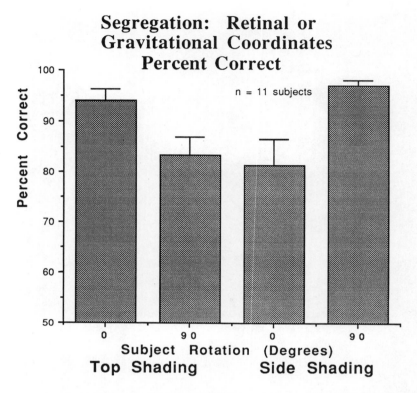

Figure 25: Results from the segregation task with shading. The subjects were much more accurate at determining whether the circle was complete or incomplete with vertical shading than with either horizontal shading or a step change in luminance. Are shading effects tied to retinal or gravitational coordinates? We explored this by having subjects view the screen either sitting up or laying down. The results of this experiment indicate that the effect depends primarily, if not exclusively, on retinal rather than gravitational cues (Ramachandran, 1988; Kleffner & Ramachandran, 1992).

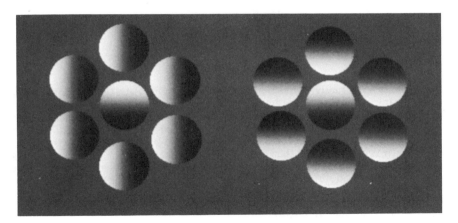

Figure 26 Demonstrates "center surround" interactions in the perception of shape from shading. The central disc in panel A is usually seen as less convex than the one in panel B. These effects are usually much more pronounced on the CRT than in the printed versions shown here.

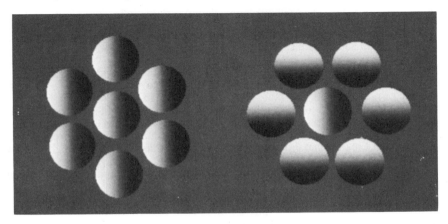

Figure 27: The central disc in A is seen as more convex than the one in B. The latter is usually seen as concave or flat.

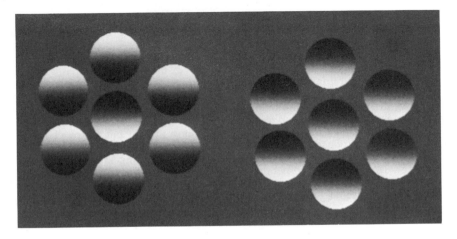

Figure 28: The central disc is more convex in B than in A.

assume that objects are usually convex. In panel B the objects in the surround are seen as unambiguously convex because of the built-in assumption that the light is shining from above. This, in turn, makes the central disc look relatively flat. In panel A on the other hand there is no conflicting information since all the discs can be assumed to be illuminated from the left. Consequently they are all seen as convex without violating the single light source assumption. The reason they are seen as convex-rather than concave - may be that there is also a tendency to assume that objects are more likely to be convex. This is not surprising given the "cohesiveness" of matter; object boundaries and surfaces in the world are more often convex than concave.

To demonstrate the 'convexity' assumption more clearly we devised the display shown in Figure 28. In this display the central disc in panel B is usually seen as more convex than the one in panel A. (43 out of 48 responses). Why?

There are two conflicting cues in panel B - the "overhead lighting" assumption which would tend to make the discs look concave vs the convexity assumption which tends to make them look like eggs. Interestingly, the convexity assumption 'wins' and subjects usually report seeing "eggs lit from below". (Notice, however, that one doesn't have to violate the single light source rule to obtain this percept.) In panel A, on the other hand, the discs in the surround are unambiguously convex since they are assumed to be lit from above. This in turn forces the central disc to be seen as concave or relatively flat since the system will accept only one light source. Clearly, the single light source rule is strong enough to override the convexity assumption in this situation.

Our next demonstration (Fig. 29) reveals a new rule that is not evident in any of our earlier demonstrations. The display depicted in Figure 29 was shown to 48 naive subjects. Most of them reported that the central disc in A was more convex than in B. (45 out of 48 trials). The reason for this difference is not obvious. Since both displays violate the single light source rule it is unclear why the central disc in A is seen as slightly more convex than in B. Perhaps what you are seeing here is an example of a 'depth contrast' effect that would be analogous to motion contrast (i.e. induced motion) - but I realize that this is merely a description, not an explanation. Notice that the discs in the surround are obviously more convex in B than in A. This might make the central disc appear less convex in B than in A; a spatial "depth contrast" effect that cannot be directly deduced from any of our earlier rules - about light sources.

Unmasking the Truth

Does the assumption of convexity play a role in the perception of more complex 3-D objects? The fact that hollow masks can be perceptually reversed to resemble a normal (convex) face has been known for a long time and has been pointed out by both Helmholtz and Gregory. Not surprisingly, such a reversal is easier to produce if the mask is illuminated from below and viewed with one eye shut to eliminate disparity cues.

What would happen if one were to illuminate the mask from above rather than from below? According to the assumption that the light source is at the top, the mask should look hollow and it should be difficult to reverse it perceptually. The importance of this assumption is evident from Figure 22 in which we had a mixture of concave and convex objects. It is virtually impossible to produce a depth reversal in this figure; i.e. one cannot "imagine" that the light is shining from below in order to generate a reversal of depth. The convex and concave objects always retain their shapes suggesting that the rule about illumination from the top is a very powerful one that cannot be over-ridden by imagery. Yet this principle appears not to be true for hollow masks (Fig. 30). When we illuminated this mask from above it did look hollow initially but we found it very easy to reverse it perceptually to produce a normal looking face, and parts of the mask that were originally convex now looked concave. Why is "imagery" effective in producing a reversal in Figure 30 but not in Figure 22? Does this have something to do with the fact that faces are familiar objects? Oddly enough, when viewing the hollow mask we could sometimes reverse the face alone with portions of the beard continuing to look hollow for a short period before they finally "caught up" with the rest of the face.

It is often assumed that effects such as these arise due to our past experience and familiarity with faces. We have observed, however, that similar depth reversals can even be obtained with

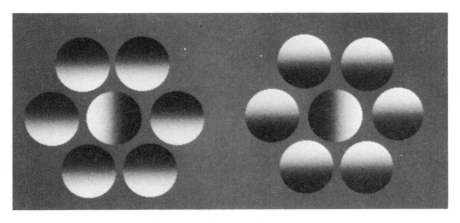

Figure 29: The central disc is usually seen as more convex in A than in B.

Figure 30: Depicts the inside of a hollow mask of Santa Claus. Notice that the face looks convex even though this requires you to "assume" that the light is shining from below.

complex 'nonsense' shapes such as abstract sculptures or convoluted lumps of clay that have several bumps and protrusions on them (Deutsch and Ramachandran, 1990). If such objects are illuminated from below and photographed, the pictures are usually seen as convex rather than concave. We may conclude, therefore, that the reluctance to see hollow faces derives from a general assumption that complex objects are usually convex rather than on specific "top-down" influence from high level semantic knowledge of faces (or other familiar objects).

Another example is shown in Figures 31. These are actually photographs of the insides of two Halloween masks illuminated from above so that they should really look hollow rather than convex. But, as in Figure 30, the visual system regards hollow objects as a highly unlikely occurrence and prefers to interpret the image as convex objects lit from below. Notice the two small circular discs between the chins of the two faces. Interestingly, even though the light is now "assumed" to come from below, the disc on the right is seen as convex and the one on the left as concave - as though they were both being lit from above. Even when the discs are directly pasted on the cheek (left mask), the one on the right is sometimes seen as a bump and the one on the left as a dimple. Perhaps the sharp outlines that delineate these from the face allow the visual system to segment them from the rest of the face so that they are regarded as separate objects illuminated from a different direction. Sure enough, when the discs are "blended" into the face by blurring their outlines (right mask) the one on the right is almost always seen as a dimple and the one on the left as a tumor or zygomatic arch, as though they were illuminated from below like the rest of the face. Thus the visual system will sometimes give up its assumption of a single light source but only if there are separate objects in the image. For different parts of a single object it is reluctant to accept more than one light source.

The reversal of a hollow mask occurs even during binocular regard, so that the visual system has to overcome conflicting stereoscopic information. The reader can verify this by an using an ordinary Halloween mask. If the mask is lit from the side, it will initially look hollow; but after several seconds of steady viewing, it will reverse and look "normal", i.e. convex rather than hollow. The implication is that the uncrossed disparity signals must either be "vetoed" or reinterpreted in the presence of other cues in the image. This vetoing can be readily demonstrated by having two LED's mounted inside the hollow mask, one at the tip of the nose and the other on the cheek. Even though the LED on the nose is not really seen as being part of the nose, it is seen to be actually *nearer* than the one on the cheek. If the room lights are switched off, on the other hand, the nose LED is instantly seen as further away than the cheek LED, as one would expect from the disparity signals. By using a stroboscopically illuminated mask we were even able to study the time-constants of the interactions between shading and stereoscopic disparity cues (Ramachandran & Gregory, 1992).

Hollow masks provide a useful technique for investigating how the visual system resolves conflicting cues. For example, we noted above that the nose LED was often seen nearer than the cheek LED. While this was *usually* the case, we found that when we went very close to the mask, the disparity signals from the LED's seemed to become dominant and one sometimes had the curious impression of a *transparent* nose with an LED seen right through it! It was as though when the disparity signals were too strong to be countermanded by the other cues on the face (shading) the visual system arrives at the "compromise" solution of seeing a transparent nose.

Is the reversal of depth seen in these displays necessarily a "global" effect that involves the entire face or can it occur separately for different parts of the face? To find out, we tried pushing one part of the face alone, e.g. the nose, inside out so that it conveyed *crossed* disparities and was no longer hollow. Curiously, we now found that although the entire face was perceptually reversed, the nose was *not* reversed; it looked like a normal convex nose! If we moved our heads from side to side, the face appeared to "follow" us but the nose appeared to move in the opposite direction. For obvious reasons the result was especially striking and comical when we used a mask of Richard Nixon (Ramachandran & Gregory, 1992). The implication of this result is that the reversal of depth seen in hollow masks does not have to involve the whole face, i.e. it does not simply involve taking all the "+" signs and changing them to "-" signs. We conclude, instead, that, even at the expense of violating the single light source rule, the visual system generates a depth reversal only for those parts of the face where it is deemed necessary.

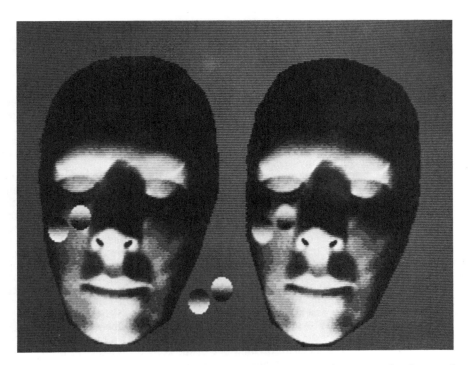

Figure 31: Hollow-mask interiors lit from above produce an eeire impression of protruding faces lit from below. In interpreting shaded images the brain usually assumes light shining from above, but here it rejects the assumption in order to interpret the images as normal, convex objects. Notice the two disks near the chin still appear as though lit from above: the right disk seems convex and the left one concave. When the disks are pasted onto the cheek (left), their depth becomes ambiguous. When blended into the cheek (right), the disks are seen as being illuminated from below, like the rest of the face.

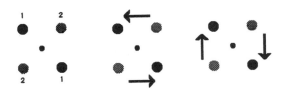

Figure 32: The basic bistable motion display used in our experiment. (Numerals indicate order of presentation.) The three possible percepts are shown (A fourth percept, anticlockwise motion, is not shown.)

VI. INTERACTIONS BETWEEN VISUAL ATTENTION AND PERCEPTION

In this section, we will consider the role of "top-down" influences such as attention and visual imagery on perception. We have begun investigating this using ambiguous apparent motion displays (Fig. 32).

Ambiguous apparent motion

Metastability is one of the most striking yet enigmatic aspects of perception. Necker cubes and other bistable figures are often used to illustrate the point that perception is really a hypothesis on the state of affairs in the world rather than a passive response to sensory stimuli. One's perception of the cube changes dramatically as the mind hesitates between alternative three-dimensional representations. Indeed, when viewing such bistable figures it is often hard to believe that something has not changed physically in the stimulus.

The stimulus used in our experiments was on the bistable apparent motion display shown in Figure 32: a matrix of four dots forming the corners of a square. It was generated on a P-4 phosphor cathode ray tube (CRT) using an Apple 2 microcomputer and viewed from a distance of 1 m. The sides of the square subtended 40 min of arc and the dots themselves were 3 min of arc in diameter. A hand-held potentiometer could be used to vary the speed alternation or stimulus onset asynchrony (SOA) continuously over a wide range: the SOA was varied by changing the duration of each of the two frames and there was no dark interval (ISI) interposed between frames. The number by each dot refers to the time at which it is presented. Dots on two diagonally opposite corners are flashed first and then switched off, followed by two dots appearing simultaneously on the remaining two corners, with the procedure repeating in a continuous cycle. The two possible percepts, are indicated in the diagram as Percept 1 (vertical oscillation) and Percept 2 (horizontal oscillation) and the display is essentially bistable just as a Necker cube is. However, we found it almost impossible to voluntarily switch from one percept to the other when the speed of alternation was higher than 3 frames per second (SOA = 350 ms). Subjects almost always experience hysteresis, that is, they tend to persist in seeing one of the percepts (either vertical or horizontal oscillation) and can switch the axis of motion only by looking away and looking back after some time has elapsed (Ramachandran & Anstis, 1983; 1986).

Another theoretically possible percept is either continuous clockwise or anticlockwise motion but at short SOAs (350-400 ms), this was impossible to achieve even through intense mental effort. However, we made the curious observation that when the SOA was made sufficiently long (>465 ms) subjects could occasionally will themselves to switch the motion axis or even to see clockwise or anticlockwise motion. Six naive subjects were asked to view the display and we then varied SOA over a wide range using a hand-held potentiometer. We began with very slow speeds so that the subject could will clockwise (or anticlockwise) motion and then gradually increased the speed of alternation. When the SOA was smaller than 465 ms, subjects reported that they could no longer hold this percept (mean of 6 readings X 6 subjects, that is, 36 readings = 465 ms; s.d. = 33). They could now only see oscillation, and the axis of oscillation could no longer be influenced at will. Furthermore, we find that if Figure 32 and its mirror-image are presented side-by-side then at slow speeds one can simultaneously "will" clockwise and anticlockwise motion suggesting that eye movements cannot explain the effect. We conclude that axis-dominance remains absolute and unambiguous at small SOAs and that it can be coupled with the effects of will and attention only at SOAs longer than 465 ms.

The subjects introspective reports were also instructive, in this regard. To switch from seeing oscillation to seeing (say) clockwise motion, for example, they would often try to "imagine" the corner dots moving continuously clockwise. Perhaps there is an upper limit to the speed at which you can imagine something moving and this might explain why only oscillation can be seen at small SOAs. If so, using bi-stable displays might provide a novel technique for exploring the time constants of obscure phenomena such as visual attention and "imagery."

Global "linking" of multiple ambiguous displays

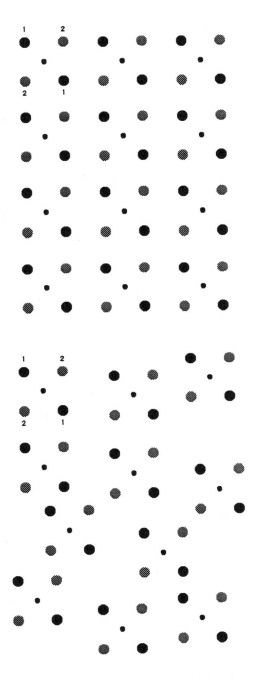

Figure 33: a) Multiple bistable displays presented simultaneously. Identical oscillations are observed in *all* the displays. b) The squares (bistable dot-pairs) are randomly positioned and the same effect is seen showing that the presence of regular arrays is not necessary. The central dot within each display is present continuously and is perceived to be static.

We have now begun to use our bistable displays to address the more general question of whether motion perception is based on a piecemeal analysis of the image or whether global field-like effects are also important as emphasized by Gestalt writers. Our strategy was to generate several bistable displays simultaneously on the screen (Fig. 33a&b). Twelve naive subjects viewing these displays reported that all the oscillating dot-pairs seemed to "lock-in" - that is, the displays always had the same motion axis. This was true even when adjacent displays were separated by several degrees. (Intriguingly, we find that the coupling is much stronger within each hemifield than across the vertical meridian.) If one of the displays changed its motion axis then all the displays invariably changed with it simultaneously. This effect suggests the presence of a global field for resolving ambiguity in bistable displays for if the different displays were being processed independently there is no *a priori* reason why their oscillations should become synchronized. The presence of such a field is interesting for two reasons. First, if several Necker cubes or other similar reversible figures are simultaneously present in the visual field then they tend to undergo reversals more or less independently of each other. This suggests that a field probably exists only for certain classes of visual displays such as the motion displays used here. Second, the effect seems to be inconsistent with the "independence-assumption" of Ullman (1979) according to which apparent motion should be based on strictly local computations and there ought to be no global field-like effects.

The role of visual attention

We wondered whether adjacent bistable displays could be "uncoupled" by changing their aspect ratios. In general, the figures remained coupled but if the aspect ratio was about 2.5:1 for each display (but in opposite directions so that one was stretched out vertically and the other was stretched out horizontally) then they uncoupled spontaneously.

An interesting effect was observed by using a display (Fig. 34) in which we had just two "stretched out" figures (i.e. two figures whose aspect ratios were 2.5:1 and 1:2.5 respectively) dispersed among six bistable ambiguous figures (which had a 1:1 aspect ratio). In this display we found that the direction of motion seen in the six ambiguous squares was strongly gated by visual attention. While fixating the central dot, if we cast our visual attention "searchlight" on the single vertically biased display then all the ambiguous displays were seen to oscillate vertically as well. On the other hand, if we paid attention to the horizontally biased display then all the ambiguous areas oscillated horizontally. Thus one could switch the perceived direction of oscillation by simply directing one's attentional spotlight to the single appropriately biased display. (Plummer and Ramachandran, 1992)

A second version of this experiment is shown in Figure 35 which is composed of several vertically biased displays interspersed among several horizontally biased displays. Notice that the 2 sets of displays are colored differently - black vs white. Also, near the center of the screen we have a bistable square (colored white) which has an aspect ratio of 1:1. Again, we found that we could pay attention to all the white spots oscillating vertically and this would cause the central bistable figure to get coupled with them spontaneously. Conversely, if we paid attention to all the horizontally oscillating black spots the central figure also oscillated horizontally, i.e. it became coupled even though it was white rather than black! Thus whatever you are paying attention to causes spontaneous coupling of the ambiguous display even though the color is not appropriate.

Occult apparent motion

The phenomena described above suggests that although the global coupling of ambiguous apparent motion displays might initially involve cooperative linking or "facilitation" between direction selective cells, the effect can be strongly modulated by effects such as attention and imagery. This was brought home to me very vividly when I discovered the phenomenon of "occult apparent motion." I have described this effect in detail elsewhere (Ramachandran & Anstis, 1983; Ramachandran, Inada & Kiama, 1985), but I will mention it here briefly since it reminds us of how little we really know about perception and how much more there is to learn.

To demonstrate this illusion we generated first Figure 33a again on the CRT screen and confirmed that all the oscillating pairs had indeed become synchronized, as in our previous

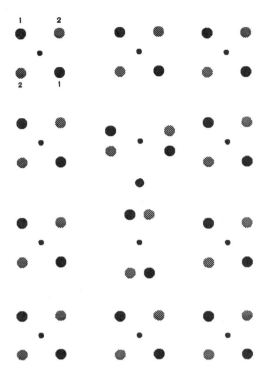

Figure 34: Demonstrates the effect of visual attention. Notice that the two displays near the center are biased by stretching them either vertically (bottom) or horizontally (top), whereas all the other displays are ambiguous (1:1 aspect ratio). Directing your attention "spotlight" to either one of these causes all the ambiguous ones to become coupled with it.

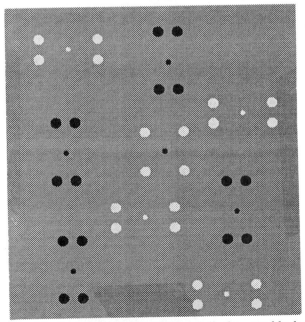

Figure 35: In this display the vertically biased displays are black and the horizontally biased ones are white. Notice that the single ambiguous display near the center (aspect ratio 1:1) is also white. This display becomes spontaneously "coupled" with the motion of the particular set of spots that one is paying attention to - either the black spots or the white spots.

experiment. We then used long strips of white tape to mask off the lower half of each horizontal row of figures. A small window was then cut in the middle so that one of the ambiguous figures alone was allowed to become fully visible. By this simple procedure we were able to create unambiguous horizontal motion throughout the surround. Not surprisingly we found that this imposed a slight tendency to see horizontal motion in the central ambiguous figure, but the effect was very small. In fact, by using voluntary effort one could easily uncouple the central ambiguous figure from the surround and decide to see vertical motion. Hence the coupling between an unambiguous and an ambiguous figure is probably much weaker than the coupling between adjacent ambiguous figures - although of course we had no simple way of measuring the strength of this coupling.

While watching this display we made a curious observation. As pointed out above we could easily uncouple the center from the surround so that vertical motion was seen. But to our astonishment we found that on some occasions when this happened the dots in the surround also moved vertically to imaginary locations behind the strips of masking tape. Since there is nothing in the retinal stimulus that would lead one to expect this, we have dubbed this illusion "occult apparent motion.". Six naive subjects were confronted with this display and they all confirmed that they could see this effect. Interestingly, the effect took time to evolve as when viewing Julesz's random-dot stereograms, and two of the subjects had to be told what to look for, that is, they did not report the effect spontaneously. We then simply masked off the central ambiguous figure and found that the illusion disappeared and was replaced by the appearance of unambiguous horizontal motion in the surround. The illusion could be seen even if the surrounding figures were several degrees (say 3°) away from the central one. It was as though the particular hypothesis the brain had selected to resolve ambiguity in the center was being automatically applied throughout the visual field. And in the absence of sufficient evidence the brain resorts to simply postulating the required dots behind the occluders.

<u>The role of attention in binding visual features</u> In the previous section we explored interactions between visual attention and apparent motion. The effects are intriguing but they tell us very little about what the <u>function</u> of focused visual attention might be in early vision. Treisman (1986) has pointed out that if different visual features such as 'color' and 'form' are initially represented separately in the visual system, the function of visual attention may be to solve the 'binding problem' i.e. it may enable the visual system to put the features back together again to create whole objects. The perceptual grouping of individual features, on the other hand, is usually assumed to be "pre-attentive".

How long does it actually take to group individual features together to form a gestalt? The oft quoted figure of 60 msec is usually based on reaction time experiments which employ briefly presented (and 'masked') displays. To take a fresh look at this problem, Diane Rogers-Ramachandran, Stuart Anstis and I created a continuous two-frame movie sequence consisting of a 'grid' of 25 spots in each frmae. In frame one, the grid was composed of alternating columns of red and green spots so that one clearly saw the spots being organized in vertical groups. In frame two, on the other hand, alternate (horizontal) rows were red and green so that the spots were seen as (horizontal) rows rather than columns. We then gradually increased the frequency of alternation of the two frames from 1 Hz to 12 Hz and found, to our surprise, that at about 4 Hz the pattern became completely scrambled and the grouping was no longer visible! This was true in spite of the fact that the alternation of the colours of the individual spots themselves could be seen clearly even at frequencies as high as 11 or 12 Hz. We conclude, therefore that the perceptual grouping based on similarity of colour takes at least 250 msec and is therefore not "pre-attentive". More recently, we have been able to show that the same principle also applies for other types of perceptual grouping such as grouping based on similarity of orientation.

SUMMARY AND CONCLUSIONS

My main purpose in writing this essay has been to emphasize some promising new lines of enquiry rather than to provide a conventional review article. Although some of the visual effects I have described can be linked to the known physiology of the visual pathways, others may require "cognitive" or "top-down" explanations of the kind I usually like to avoid.

In Section I, I described a new class of stimuli that selectively stimulate a fast sign-invariant pathway in the visual system. Further experiments are needed to establish clearly whether this pathway is indeed the magnocellular pathway of physiologists. If so, it might provide a powerful psychophysical probe for isolating and studying visual functions that are mediated by this pathway.

In Section II, I described the "filling in" of artificial scotomas and blind spots. Many philosophers and psychologists in the past have argued that we don't really "fill in" the blind spot - you simply ignore what's out there. Our experiment demonstrating pop-out flatly contradicts this view. How can something you ignore actually *pop-out* at you? This experiment clearly suggests that you really do "fill in" the blind spot, if the phrase filling has any meaning at all. Also, for filling in of artificial scotomas, we can now provide a straightforward physiological explanation. The effect probably arises from the kinds of short-term receptive field changes observed recently by Gilbert and Wiesel. I should point out, by the way, that it is not often the case in our field that you can discover a phenomenon that has an immediate impact on both physiology and philosophy. Results in psychology usually take several decades to trickle down to physiologists and results in science usually take centuries to change the thinking of philosophers!

Section IV is concerned with the time course of seeing illusory contours. Although perception is usually assumed to occur in stages (Marr, 1982) there have been very few experiments demonstrating this empirically. The experiment we describe provides a novel approach to this problem since it allows us to "dissect" the different stages of processing. Since illusory contours are signalled by cells in area 18, we suggest that the "vetoing" depends on inhibitory feedback from higher visual areas, whereas the enhancement of the contours may be based on "resonance" resulting from positive feedback between the different extra-striate areas. This line of reasoning is consistent with the recent speculations of Edelman (1991) and his co-workers.

In Section V we present several demonstrations which suggest that the extraction of shape from shading probably occurs fairly early in visual processing. One especially surprising result is that the assumption of overhead lighting depends on retinal rather than gravitational (or "world centered") coordinates. This seems counterintuitive, since it implies that when you tilt your head the visual system "assumes" that the sun is stuck to your head! Why has such a stupid assumption been built into our visual systems? After all we do tilt our heads occasionally and it would therefore be much more sensible to compute shape from shading *after* making the appropriate corrections for head tilt. The answer, of course, has to do with *speed*. It may indeed be more appropriate to make the correction for head tilt but in real life situations there simply may not be enough time to perform the required correction. And since most of the time we do keep our heads upright the system can get away with the "short-cut" of using retinal rather than gravitational coordinations. The point is that this short-cut worked often enough for your arboreal ancestors to survive and leave babies and, consequently, there was no selection pressure to evolve anything more sophisticated or time-consuming.

My physiological and neuroanatomical colleagues often ask me why I study illusions. Why not study "reality" instead? My answer would be that illusions have the same aesthetic appeal to a psychophysicist that Drosophila mutants such as *Antennepedia* have to the developmental biologist. They are bizarre and beautiful but, hopefully, they will also allow us to peek inside the black box and tell us something about the underlying mechanisms.

V.S.R. was supported by grants from the Air Force Office of Scientific Research (No. 89-0414) and the Office of Naval Research (ONR Grant No.N00014-91-J-1735). We thank Chandramani Ramachandran, Richard Gregory, John Allman, Francis Crick, Terry Sejnowski, and Harold Pashler for stimulating discussions. The authors' mailing address is Department of Psychology, 0109, University of California, San Diego, La Jolla, CA 92093-0109.

REFERENCES

Blakemore, C. (1973) 'The Baffled Brain', in R.L. Gregory and E.H. Gombrich, Eds. *Illusion in Nature and Art.* Charles Scribner & Sons; New York.

Brewster, D. (1832). *Lectures on Natural Magic*; John Murray, London.

Dennet, D. (1991) *Consciousness Explained.* Random House, New York.

Deutsch, J.A. and Ramachandran, V.S. (1990) Binocular depth reversals despite familiarity cues; an artifact? Science, **249**, 565-566.

Edelman, G.M. (1989) *The Remembered Present*, Basic Books Inc., New York.

Gilbert, C. & Wiesel, T.N. (1991). Short and long-term changes in receptive field size and position following following focal retinal lesions. *Society for Neuroscience Abstracts*, 1090.

Grossberg, S. and Mingolla, E. (1985) *Psychological Reviews*, **92**, 173-211.

Gregory, R. (1977) *Eye and Brain*, Wiedenfeld and Nicholson, London.

Julesz, B. (1971) *Foundations of Cyclopean Perception*, University of Chicago Press, Chicago.

Kleffner, D. and Ramachandran, V.S. (1992) On the perception of shape from shading. *Perception and Psychophysics*, in press.

Land, E.H. (1983) Recent advances in Retinex theory and some implications for cortical computations: color vision and the natural image. *Proceedings of the National Academy of Sciences, USA,* **80**, 5163-5169.

Livingstone, M., and Galaburda, A. (1991), in press, *Proceedings of the National Academy of Sciences.*

Livingstone, M. and Hubel, D.H. (1987). Psychophysical evidence for separate channels for the perception of form, colour, movement, and depth. *Journal of Neuroscience*, 7, 3416-3486.

Marr, D. (1982). *Vision.* Freeman, San Francisco.

Nakayama, K. and Loomis, J.M. (1974). Optical velocity patterns; celocity sensitive neurons and space perception: a hypothesis. *Perception*, 3, 63-80.

Nakayama, K. and Shimojo, S. (1992). Visual surface perception. *Science*, in press.

Plummer, D.J. and Ramachandran, V.S. (1992) The role of visual attention in perceiving apparent motion, in preparation.

Quigley, H.A., Addicks, E.M., and Green, W.R. (1982) Optic nerve damage in human glaucoma. *Archives of Opthamology*, **100**, 135-146.

Ramachandran, V.S. (1985) Apparent motion of subjective surfaces. *Perception*, **14**, 127-134.

Ramachandran, V.S. (1986) Illusory contours capture stereopsis and apparent motion. *Perception and Psychophysics*, 39, 361-373.

Ramachandran, V.S. (1988a). Perception of shape from shading. *Nature*, **331**, 133-136.

Ramachandran, V.S. (1988b). Perception of depth from shading. *Scientific American*, **269**, 76-83.

Ramachandran, V.S. (1989a). Talk given to NEI Symposium on Vision, Art, and the Brain, Bristol, U.K. Abstracts published in *Perception*, **19**, 273 (1990).

Ramachandran, V.S. (1989b). Visual perception in people and machines. Presedential special lecture given at the annual meeting of the Society for Neuroscience.

Ramachandran, V.S. (1991a). Visual perception in people and machines. In A. Blake and T. Troscianko, Eds. AI and the Eye. J. Wiley and Sons, Bristol.

Ramachandran, V.S. (1991b). 2D or not 2D; that is the question. In R.L. Gregory, P. Heard, and J. Harris, Eds. ' The Artful Brain', Oxford University Press, Oxford. (in press)

Ramachandran, V.S. (1992a). Filling in the blind spot, *Nature*, **355**,

Ramachandran, V.S. (1992b). Filling in gaps in perception. *Scientific American*, in press.

Ramachandran, V.S. and Anstis, S.M. (1983). Perceptual organisation in moving displays. *Nature*, **304**, 529-531.
Ramachandran, V.S. and Anstis, S.M. (1986). Perception of apparent motion. *Scientific American*, **254**, 102-109.
Ramachandran, V.S. and Anstis, S.M. (1987). Motion capture causes illusory displacements of Kinetic Edges. The European Conference on Visual Perception, Bristol, U.K.
Ramachandran, V.S., Cobb, S., & Rogers-Ramachandran, D. (1988). Recovering 3-D structure from motion: some new constraints. *Perception and Psychophysics*, **44**, 390-393.
Ramachandran, V.S. and Gregory, R.L. (1978). Does colour privide an input to human motion perception? *Nature (London)*, **275**, 55-56.
Ramachandran, V.S. and Gregory, R.L. (1991). Perceptual filling in of artificially induced scotomas in human vision. *Nature*, **350**, 699-702.
Ramachandran, V.S. and Gregory, R.L. (1992). 'Unmasking the Truth'- in preparation.
Ramachandran, V.S., Inada, V., and Kiama, G. (1985). Perception of illusory occlusion in apparent motion. *Vision research*.
Ramachandran, V.S., Rao, V.M., and Vidyasagar, T.R. (1973). Apparent motion with subjective contours. *Vision Research*, **13**, 1399-1401.
Reynolds, R. (1981). Perception of illusory contours as a function of processing time. *Perception*, **10**, 107-115.
Rogers-Ramachandran, D. and Ramachandran, V.S. (1991). Phantom contours; a new class of stimuli that selectively activate the magnocellular pathway in man. Association for Research in Vision and Opthamology abstracts.
Todd, J.T. and Mingolla, E. (1983). Perception of surface curvature and direction of illumination from patterns of shading. *Journal of Experimental Psychology: Human Performance*, **9(4)**, 583-595.
Treisman, A. (1986). Features and objects in visual processing. *Scientific American*, **255**, 114-126.
Ullman, S. (1979). *The Interpretation of Visual Motion*. MIT Press, Cambridge.
Von der Heydt, R., Peterhans, E. & Baumgartner, G. (1984) Illusory contours and cortical neural responses. *Science*, **224**, 1260-1261.
Yarbus, A.L. (1967). Eye movement and vision. (English translation by Haigh, B.; Ed. Riggs, L.A.). Plenum; New York.
Yonas, A., Kuskowski, M., & Sternfels, S. (1979). The role of frames of reference in the development of responsiveness to shading. *Child Development*, **50(2)**, 495-500.

THE VISUAL PERCEPTION OF 3-DIMENSIONAL FORM

J. Farley Norman and James T. Todd[1]
Brandeis University

1. Introduction

One of the most perplexing phenomena in the study of vision is the ability of observers to determine an object's 3-dimensional structure from patterns of light that project onto the retina. Indeed, were it not for the facts of our day-to-day experience, it would be tempting to conclude that the perception of 3-dimensional form is computationally impossible, since the properties of optical stimulation have so little in common with properties of real objects encountered in nature. Whereas real objects exist in 3-dimensional space and are composed of tangible substances such as earth, metal, or flesh, an optical image of an object is confined to a 2-dimensional projection surface and consists of nothing more than flickering patterns of light. Nevertheless, for many animals including humans, these seemingly uninterpretable patterns of light are the primary source of sensory information about the layout of objects and surfaces in the surrounding environment.

Previous research has identified several different properties of optical structure from which an object's 3-dimensional form can be perceptually specified. Some of these properties -- the so-called pictorial depth cues -- are available within individual static images. Consider, for example, the patterns of image contours presented in Figure 1. The upper left panel of this figure shows a pattern of converging line segments, which is perceived as a ground plane receding in depth; the upper right panel shows a pattern of connectivity among parallel contours, which is perceived as two solid rectangular objects, one resting on top of the other; and the lower panel shows a pattern of curved contours, which is perceived as a smooth surface. Other perceptually informative properties of individual images that have been studied extensively include gradients of shading or texture (e.g., see Todd and Mingolla, 1983; Mingolla and Todd, 1986; Todd and Akerstrom, 1987).

Additional information about an object's 3-dimensional form can also be provided by the systematic transformations among a sequence of multiple images. For example, when an observer moves within a cluttered environment the texture elements on visible surfaces move projectively at different velocities (i.e., motion parallax) due to their different depths relative to the observer. Binocular

---

[1] The research described in this manuscript was supported by grants #89-0016 from the Air Force Office of Scientific Research and #BNS-8908426 from the National Science Foundation, the Office of Naval Research and the Air Force Office of Scientific Research to James Todd.

parallax occurs whenever an observer views an object with both eyes simultaneously. Because each eye views the world from a slightly different vantage point, texture elements at different depths project to different relative positions (binocular disparity) within each retinal image. Other optical transformations that can provide information about 3-dimensional form include accretions or deletions of texture when one surface occludes another, or the deformations of shading that occur when an object moves relative to the direction of illumination.

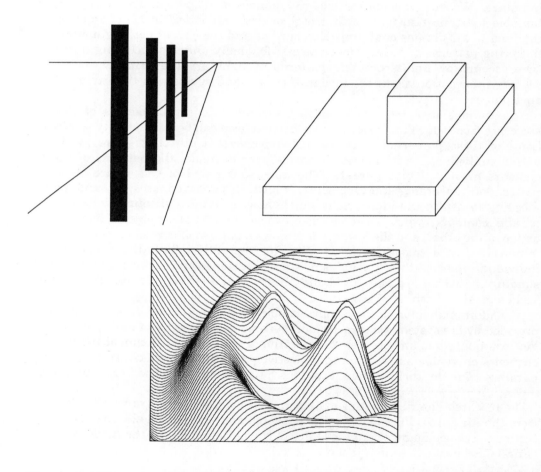

Figure 1. Three examples of how static configurations of image contours can provide perceptually compelling information about an object's 3-dimensional structure.

All of these different properties of optical structure have been studied extensively within controlled laboratory experiments, and the results show clearly that they can all produce compelling perceptions of 3-dimensional form. What is not clear, however, is how these perceptions relate to the actual structure of the physical environment. The mere fact that observers are able to perceive 3-dimensional form does not reveal the specific parameters by which visible objects are perceptually represented. Thus, in an effort to shed new light on this issue, the present article will examine some alternative geometric frameworks for representing shape, and it will review the available psychophysical evidence to see which of these frameworks are most similar to the properties of human perception.

2. What is shape ?

As it has evolved from ancient times, the concept of space in modern science has become increasingly abstract and farther removed from our intuitive beliefs derived from experience. In speculating about the properties of physical space, for example, some theorists have argued that the fabric of space can be deformed by the presence of large masses, and that these deformations are responsible for the phenomenon of gravity. Imagine taking a glass marble and rolling it across a smooth surface. When the rolling marble encounters a hill or a valley, its movements will be constrained by the overall shape of the surrounding terrain. In an analogous fashion, the large mass of the sun deforms its surrounding space, which constrains the earth to orbit in an elliptical trajectory.

The nature of visual space is at present not adequately known. We perceive a compelling 3-dimensional environment filled with many differently-shaped objects, despite the obvious fact that the spatio-temporal optical patterns that give rise to this perception are only 2-dimensional. A more complete understanding of the nature of visual space is critical for the study of solid shape, because the structure of this space determines the measurable properties of objects embedded within it. Let us now consider some of the possible alternatives.

Throughout the literature on human perception, classical euclidean geometry is by far the most common framework for describing the structure of the environment. The defining characteristic of a euclidean space is that it possesses a distance metric, such that the absolute distance between two points $\mathbf{a}$ and $\mathbf{b}$ having Cartesian coordinates $(\mathbf{a}_1, \mathbf{a}_2, \mathbf{a}_3)$ and $(\mathbf{b}_1, \mathbf{b}_2, \mathbf{b}_3)$ is $\sqrt{(\mathbf{a}_1 - \mathbf{b}_1)^2 + (\mathbf{a}_2 - \mathbf{b}_2)^2 + (\mathbf{a}_3 - \mathbf{b}_3)^2}$. Since a euclidean space is isotropic, we can measure distances between any pair of arbitrary points. Thus, if human observers are able to perceive the euclidean metric structure of the environment, then they ought to be able to accurately compare distances (or angles) of line segments oriented in different directions according to the pythagorean theorem.

It is important to recognize that the euclidean distance metric is not universally applicable in all contexts. Consider, for example, the structure of a simple triangle. One of the fundamental theorems of euclidean geometry is that the sum of the three angles of a triangle must be exactly 180 degrees. This will always be true if the triangle is drawn on a flat surface. However, if a triangle is drawn on a sphere, then the sum of its three angles will always be greater than 180 degrees, and if a triangle is drawn on a saddle then the sum of its angles will always be less than 180 degrees. Beginning in the nineteenth century, results such as these led Gauss and others to conclude that the constraints imposed by euclidean geometry might not always be appropriate for describing the structure of the environment. In order to deal with curved surfaces, they discovered, it is necessary to adopt alternative noneuclidean distance metrics for which the Pythagorean theorem is invalid.

It is also possible to define even more abstract spaces. For example, in affine geometry the distance metric is allowed to vary in different directions -- i.e., it is anisotropic. Because distance intervals can only be compared when they are in the same direction, one cannot determine either distances or angles between arbitrary points located within an affine space. Surprisingly, however, this limitation does not seriously handicap the study of shape. As described by Snapper and Troyer (1971) : "affine geometry is what remains after practically all ability to measure length, area, angles, etc., has been removed from Euclidean geometry. One might think that affine geometry is a poverty-stricken subject. On the contrary, affine geometry is quite rich".

Other types of representation are possible which have no distance metric at all. For example, many of the important properties of a smoothly curved surface can be adequately described by the depth order relations among neighboring points, without including any information about depth magnitudes. Objects can also be described by their patterns of connectivity, or as a collection of categorically distinct parts.

There are many possible geometries for describing an object's 3-dimensional structure. As was first noticed by the German mathematician Felix Klein in 1872, these geometries can be organized in a hierarchy, based on the different transformations they allow, and the structural properties that remain invariant under those transformations. Euclidean geometry allows arbitrary translations and rotations, which preserve the distance between any pair of points on an object. Affine geometry allows arbitrary stretching transformations, which do not leave distance invariant, but do preserve a wide variety of other properties, such as the sign of Gaussian curvature of a surface or the parallelism of a pair of line segments. Similarly, with more general projective transformations, a conic section will remain a conic section and a collinear set of points will remain collinear. Even when an object is subjected to arbitrary smooth deformations, some of its properties such as the pattern of connectivity among neighboring points will still remain invariant.

3. The geometry of human vision

Which of these geometries is most relevant to the visual perception of

3-dimensional form? In the discussion that follows we shall consider several possible aspects of an object's structure, and we shall review the available psychophysical evidence about their relative perceptual salience.

A. Euclidean Structure

In order to obtain an accurate euclidean representation of the environment, it would first be necessary to somehow determine the relative depths of every visible point from the available patterns of optical stimulation (e.g., see Marr and Poggio, 1979; Marr, 1982). As we shall see, some types of optical information are theoretically sufficient to obtain an accurate depth map while others are not.

Consider, for example, the phenomenon of stereopsis. It is well known that human observers can perceive vivid and compelling 3-dimensional shapes defined by variations in binocular disparity. It is important to keep in mind, however, that it is not possible, even in principle, to uniquely determine euclidean structure from binocular optical patterns based solely on variations in horizontal disparity. In order to recover the 3-dimensional distance between a pair of points on the basis of their disparity, one must have accurate information about the distance to the observer's fixation point. This information about viewing distance could come from some optical source other than binocular horizontal disparity or from some non-optical, extra-retinal source like the state of convergence of the two eyes.

Figure 2 illustrates Panum's limiting case, the simplest possible stereoscopic situation. An observer fixates F and views two points separated in depth. In the left eye, the two points project to the same retinal location; while in the right eye, they project to different retinal locations (disparity) as a result of the points' differing depths. Since the binocular disparity δ is an angle, the actual depth difference D cannot be determined on the basis of the optical patterns alone. In order to recover D, one must know the viewing distance from the observer to F. The key point is that there is no one-to-one relationship between horizontal retinal disparity and depth. Any given disparity δ could result from any physical depth difference depending on the viewing distance. In our example, both viewing situations have the same disparity δ, yet the physical depth interval corresponding to that disparity is much larger for the viewing situation illustrated in the left half of the figure due to the larger viewing distance.

Longuet-Higgins (1982) and Mayhew and Longuet-Higgins (1982) have shown that the *vertical* disparity that exists for all environmental points lying off the horizontal meridian could be used to obtain the necessary information about viewing distance. However, Fox, Cormack, and Norman (1987) manipulated the magnitude of vertical disparity within line-element stereograms and found that variations in vertical disparity had no effect on the perceived depth intervals resulting from a given horizontal disparity.

The convergence angle of the two eyes could also be used as a possible source of information about viewing distance. Although Foley (1980) has shown that manipulations of convergence can have measurable effects on stereoscopically defined figures, it is important to keep in mind that the eyes are essentially parallel for viewing distances over two meters while stereopsis occurs over much larger viewing distances involving hundreds of meters (Cormack, 1984). Given that a pattern of binocular disparities covaries with both an object's physical form and its position relative to an observer, stereopsis appears to be ill-suited for delivering information about euclidean relations in 3-dimensional space. Indeed, the invariance of perceived binocular shape under changes in viewing position and distance led Julesz (1971, p. 290) to conclude that : "for stereopsis one must generalize the metric of space from a rigid Euclidean one to a less rigid affine or topological space" (see also Luneberg, 1947, 1950).

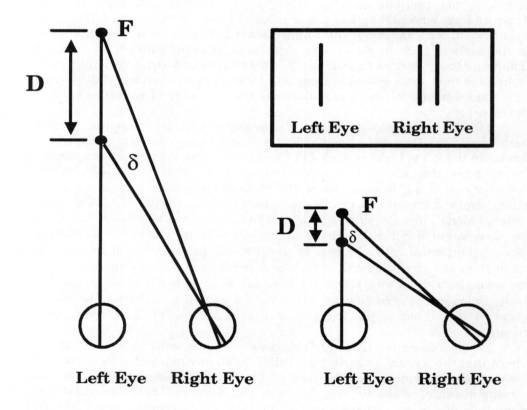

Figure 2. A schematic illustration of Panum's limiting case used to show that any given binocular disparity δ can be produced by any physical depth interval depending on the observer's viewing distance to the fixation point.

Although perception of 3-dimensional form from stereopsis may be inherently limited, perhaps there are other sources of information that are potentially more reliable. Of all the different aspects of optical stimulation that are known to influence observers' perceptions of 3-dimensional form, motion is the one that is most likely to provide perceptually useful information about euclidean metric structure. During the past decade, there have been numerous theoretical analyses of how it is possible to compute an object's structure from motion, provided that certain minimal conditions are satisfied. Most of these analyses are designed to be used with a discrete sequence of orthographic projections of an arbitrary configuration of points rotating in depth about an arbitrary axis. Within this context, it can be proven mathematically that there will always be a unique rigid interpretation for any apparent motion sequence that contains at least three views of four or more noncoplanar points (see Bennett, Hoffman, Nicola, and Prakash, 1989; Huang and Lee, 1989; Ullman, 1979). These conditions are both necessary and sufficient. For arbitrary configurations that contain fewer than three views or fewer than four points, the 3-dimensional structure will be mathematically ambiguous with an infinity of possible rigid interpretations.

During the past several years, however, there has been a growing amount of evidence that these theoretical limits may have surprisingly little relevance to actual human vision. Of particular importance in this regard are the recent findings from several different laboratories that 2-frame apparent motion sequences presented in alternation provide sufficient information to obtain compelling kinetic depth effects and to accurately discriminate between different 3-dimensional structures (Braunstein, Hoffman, and Pollick, 1990; Braunstein, Hoffman, Shapiro, Andersen, and Bennett, 1987; Doner, Lappin, and Perfetto, 1984; Lappin, Doner, and Kottas, 1980; Todd, Akerstrom, Reichel and Hayes, 1988; Todd and Bressan, 1990). Similar results can also be obtained using longer length sequences of scintillating random dot surfaces for which no dot is allowed to survive for more than two successive frames (Dosher, Landy, and Sperling, 1990; Norman, 1990; Todd, 1985).

Since euclidean depths and orientations cannot in principle be determined from 2-frame apparent motion sequences under orthographic projection, and human observers can perform accurately on tasks where the motion sequences are limited to two views, it seems reasonable to conclude that observers performance on these tasks cannot be based on a computational analysis of euclidean structure from motion. It would appear from this finding that observers are able to make use of other, more abstract forms of perceptual representation when presented with minimal amounts of information, but are they capable of perceiving euclidean structure when sufficient information is available to support such an analysis? It is important to keep in mind when considering this issue that the defining characteristic of euclidean structure that distinguishes it from other possible geometries is the existence of an isotropic distance metric. Thus, if there are any conditions in which observers can accurately discriminate lengths and angles of line segments oriented in different directions, then, by definition, their knowledge of 3-dimensional structure in those conditions must be euclidean.

Surprisingly, even though this is the defining characteristic of euclidean geometry, there have been relatively few experiments in which observers were required to make explicit judgements about isotropic metric structure. One such experiment has recently been performed by Todd and Bressan (1990). Observers in their study were asked to discriminate the relative 3-dimensional lengths or angles between moving line segments, whose relative orientations were carefully controlled so that above chance performance could not be achieved based solely on the projected lengths or projected angles depicted in each display. Performance on these tasks was extremely poor relative to other types of sensory discrimination. Weber fractions for the length and angle judgements were 25 and 50 percent, respectively. Moreover, although the overall level of performance was above chance, there were no significant improvements as the number of distinct frames in an apparent motion sequence was increased from two to eight. Thus, whatever information was used for performing these tasks, it was fully available within 2-frame displays for which an accurate analysis of euclidean metric structure was computationally impossible.

B. Affine Structure

If not euclidean structure, then what can observers perceive from the minimal amounts of information provided by 2-frame apparent motion sequences or stereograms under orthographic projection? Recent analyses by Koenderink and van Doorn (1991) and Todd and Bressan (1990) have shown that this information is mathematically sufficient to determine an object's structure up to an affine stretching transformation along the line of sight (see also Bennett et al, 1989). Although an object's euclidean structure cannot be uniquely specified from such minimal amounts of information, it is nonetheless severely constrained.

There are a wide variety of object properties that can be reliably detected based solely on an analysis of affine structure. For example, it is possible with this analysis to determine the metric length ratio between any pair of parallel line segments; to perform various nominal categorizations, such as distinguishing between planar and nonplanar configurations; and to accurately discriminate structural differences between any pair of objects that cannot be made congruent by an affine stretching transformation along the line of sight. It is also interesting to note in this regard that an analysis of affine structure from 2-frame displays is sufficiently powerful to perform most of the existing psychophysical tasks that have been employed previously to study observers' perceptions of structure from motion or stereopsis, including judgements of rigidity, or coherence, discriminations of rigid from nonrigid motion, judgements of ordinal depth relations, and the discrimination or identification of complex 3-dimensional forms.

In a recent series of experiments, Todd and Bressan (1990) and Todd and Norman (1991) have examined the accuracy of observer's judgements for several different aspects of a moving object's 3-dimensional form. The results reveal that performance is quite poor for tasks that require an analysis of euclidean metric structure, but that observers' judgements can be extremely accurate for tasks that are mathematically possible based on an analysis of affine structure. In addition, there is little or no improvement on any of these tasks as the number of distinct frames in an apparent motion sequence is increased beyond two.

It is important to keep in mind that a 2-frame apparent motion sequence or stereogram under orthographic projection can only provide sufficient information to determine an object's structure up to an affine stretching transformation along the line of sight. Thus, this latter finding suggests a surprising prediction: Consider an extended apparent motion sequence of an object rotating in depth that is stretched or compressed along the line of sight at each frame transition. The cumulative effects of these stretching transformations would be carried along by the rotation, resulting in potentially large deformations of the object's structure. However, if the human visual system is restricted to an analysis of first order displacements between 2-frame sequences, as we have suggested above, then this particular type of deformation should be perceptually undetectable, since every successive 2-frame sequence would have a possible rigid interpretation.

This prediction has been confirmed empirically in a recent series of experiments by Norman and Todd (1991). When a rotating object is stretched or compressed along the line of sight, it appears indistinguishable from a perfectly rigid object whose rate of rotation is accelerated or decelerated. Norman and Todd have also performed computer simulations to demonstrate that these different transformations can be distinguished by analyses of euclidean structure from motion that are able to integrate information over three or more views. A typical pattern of results from one such algorithm by Hoffman and Bennett (1986) is shown in Figure 3. From the projected positions of a set of points rotating in depth about a fixed axis in the image plane, this algorithm computes the radius of each point relative to the axis of rotation. The upper curve in Figure 3 shows the computed radius over a 100 frame sequence for a single point within a rigid configuration whose angular velocity varies sinusoidally over time. The computed radius in this case remains perfectly constant, indicating that the object's motion is rigid. The lower curve in this figure shows the output from the same algorithm for a nonrigid configuration that rotates at a constant velocity, but is sinusoidally stretched along the line of sight as it rotates. Note in this case that the computed radius varies over time indicating the object's motion is nonrigid. It is clear that these different transformations could easily be distinguished using existing algorithms for computing euclidean structure from motion. Thus, the fact that they are perceptually identical provides especially strong evidence that human observers may be restricted to a more abstract analysis of affine structure.

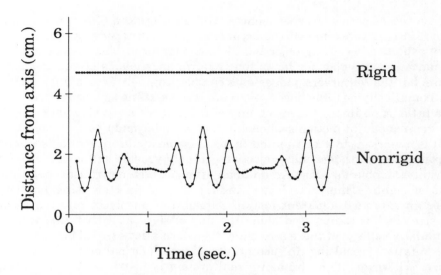

Figure 3. The distance of two points from their axes of rotation as a function of time. Each trajectory was computed from the optical projection of a moving configuration using an algorithm developed by Hoffman and Bennett (1986). The point represented by the upper curve was part of a rigid configuration, whose angular velocity varied sinusoidally over time. The point represented by the lower curve, in contrast, was part of a nonrigid configuration that was stretched sinusoidally along the line of sight as it rotated. Although the differences between these two trajectories are easily detectable using an analysis of euclidean structure from motion, they cannot be detected by actual human observers.

C. Ordinal Structure

Whereas an affine representation of 3-dimensional form retains some rudimentary information about ratiometric distances in any given direction, it is also possible to describe many of the essential properties of an object's structure without any distance metric at all. Gibson (1950) argued that much of our perceptual awareness of the environment is based on simple order relations that can be described in terms of "greater than" or "less than". More recently, Todd and Reichel (1989) have suggested that an observer's knowledge of smoothly curved surfaces can often involve a form of ordinal representation, in which neighboring surface regions are labeled in terms of which region is closer to the point of observation without specifying how much closer.

It is important to recognize that the order relations in this proposed representation are only defined for adjacent regions within an arbitrarily small neighborhood. This has some important consequences. Suppose, for example, that we wish to determine the ordinal depth relation between two visible surface regions R_1 and R_n that are not locally adjacent to one another. Using an ordinal representation, the relative depth of these regions can only be determined if there is a continuous chain of intervening regions that are ordinally transitive (i.e., if $R_1 < R_2 < R_3 ... < R_n$, then $R_1 < R_n$). If this restriction is violated (i.e., $R_1 < R_2 < R_3 ... > R_n$), then the relative depths of R_1 and R_n cannot be determined from an ordinal representation without providing additional information. Todd and Reichel (1989) have demonstrated psychophysically that ordinal transitivity is important for the perception of smoothly curved surfaces. That is to say, when observers are required to judge the relative depths of ordinally transitive surface regions, their responses are more accurate and they have faster reaction times, than when similar judgements are performed for ordinally intransitive surface regions.

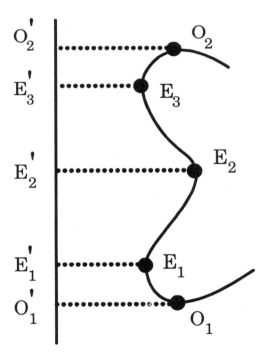

Figure 4. The ordinal structure of any surface is completely determined by the optical projections of its occlusion points and depth extrema. In this particular example there are two occlusion points O_1 and O_2, and three depth extrema E_1, E_2, and E_3. Their corresponding optical projections in the image plane are labeled O_1', O_2', E_1', E_2', and E_3', respectively.

To better appreciate the precise nature of ordinal representations it is useful to consider a planar cross-section of a smoothly curved surface as shown in Figure 4. Note in the figure that there are two occlusion points O_1 and O_2. It can be demonstrated theoretically (see Koenderink and van Doorn, 1976, 1982) that as we move from an occlusion point in an attached region, the ordinal depth of a surface relative to the image plane must decrease monotonically until a depth minimum is reached. Thus, occlusion contours provide potential information about the ordinal structure of attached surface regions in their immediate local neighborhood, and there is considerable evidence to suggest that human observers rely heavily on this information for the visual perception of 3-dimensional form (see Reichel and Todd, 1990; Todd and Reichel, 1989). A complete ordinal representation cannot be achieved, however, without also identifying the depth extrema (i.e., the depth maxima and minima), which define the boundaries of ordinal transitivity where a monotonic depth change switches from positive to negative or vice versa. The optical projections of these depth extrema together with those of the occlusion points are both necessary and sufficient for visually specifying the complete ordinal structure of any surface cross-section.

D. Topological Structure

Some aspects of an object's structure can be adequately characterized using even more abstract representations, in which the concept of distance is abandoned altogether. Consider the topological structure formed by the pattern of connectivity among the vertices of a polyhedron or the neighborhood relations among identifiable points on a continuous surface. If we allow objects to be smoothly deformed without tearing by arbitrary combinations of bending and stretching transformations, these connectivity and neighborhood relations will remain invariant. Solid objects can be distinguished topologically by the number of holes they contain. Within this framework, a doughnut and a coffee mug are topologically equivalent because they have the same number of holes and can therefore be deformed into one another without tearing. A doughnut and a potato, on the other hand, are topologically different because they do not have the same number of holes.

Is topological structure of any relevance to human perception? Consider the sequence of projected silhouettes of an object rotating in 3-dimensional space depicted in Figure 5. Note in the first frame of this sequence that the object's silhouette is bounded by a single connected contour. Between frames two and three, however, the pattern undergoes a qualitative change in which the emergence of a hole causes the silhouette to be bounded by two different contours that are not connected to one another. Once this hole is revealed, it provides potentially useful information about the depicted object's 3-dimensional structure -- i.e., if an object's silhouette contains a hole, then the object itself must also contain a hole.

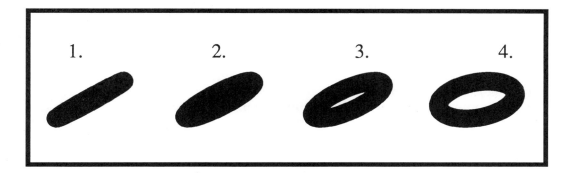

Figure 5. Four successive views of the projected silhouette of a rotating torus. Note the catastrophe between views two and three, which indicate the presence of a hole.

There have been several recent psychophysical experiments designed to demonstrate the importance of topological structure in human vision. For example, Chen (1982) has shown that topologically equivalent forms, such as a filled circle and a filled triangle, are more difficult to discriminate at brief exposure durations than are topologically different forms, such as a filled circle and a circular annulus, whose bounding contours have identical shapes. Topologically equivalent forms also produce stronger perceptions of apparent motion when presented sequentially over time (Chen, 1985; Prazdny, 1986). These findings provide strong evidence that human observers are indeed sensitive to topological relations at a relatively early stage of visual processing.

E. Nominal or Categorical Structure

Another way of representing 3-dimensional form that does not require the concept of distance is to decompose an object into a relatively small set of categorically distinct parts. There have been several variations of this approach described in the literature. For example, one possible strategy proposed by Koenderink and van Doorn (1976, 1980, 1982) is to decompose a surface into bounded regions of positive (elliptic), negative (hyperbolic), or zero (parabolic) Gaussian curvature (see also Koenderink, 1984; Richards, Koenderink, and Hoffman, 1987). These authors have demonstrated mathematically that the local Gaussian curvature of a surface can be optically specified by certain types of image features such as smooth occlusion contours and singular points within the field of image intensities. Objects can also be divided into parts at loci of negative minima along lines of principle curvature (Beusmans, Hoffman, and Bennett,

1987; Hoffman and Richards, 1984) or by decomposing their boundaries into the largest convex surface patches (Vaina and Zlateva, 1990).

With respect to the study of actual human perception, there is a growing amount of evidence that certain types of object recognition may be primarily dependent on part-based representations. Biederman (1987) has argued that most objects in the environment can be adequately represented to achieve recognition using a limited set of volumetric primitives, called geons, which are connected to one another in simple combinations, in much the same way that words can be composed from a relatively small alphabet of phonemes. The optical information from which these geons are perceptually specified is assumed to be based on easily measurable properties of image contours, such as the presence or absence of curvature, parallelism or symmetry, and the cotermination of contours at vertices. Biederman and his colleagues have conducted numerous empirical studies of object recognition which have generally supported the psychophysical validity of this analysis. The results have revealed that only two or three geons are usually sufficient to allow rapid and accurate object recognition, and that performance remains surprisingly unimpaired even when a test image is systematically transformed by changing its scale, deleting a large portion of its contours, or by rotating the depicted object in depth.

4. Discussion

Based on the available evidence, we believe it is the case that there is no one type of visual representation that can adequately account for observers' perceptions of 3-dimensional form. Consider, for example, the apparent usefulness of part-based representations for object recognition. The primary selling point for a nominal description of 3-dimensional structure is its stability over changes in viewing position. When an object moves relative to the observer -- even when the motion is not perfectly rigid as in a human gait -- the decomposition of its structure into parts remains largely invariant. The primary disadvantage of a nominal representation is its lack of precision. Part-based descriptions are invariant over change, because they typically do not encode the size, shape, or orientation of each part. This makes them incapable, however, of describing any variations along these dimensions, unless they are supplemented with some additional form of metric representation.

In some respects, the perceptual performance of human observers in judging various aspects of an object's 3-dimensional form seems to have much in common with the hierarchy of geometries proposed by Felix Klein. In a speech at Erlangen University in 1872, Klein argued that different geometries can be stratified by the properties of objects that are preserved by different types of geometric transformations. There is a growing amount of evidence to suggest that human vision involves a similar type of stratification in which the most perceptually salient aspects of an object's structure are those that remain invariant over the largest number of possible transformations.

Such findings indicate that visual knowledge of 3-dimensional form may exist at multiple levels of description, and that the specific type of representation required for any given task is dependent on the particular judgement an observer is asked to perform. The most difficult tasks are those that require a knowledge of metric structure. Ordinal judgements, in contrast, are performed significantly faster and significantly more accurately (Todd and Reichel, 1989), and those measures of performance can be improved still further for tasks such as object recognition involving topological or nominal judgements (e.g., see Biederman, 1987).

Ironically, although many theorists assume that the primary goal of human perception is to achieve an accurate euclidean representation of visible objects in the surrounding environment, there are good reasons to question whether human observers are able to perceive euclidean metric structure under any circumstances. For example, recent research by Todd and Bressan (1990) and Todd and Norman (1991) has shown that most tasks employed to measure observers perceptions of 3-dimensional form can be performed reliably based solely on an analysis of affine structure, and that tasks which specifically require a knowledge of euclidean structure typically result in dramatically impaired performance. All of this suggests that the geometry of perceived 3-dimensional form may be much more abstract than is generally taken for granted, and that our common intuitions about the importance of metric structure may have surprisingly little relevance to the processes of human perception.

REFERENCES

Bennett, B. M., Hoffman, D. D., Nicola, J. E., and Prakash, C. (1989). Structure from two orthographic views of rigid motion. *Journal of the Optical Society of America A*, **6**, 1052-1069.

Beusmans, J. M. H., Hoffman, D. D., and Bennett, B. M. (1987). Description of solid shape and its inference from occluding contours. *Journal of the Optical Society of America A*, **4**, 1155-1167.

Biederman, I. (1987). Recognition-by-Components: A theory of human image understanding. *Psychological Review*, **94**, 115-147.

Braunstein, M. L., Hoffman, D. D., Shapiro, L. R., Andersen, G. J., and Bennett, B. M. (1987). Minimum points and views for the recovery of three-dimensional structure. *Journal of Experimental Psychology: Human Perception and Performance*, **13**, 335-343.

Braunstein, M. L., Hoffman, D. D. and Pollick, F. E. (1990). Discriminating rigid from nonrigid motion. *Perception and Psychophysics*, **47**, 205-214.

Chen, L. (1982). Topological structure in visual perception. *Science*, **218**, 699-700.

Chen, L. (1985). Topological structure in the perception of apparent motion. *Perception*, **14**, 197-208.

Cormack, R. H. (1984). Stereoscopic depth perception at far viewing distances. *Perception and Psychophysics*, **35**, 423-428.

Doner, J., Lappin, J. S., and Perfetto, G. (1984). Detection of three- dimensional structure in moving optical patterns. *Journal of Experimental Psychology: Human Perception and Performance*, **10**, 1-11.

Dosher, B. A., Landy, M. S. and Sperling, G. (1990). Kinetic depth effect and optic flow - I. 3-D shape from fourier motion. *Vision Research*, **29**, 1789-1814.

Foley, J. M. (1980). Binocular distance perception. *Psychological Review*, **87**, 411-434.

Fox, R., Cormack, L. and Norman, F. (1987). The effect of vertical disparity on depth scaling. *Investigative Ophthalmology and Visual Science (Suppl.)*, **28**, 293.

Gibson, J. J. (1950). **The perception of the visual world**. Boston, MA : Houghton Mifflin.

Hoffman, D. D., and Richards, W. A. (1984). Parts of recognition. *Cognition*, **18**, 65-96.

Hoffman, D. and Bennett, B. (1986). The computation of structure from fixed axis motion: Rigid structures. *Biological Cybernetics*, **54**, 1-13.

Huang, T. S. and Lee, C. H. (1989). Motion and structure from orthographic projections. *IEEE Transactions on Pattern Analysis and Machine Intelligence*, **11**, 536-540.

Julesz, B. (1971). **Foundations of cyclopean perception.** Chicago, IL : University of Chicago Press.

Koenderink, J. J. (1984). What does the occluding contour tell us about solid shape? *Perception*, **13**, 321-330.

Koenderink, J. J., and van Doorn, A. J. (1976). The singularities of the visual mapping. *Biological Cybernetics*, **24**, 51-59.

Koenderink, J. J., and van Doorn, A. J. (1980). Photometric invariants related to solid shape. *Optica Acta*, **27**, 981-996.

Koenderink, J. J., and van Doorn, A. J. (1982). The shape of smooth objects and the way contours end. *Perception*, **11**, 129-137.

Koenderink, J. J. and van Doorn, A. J. (1991). Affine structure from motion. *Journal of the Optical Society of America A*, **8**, 377-385.

Lappin, J. S., Doner, J. F., and Kottas, B. L. (1980). Minimal conditions for the visual detection of structure and motion in three dimensions. *Science*, **209**, 717-719.

Longuet-Higgins, H. C. (1982). The role of the vertical dimension in stereoscopic vision. *Perception*, **11**, 377-386.

Luneberg, R. K. (1947). **Mathematical analysis of binocular vision.** Princeton, New Jersey : Princeton University Press.

Luneberg, R. K. (1950). The metric of binocular visual space.*Journal of the Optical Society of America*, **40**, 627-642.

Marr, D. and Poggio, T. (1979). A computational theory of human stereo vision. *Proceedings of the Royal Society of London B*, **204**, 301-328.

Marr, D. (1982). **Vision**. San Francisco: W. H. Freeman and Co.

Mayhew, J. E. W. and Longuet-Higgins, H. C. (1982). A computational model of binocular depth perception. *Nature*, **297**, 376-378.

Mingolla, E., and Todd, J. T. (1986). Perception of solid shape from shading. *Biological Cybernetics*, **53**, 137-151.

Norman, J. F. (1990). The perception of curved surfaces defined by optical motion. Ph.D. Dissertation, Vanderbilt University.

Norman, J. F., and Todd, J. T. (1991) The perception of rigid motion for affine distortions in depth. *Investigative Ophthalmology and Visual Science (Suppl.)*, **32**, 958.

Prazdny, K. (1986). What variables control (long-range) apparent motion? *Perception*, **15**, 37-40.

Reichel, F. D., and Todd, J. T. (1990). Perceived depth inversion of smoothly curved surfaces due to image orientation. *Journal of Experimental Psychology: Human Perception and Performance*, **16**, 653-664.

Richards, W. A., Koenderink, J. J., and Hoffman, D. D. (1987). Inferring three-dimensional shapes from two-dimensional silhouettes. *Journal of the Optical Society of America A*, **4**, 1168-1175.

Snapper, E. and Troyer, R. (1971). **Metric affine geometry**. New York : Academic Press.

Todd, J. T. (1985). The perception of structure from motion: Is projective correspondence of moving elements a necessary condition? *Journal of Experimental Psychology: Human Perception and Performance*, **11**, 689-710.

Todd, J. T., and Mingolla, E. (1983). Perception of surface curvature and direction of illumination from patterns of shading. *Journal of Experimental Psychology: Human Perception and Performance*, **4**, 583-595.

Todd, J. T., and Akerstrom, R. A. (1987). Perception of three-dimensional form from patterns of optical texture. *Journal of Experimental Psychology: Human Perception and Performance*, **13**, 242-255.

Todd, J. T., Akerstrom, R. A., Reichel, F. D. and Hayes, W. (1988). Apparent rotation in three-dimensional space: Effects of temporal, spatial, and structural factors. *Perception and Psychophysics*, **43**, 179-188.

Todd, J. T. and Reichel, F. D. (1989). Ordinal structure in the visual perception and cognition of smoothly curved surfaces. *Psychological Review*, **96**, 643-657.

Todd, J. T. and Bressan, P. (1990). The perception of 3-dimensional affine structure from minimal apparent motion sequences. *Perception and Psychophysics*, **48**, 419-430.

Todd, J. T. and Norman, J. F. (1991). The visual perception of smoothly curved surfaces from minimal apparent motion sequences. *Perception and Psychophysics*, **50**, 509-523.

Ullman, S. (1979). The interpretation of structure from motion. *Proceedings of the Royal Society of London B*, **203**, 405-426.

Vaina, L. M., and Zlateva, S. D. (1990). The largest convex patches: A boundary-based method for obtaining object parts. *Biological Cybernetics*, **62**, 225-236.

A New Approach to Shape from Shading

Pierre Breton[1], Lee A. Iverson[1], Michael S. Langer[1], Steven W. Zucker[1,2]

[1] McGill University, Research Center for Intelligent Machine, 3480 rue Université, Montréal, Québec, Canada, H3A 2A7, e-mail: zucker@mcrcim.mcgill.edu
[2] Fellow, Canadian Institute for Advanced Research

Abstract. The classical approach to shape from shading problems is to find a numerical solution of the image irradiance partial differential equation. It is always assumed that the parameters of this equation (the light source direction and surface albedo) can be estimated in advance. For images which contain shadows and occluding contours, this decoupling of problems is artificial. We develop a new approach to solving these equations. It is based on modern differential geometry, and solves for light source, surface shape, and material changes concurrently. Local scene elements (scenels) are estimated from the shading flow field, and smoothness, material, and light source compatibility conditions resolve them into consistent scene descriptions. Shadows and related difficulties for the classical approach are discussed.

1 Introduction

The shape from shading problem is classical in vision; E. Mach (1866) was perhaps the first to formulate a formal relationship between image (Ratliff, 1965) and scene domains, and to capture their inter-relationships in a partial differential equation. Horn set the modern approach by focusing on the solution of such equations by classical and numerical techniques (Horn, 1975; Ikeuchi and Horn, 1981; Horn and Brooks, 1986), and others have built upon it (Nayar et al., 1991; Pentland, 1990). Nevertheless, problems remain which are not naturally treated in the classical sense, especially those related to discontinuities and shadows. We present a new approach to the shape from shading problem motivated by modern notions of fibre bundles in differential geometry (Spivak, 1979). The global shape from shading problem is posed as a coupled collection of "local" problems, each of which attempts to find that local scene element (or scenel[3]) that captures the local photometry, and which are then coupled together to form global piecewise smooth solutions.

The paper is written in a discursive style to convey a sense of the "picture" behind our approach rather than the formal treatment. We begin with an overview of the classical formulation, then proceed to define the structure of a "scenel" and our new conceptualization.

1.1 The "Classical" Shape from Shading Problem

We take the classical setting in computer vision for shape from shading to be the following: a point light source at infinity uniformly illuminates a smooth matte surface of constant albedo whose image is formed by orthographic projection.

The matte surface is traditionally modeled with Lambert's reflectance function so the image irradiance equation is

$$I(x,y) = \rho\lambda \mathbf{L} \cdot \mathbf{N}(x,y)$$

where $I(x,y)$ is the intensity of an image point (x,y); ρ, the albedo of the surface, i.e. the fraction of the shining light which is reflected; λ, the illumination, i.e. the amount of shining light; $\mathbf{L}$, the light source direction; $\mathbf{N}(x,y)$, the normal at the surface point corresponding to an image point (x,y).

The literature cited above describes the various attempts to solve this (or a closely related) problem[4] from first principles. We emphasize, however, that, to make these approaches tractable, certain parameters are assumed known (e.g. typically ρ, λ and $\mathbf{L}$). Operationally this decouples problems; e.g., it decouples the shape from shading problem from light source estimation problems (Pentland, 1982).

1.2 Piecewise Smooth Shape from Shading Problems

We submit that such decoupling, while appropriate for certain highly engineered situations, is not always necessary; moreover, it can make shading analysis impotent precisely when it should be useful. For example, a human observer confronted with a static, monocular view of a scene will succeed in obtaining some estimate of the shapes of the surfaces within it even when some of the classical setting's constraints are relaxed. The presence of a shadow, a diffuse light source, or even a patterned surface does not necessarily interfere with our ability to recover shape from shading. Thus the

---
[3] *cf.* Pixel, voxel, ... scenel.
[4] Other reflectance functions have been studied, e.g. for glossy surfaces or the moon.

classical constraints can be relaxed in principle; but how far, and once relaxed by what mechanism can solutions be found? These are precisely the questions with which we shall be concerned.

We retain the basic assumption that smooth variation in intensity is entirely due to smooth variation in surface orientation; thus:

$$\nabla I(x,y) = \rho \lambda \mathbf{L} \cdot \nabla \mathbf{N}(x,y) \ . \tag{1}$$

But we diverge from the classical model in two ways. First, we develop an approach that handles light source and surface properties concurrently; neither problem must be solved "before" the other. Second, we allow for discontinuities. Geometric discontinuities (in curvature, orientation, and depth) are unavoidable and their projection into the image has widely recognized importance (Kœnderink, 1984; Kœnderink and Van Doorn, 1982; Biederman, 1987). We will therefore assume that the scene is composed of piecewise smooth surfaces.[5]

Since we are requiring that all smooth variation in the image intensity arises from variation in surface normal, we also assume that the albedo is constant on each smooth surface region.

When several light sources (point or non-point sources) are present, one can consider instead a single equivalent light source under the condition that all light sources are visible from the lambertian surface patch. As long as this equivalent light source is constant over a surface patch, (1) will be valid.

Shadow boundaries are problematic since their effect is to change the direction and magnitude of the equivalent light source. These problems are discussed in Sect. 5. Another problem is the mutual illumination between bright surfaces which can be considered as nearby large light sources. However, (1) is only valid when light sources are far from the surface.

In summary, we attribute all smooth variation in intensity to smooth variation in orientation of surface elements, and thus to surface shape. Thus, in general, any shape from shading process should reconstruct shape information identically for a photograph (where the albedo varies continuously), a projected slide (where the illumination on the screen varies continuously) and the scene itself.

---

[5] Since the reflectance function depends on the existence of a differentiable surface normal, allowing surfaces that are nowhere smooth is clearly inappropriate.

2 Shape from Shading as a Coupled Family of Local Problems: Outline of the Scenel Bundle Approach

The key idea underlying our approach is to consider the shape from shading problem as a coupled family of local problems. Each of these is a "micro"-version of the shape from shading problem in which a lighting and surface model interact to produce the locally observed shading structure. We call each of these different models a scene element, or *scenel*, and, since many different scenels may be consistent with the local image structure, utilize fibre bundles to provide a framework to couple them together.

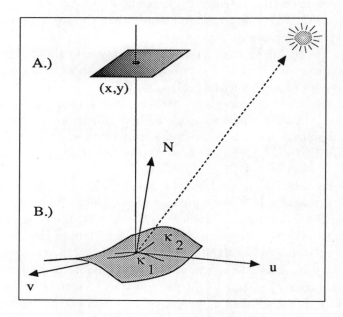

Fig. 1. Depiction of an abstract scene element, or scenel, corresponding to an image patch (A). The scenel (B) consists of a surface patch, described by its image coordinates, surface normal, and curvature. Its material properties (albedo) are also represented. Finally, a virtual light source completes the photometry.

The notion of FIBRE BUNDLES is fundamental to modern differential geometry (Husemoller, 1966). A fibre bundle consists of a triple (E, π, M), where E is called the total space, π is a projection operator, and M is the base space. For each point $p \in M$, the subset $\pi^{-1}(p) \subset E$ is called the FIBRE over p. Denoting the fibre F, an example is the product bundle $(M \times F, \pi, M)$, which illustrates how the total space can be viewed as the base manifold crossed with the fibre space. While this construction is quite abstract, we use it in the manner shown in Figs. 1, 2 and 5. Essentially it allows us to consider the global shape from shading problem as a coupled family of local problems.

We take the image manifold as the base space, and consider the photometry for each point on it. Locally, in a neighbourhood around the point (x, y), the shading information can be described by a combination of local light source and surface property values (including curvature, albedo, etc.). Each of these defines a scenel (Fig. 1), and the space of all possible scenels defines the fibre over that point. Together, the collection of scenel fibres defines the scenel bundle (Fig. 2).

We seek a solution of the shape from shading problem as connected sets of scenels in which neighbours are consistent; such a solution is called a CROSS SECTION through the scenel bundle. Formally, a cross section of a bundle (E, π, M) is a map $s: M \to E$ such that $\pi s = 1_M$. In other words, a cross section assigns a member of each fibre to each position in the manifold.

A sub-bundle of the scenel bundle is the TANGENT BUNDLE, in which the fibres consist of the tangent spaces at each point and sections correspond to vector fields; Sander and Zucker (1990) previously used this bundle in their study of inferring principle direction fields on surfaces.

We define what is meant by consistency shortly; first, we introduce the shading flow field as our initial data.

2.1 The Shading Flow Field as Initial Data

Observe that a sensitivity issue arises in the scenel framework; spatial quantization of the image induces a quantization of the scene domain. Analogously to the manner in which integer solutions are not always possible for algebraic equations, we begin with "quantized" initial data as well. In particular, we derive our initial estimates from *the shading flow field* instead of directly from the intensity image. This field is the first order differential structure of the intensity image expressed as the isoluminance direction and gradient magnitude (Fig. 3(a)); we supplement it with the intensity "edge" image (Fig. 3(b)). We suggest that dealing with uncertainties at the level of the shading flow field will expose more of the natural spatial consistency of the intensity variation, and will thus lead to more robust processing than the raw intensities. The shading flow field ideas are related to Kœnderink's isophotes (Kœnderink and Van Doorn, 1980).

Traditionally, the gradient of an image is computed by estimating direc-

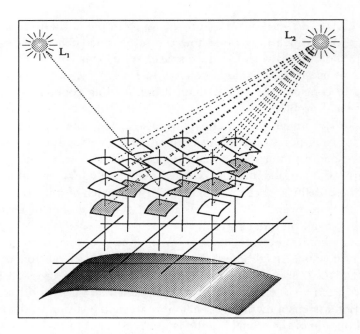

Fig. 2. Depiction of a Scenel Bundle over an image. At each point in the image there are many possible scene elements, or scenels. Each of these scenels is depicted along a fibre, or vertical space above each image coordinate. The union of scenel fibres over the entire image is called a scenel bundle. The shape from shading problem is formulated as determining sections through the scenel bundle. Such a section is depicted by the shaded scenels, and represents a horizontal slice across the bundle. Scenel participation in a horizontal section is governed by surface smoothness and material and light source constancy constraints.

tional derivatives by

$$\frac{\partial I}{\partial x} \approx I * G'_\sigma(x) G_\sigma(y) ,$$
$$\frac{\partial I}{\partial y} \approx I * G_\sigma(x) G'_\sigma(y) .$$

The gradient estimate follows immediately, and the isoluminance direction is simply perpendicular to the gradient. The one limitation of this approach is that (depending on the magnitude of σ) it always infers a smooth gradient field, even when the underlying image is non-continuous. We have investigated methods of obtaining stable, discontinuous shading flow fields using logical/linear operators (Iverson and Zucker, 1990).

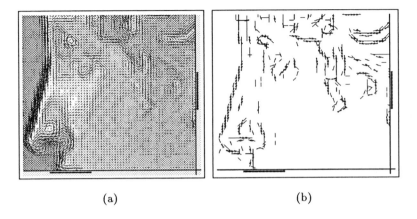

(a) (b)

Fig. 3. Typical shading flow (a) and edge (b) fields The left shading flow field depicts the first order differential structure of the image intensities, and the right is an edge map. Our shading analysis is based on these data, and not on the raw image intensities. This is to focus on the geometry of shape from shading analysis, and perhaps to capture something implicit in the biological approach.

Our motivation for starting from the shading flow field is also biological. We take shading analysis to be an inherently geometric process, and hence handled within the same cortical systems that provide orientation selection and texture flow analysis. Shading flow is simply a natural extension.

2.2 Constraints between Local Scenels

The coupling between the local scenel problems dictates a consistency relationship over them, and derives from three principle considerations:

1. A SURFACE SMOOTHNESS CONSTRAINT, which states that the surface normal and curvatures must vary according to a Lipschitz condition between pairs of scenels which project to neighbouring points in the image domain. This notion is subtle to implement, because it involves comparison of normal vectors following parallel transport to the proper position (see (Sander and Zucker, 1990)).

2. A SURFACE MATERIAL CONSTRAINT, which states that the surface material (albedo and reflectance) is constant between pairs of scenels which project to neighbouring points in the image domain.

3. A LIGHT SOURCE CONSTRAINT, which states that the virtual light source is constant for pairs of scenels which project to neighbouring points in the image domain.

Viewed globally, the solution we seek consists of sections in which a single (equivalent) light source illuminates a collection of surface patches with constant material properties but whose shape properties vary smoothly. The above constraints are embedded into a functional, and consistent sections through the scenel bundle are stationary points of this functional. More specifically, the constraints are expressed as compatibility relationships between pairs of neighbouring estimates within a relaxation labelling process. As background, we next sketch the framework of relaxation labelling.

2.3 Introduction to Relaxation Labelling

Relaxation labelling is an inference procedure for selecting labels attached to a graph according to optimal (symmetric) or variational (asymmetric) principles. Think of nodes in the graph as scenels, edges in the graph as links connecting "nearby" scenels with weights representing their compatibility. Formally, let $i \in \mathcal{I}$ denote the discrete scenel i, and let $l \in \mathcal{L}_i$ denote the set of labels for scenel i. The labels at each position are ordered according to the measure $p_i(l)$ such that $0 \leq p_i(l) \leq 1$ and $\sum_{l \in \mathcal{L}_i} p_i(l) = 1 \ \forall i$. (In biological terms, think of $p_i(l)$ as the firing rate for a neuron coding label l for scenel i.) Compatibility functions $r_{i,j}(l, l')$ are defined between label l at position i and label l' at position j such that increasingly positive values represent stronger compatibility. The abstract network structure is obtained from the support S that label l obtains from the labelling of it's neighbours $N(i)$; in symbols,

$$S_i(l) = \sum_{j \in N(i)} \sum_{l' \in \mathcal{L}_j} r_{i,j}(l, l') p_j(l') \ .$$

The final labelling is selected such that it maximizes the average local support

$$A(p) = \sum_{i \in I} S_i(l) p_i(l) = \sum_{i \in \mathcal{I}} \sum_{j \in N(i)} \sum_{l' \in \mathcal{L}_j} r_{i,j}(l, l') p_j(l') p_i(l) \ .$$

Such a labelling is said to be consistent (Hummel and Zucker, 1983). We now simply remark that such "computational energy" forms have become common in neural networks, and observe that Hopfield (Hopfield, 1984) networks are a special case, as are polymatrix games, under certain conditions (Miller and Zucker, 1991).

2.4 Overview of the Paper

The paper is organized as follows from this point on. We first formally define a scene element. We then show how to find those scene elements that are consistent with the shading flow field, and solve the forward problem of calculating the shading flow field expected for each scene element. The remainder of the paper is concerned with the inverse problems of inferring those scenels that are consistent with the shading flow field. The subtle interaction between expressing the scenel variables and the relaxation compatibilities is described, and scene element ambiguity addressed. An advantage of our technique is that, since both surface and lighting geometry are estimated, different types of shadow and illumination discontinuities can be handled; these are discussed in the Sect. 5.

3 Initial Estimation of Scene Attributes

We adopt a coarse coding of surface and light source attributes by quantizing the range of values of each attribute. Each scene element is defined by an assignment of a value to each of the scene attributes. The set of scene elements is viewed as a set of existence hypotheses of a surface patch of a fixed shape and orientation at a fixed image position, illuminated from a fixed direction with a fixed product of albedo and illumination.

3.1 The Scene Element

The attributes we consider are

1. IMAGE POSITION: The image pixels are themselves a set of discrete values of x and y position.

2. VIRTUAL ILLUMINANT DIRECTION: We take the light sources in the scene as a set of M distant point sources[6], $\{\lambda_{(i)} \mathbf{L}_{(i)} : 1 \leq i \leq M\}$ where $\lambda_{(i)}$ is the intensity and $\mathbf{L}_{(i)}$ is a unit vector. Let $V_i(x,y)$ be a binary "View" function such that $V_i(x,y) = 1$ if and only if light source $\mathbf{L}_{(i)}$ directly illuminates the surface element corresponding to pixel (x,y).

   We define the *virtual point source* at (x,y) by the two attributes λ and $\mathbf{L}$ such that
   $$\lambda \mathbf{L}(x,y) \equiv \sum_i \lambda_{(i)} \mathbf{L}_{(i)} V_i(x,y) \ . \tag{2}$$

---

[6] M would be quite large in the case of a diffuse source

This virtual point source is constant for any neighbourhood in which all the V_i functions are constant. In such a neighbourhood, surface luminance satisfies
$$I(x,y) = \rho\lambda \mathbf{L} \cdot \mathbf{N}(x,y) \; .$$

The possible virtual light source directions map onto a unit sphere. We sample this sphere as uniformly as possible to get a discrete set of virtual illuminant directions. In viewer-centered coordinates, this unit vector is given as
$$\mathbf{L} = (L_x, L_y, L_z) \; .$$

3. MATERIAL PROPERTIES or the product $\rho\lambda$: We need only consider the product of the albedo and the illuminance (see (1)). Usually the imaging process normalizes to some maximum value, so we can assume a range between zero and one, and discretely sample this range.

4. SURFACE SHAPE DESCRIPTORS: The two principal curvatures (κ_1, κ_2) describe the shape up to rotation. Two angles slant σ and tilt τ are needed to describe the surface tangent plane orientation with respect to the viewer's coordinate frame. An additional angle ϕ is needed to describe the principal direction of the Darboux frame in the surface tangent plane.

   (a) The two angles needed to orient the surface tangent plane in space describe the surface normal
   $$\mathbf{N} = (N_x, N_y, N_z) = (\cos\tau \sin\sigma, \sin\tau \sin\sigma, \cos\sigma) \; .$$

   The set of all such normals form a unit sphere. The surface of this sphere is sample as uniformly as possible to derive a discrete set of normals. Of these, only the ones in the hemisphere facing the viewer are used; the others are not visible; see Fig. 4.

   (b) The two principal curvatures (κ_1, κ_2) are mapped into a curveness measure
   $$c = \max(|\kappa_1|, |\kappa_2|)$$
   and a shape index measure
   $$s = \cos^{-1}\left(\frac{\kappa_1 + \kappa_2}{2c}\right) \; .$$

   These are analogous to K
   oenderink's curveness and shape index (Kœnderink, 1990), with the

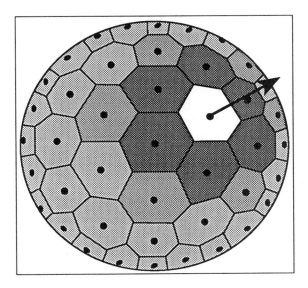

Fig. 4. The shape of surface patches is represented as a function of the principle curvatures mapped onto an abstract sphere through "curveness" and "shape index" measures (see text). Nearby positions on the sphere indicate smooth changes in either the shape or orientation of a scenel. This representation on the sphere facilitates the definition of scenel compatibilities later in the text.

choice of norm for the curveness and the spreading function for the shape index modified slightly.

The angles 2ϕ and s are, respectively, the longitude and latitude of a spherical coordinate system covering shape variation in the tangent plane. As we stated previously, the angle ϕ represents the principal direction, while s values 0 and π represent umbilic surfaces where principal directions are not defined. Any smooth curve on the $(2\phi, s)$ sphere represents a smooth deformation or rotation of the surface. We define a unit vector K as follows

$$\mathbf{K} = (K_1, K_2, K_3) = (\cos 2\phi \sin s, \sin 2\phi \sin s, \cos s) \ .$$

Thus by sampling the surface of the sphere uniformly we derive a discrete set of parameters which cover all variations of smooth, oriented shape in the tangent plane. Augmenting this with the curveness index provides a complete, discretely sampled shape descriptor.

Given the discrete sampling of the scene attributes as defined above, we derive a set of scenel labels

$$\mathcal{I} = \{x, y, \rho\lambda, \mathbf{L}, \mathbf{N}, c, \mathbf{K}\}$$

which represent all potential assignments of these scene attributes. Thus each i represents the hypothesis that the scene can be locally described by the scenel $(x_i, y_i, (\rho\lambda)_i, \mathbf{L}_i, \mathbf{N}_i, c_i, \mathbf{K}_i)$. To relate this to the *relaxation labelling* paradigm (Hummel and Zucker, 1983), we distribute a measure p_i over each scenel i representing confirmation of the hypothesis. The first step is to obtain an initial estimate for the confidence measure p_i from the shading flow field. The second is to extract locally consistent sets of hypotheses by relaxation labelling. The third step is to prune these sets by imposing appropriate boundary conditions. This is a large scale parallel computation, and we are currently implementing it on a massively parallel machine (MasPar MP-1).

3.2 The Expected Shading Flow Field

For each scenel i, we need an initial estimate for the associated weight p_i. We take this weight to reflect the match between the local properties of the shading flow field and the EXPECTED VALUES for the scene element i.

The EXPECTED SHADING FLOW FIELD is obtained by computing the light intensity gradient for a surface locally described by the scenel i. The paraboloid is an arbitrarily curved surface with the local parametric form

$$(u, v, w) \text{ where } w = -\frac{\kappa_1 u^2 + \kappa_2 v^2}{2}.$$

The u and v axes correspond to the two principal directions. The normal to the surface at a point (u, v) is given by:

$$\begin{aligned}\mathbf{N}(u, v) &= \Big(N_u(u, v), N_v(u, v), N_w(u, v)\Big) \\ &= \left(\frac{\kappa_1 u}{(\kappa_1^2 u^2 + \kappa_2^2 v^2 + 1)^{\frac{1}{2}}}, \frac{\kappa_2 v}{(\kappa_1^2 u^2 + \kappa_2^2 v^2 + 1)^{\frac{1}{2}}}, \frac{1}{(\kappa_1^2 u^2 + \kappa_2^2 v^2 + 1)^{\frac{1}{2}}}\right).\end{aligned}$$

Two rotations relate the viewer's coordinate frame to the paraboloid local frame. The first rotation takes care of the surface orientation and the second, of the principal directions. In matrix form

$$\mathbf{M} = \mathbf{M}_n \mathbf{M}_c$$

where

$$\mathbf{M}_n = \begin{pmatrix} \cos^2 \tau \cos \sigma + \sin^2 \tau & \sin \tau \cos \tau \cos \sigma - \sin \tau \cos \tau & \cos \tau \sin \sigma \\ \sin \tau \cos \tau \cos \sigma - \sin \tau \cos \tau & \sin^2 \tau \cos \sigma + \cos^2 \tau & \sin \tau \sin \sigma \\ -\cos \tau \sin \sigma & -\sin \tau \sin \sigma & \cos \sigma \end{pmatrix}$$

and
$$\mathbf{M}_c = \begin{pmatrix} \cos\phi & -\sin\phi & 0 \\ \sin\phi & \cos\phi & 0 \\ 0 & 0 & 1 \end{pmatrix} .$$

Since the inverse of the matrix $\mathbf{M}$ is simply its transpose, the light source expressed in the local surface coordinate is:

$$(L_u, L_v, L_w) = \mathbf{M}^t (L_x, L_y, L_z) .$$

Therefore the surface luminance can be computed.

$$I(u,v) = \rho\lambda \mathbf{L} \cdot \mathbf{N}(u,v) = \rho\lambda (L_u N_u(u,v) + L_v N_v(u,v) + L_w N_w(u,v)) .$$

Thus we can derive the local properties of the expected flow field. The orientation of the isoluminance line:

$$\theta_i = \tan^{-1}\left(\frac{L_u \kappa_1 M_{21} + L_v \kappa_2 M_{22}}{L_u \kappa_1 M_{11} + L_v \kappa_2 M_{12}}\right) .$$

The gradient magnitude:

$$|\nabla I(x,y)| = \rho\lambda \left| \frac{\partial(u,v)}{\partial(x,y)} (\kappa_1 L_u, \kappa_2 L_v) \right| .$$

The curvature of the isoluminance line:

$$\kappa_i = \frac{2\kappa_1^2 \kappa_2^2 L_w (L_u^2 + L_v^2) \det(M^{33}) - \kappa_1 \kappa_2 (\kappa_2 L_v \det(M^{32}) + \kappa_1 L_u \det(M^{31}))}{((M_{11}\kappa_2 L_v - M_{12}\kappa_1 L_u)^2 + (M_{21}\kappa_2 L_v - M_{22}\kappa_1 L_u)^2)^{\frac{3}{2}}} .$$

where

$$\det(M^{33}) = (M_{12}M_{21} - M_{11}M_{22}) ,$$
$$\det(M^{32}) = (M_{13}M_{21} - M_{23}M_{11}) ,$$
$$\det(M^{31}) = (M_{23}M_{12} - M_{13}M_{22}) .$$

The initial weight p_i is a decreasing function of some distance measure between the observed flow field in the neighbourhood of (x_i, y_i) and the expected flow field for scenel i.

$$p_i = G(\theta_i - \theta_{obs.}) \cdot G(|\nabla I|_i - |\nabla I|_{obs.}) \cdot G(\kappa_i - \kappa_{obs.})$$

where $\theta_{obs.}$, $|\nabla I|_{obs.}$ and $\kappa_{obs.}$ are extracted from the initial flow field.

3.3 Scene Element Ambiguity

It should be clear that a local shading flow field will not select a unique scenel. In general, for an arbitrary shading flow field, several scene elements will be assigned a significant weight for each image position. Identical flow fields can be generated by surfaces of different shapes because of

- Intrinsic ambiguity. For example, the cases of convex (κ_1, κ_2), concave $(-\kappa_1, -\kappa_2)$, hyperbolic $(-\kappa_1, \kappa_2)$ or $(\kappa_1, -\kappa_2)$ surfaces facing the viewer and the light source all have the same intensity profile.

- Accidental correspondence between light source and surface orientations. For example, a surface with $\kappa_1 = \kappa_2$ will generate concentric circular isoluminance lines in the plane facing the light source. If the curvature is small, the isoluminance lines projected in the image plane form concentric ellipses. Such isoluminance lines can also be seen when an elliptic surface faces the viewer and the light source.

The example of elliptic isoluminance lines is particulary revealing. Such lines could be due to the shading of a spherical patch directly facing the light source but slanted away from the viewer; or it could be due to the shading of an elliptical (convex or concave) patch directly facing the light source and the viewer. Here, two phenomena are closely coupled: the formation of the isoluminance lines on the surface (related to the slant of the surface with respect to the light source) and the projection on the image plane (related to the slant of the surface with respect to the viewer).

However, these local ambiguities are not always accompanied by global ambiguities. Although, as Mach observed in 1866, "many curved surfaces may correspond to one light surface even if they are illuminated in the same manner" (Ratliff, 1965), these global ambiguities are often finite (Oliensis, 1991). In the Sect. 4, we propose a relaxation labelling process to disambiguate the local surface geometry.

4 The relaxation labelling process

Recall that we are imposing three constraints on our surfaces: locally constant albedo, locally constant lighting conditions, and locally smooth geometry. For the relaxation labelling process, these translate into the following:

- A scenel j is *compatible* with the scenel i if they have the same constant $\rho\lambda$ and the same virtual illuminance direction **L**, and if scenel j's surface descriptors fall on scenel i's extrapolated surface at the corresponding relative position.

Shape from Shading

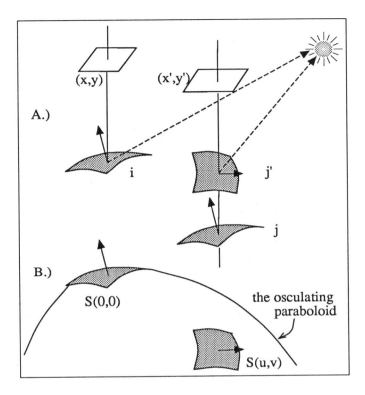

Fig. 5. Illustration of the compatibility relationship for scenel consistency. Two scenels are shown on the fibre at image location (x', y'), and are evaluated against the scenel (i) at (x, y). The surface represented in scenel$_{x,y}$ is modeled by the osculating paraboloid, and extended to (x', y'). It is now clear that one scenel (j') at (x', y') is consistent, because its surface patch lies on this paraboloid and light source and albedo agree. The other scenel (j) is inconsistent, because its surface does not match the extended paraboloid. Such osculating paraboloids are used to simulate the parallel transport of scenel$_{x',y'}$ onto scenel$_{x,y}$.

- A scenel j is *incompatible* with the scenel i if they have the same constant $\rho\lambda$ and the same virtual illuminance direction, and if there exists another scenel j', neighbouring scenel j along the fibre, that better fits the extrapolated surface from scenel i than scenel j. Observe that this incompatibility serves to localize information along each fibre.
- otherwise a scenel j is *unrelated* to the scenel i.

Using these guiding principles, we assign a value to the compatibility r_{ij} between two scenels i and j. This compatibility will be positive for compatible hypotheses, negative for incompatible hypotheses, and zero otherwise. In general, variation in r_{ij} is assumed to be smoothly varying between nearby points in the parameter space $\mathcal{I}$. The process is illustrated in Fig. 5.

Consider the unique paraboloid $S_i(u,v)$ such that the neighbourhood of $S_i(0,0)$ is described by the surface parameters of scenel i. The compatibility of a scenel j with i (r_{ij}) is then defined in terms of the relationship between scenel j and the point $S_i(u,v) \in S_i$ which is "closest to" scenel j. This operation is equivalent to the minimization of the distance measure

$$(x_j - x_i^*(u,v))^2 + (y_j - y_i^*(u,v))^2 + \left(\frac{\cos^{-1}(\mathbf{N}_j \cdot \mathbf{N}_i^*(u,v))}{\Delta \gamma_N}\right)^2$$
$$+ \left(\frac{\cos^{-1}(\mathbf{K}_j \cdot \mathbf{K}_i^*(u,v))}{\Delta \gamma_K}\right)^2 + \left(\frac{(c_j - c_i^*(u,v))}{\Delta c}\right)^2$$

over (u,v). Here, $\Delta \gamma_N, \Delta \gamma_K, \Delta c$ are the distance between neighbouring scenels for the given attribute and $(x_i^*, y_i^*, \mathbf{N}_i^*, \mathbf{K}_i^*, c_i^*)$ are the surface descriptors for the point $S_i(u,v)$.

So by physical consideration, the compatibility between scenel i and and scenel j can be expressed as

$$r_{ij} = \delta_{\rho\lambda_i,\rho\lambda_j} \cdot \delta_{\mathbf{L}_i,\mathbf{L}_j} \cdot \mathrm{G}\left(\sqrt{(x_j - x_i^*)^2 + (y_j - y_i^*)^2}\right) \cdot -\mathrm{G}''_{\Delta\gamma_N}\left(\cos^{-1}(\mathbf{N}_j - \mathbf{N}_i^*)\right)$$
$$\cdot -\mathrm{G}''_{\Delta\gamma_K}\left(\cos^{-1}(\mathbf{K}_j - \mathbf{K}_i^*)\right) \cdot \mathrm{G}_{\Delta c}(c_j - c_i^*)$$

where $\mathrm{G}_\sigma(x)$ is the Gaussian, and $\mathrm{G}''_\sigma(x)$ is its second derivative.

Notice that these values depend only on the relationship between scenel i and scenel j, which are fixed and constant throughout the computation. Therefore, these compatibilities can be calculated once and then stored in either a lookup table or as the weights in some sort of network.

5 A Case Study of Discontinuities

We have described a framework for computing sections of the scenel fibre bundle that are constrained to have a constant virtual light source, constant albedo, and smooth surface geometry. Any discontinuity in these attributes will demarcate a section boundary. Discontinuities can arise in many ways: the albedo can change suddenly along a smooth surface; a shadow can be cast across a smooth surface; the surface normal can change abruptly along a contour. In all three of these cases there will be a discontinuity in the image intensity. In this Sect., we address the question of how scenel discontinuities are manifested as discontinuities in the image. We restrict our attention to two types of image discontinuities : intensity discontinuities (edges) and shading flow discontinuities.

Shape from Shading 127

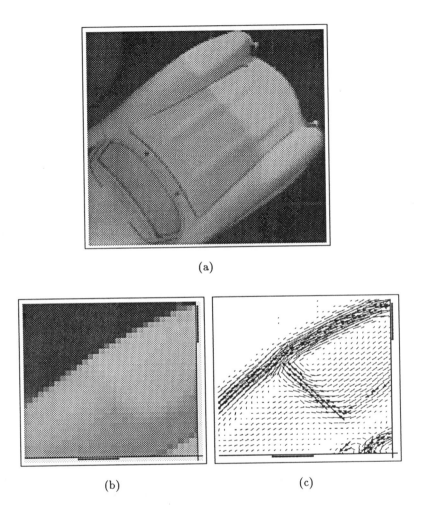

Fig. 6. An illustration of shadow boundaries and how they interact with flow fields. In (a) a shadow is cast across the hood of a car. In (b) we show a subimage of the cast shadow on the left fender, and in (c) we show the shading flow field (represented as a direction field with no arrowheads) and the intensity edges (represented as short arrows; the dark side of the edge is to the left of the arrow). Observe how the shading field remains continuous (in fact, virtually constant) across the intensity edge. This hold because the surface is cylindrical, and the shading flow field is parallel to the axis of the cylinder.

5.1 Illumination Discontinuities (Shadows)

Shadows are produced by variations in the virtual point source along a single surface patch (recall Sect.3.1 for the definition of $\lambda \mathbf{L}$). These variations are the direct result of a discontinuity in some $V_i(x, y)$. We say that there is a shadow boundary along this discontinuity. The image intensity cannot satisfy our model (1) in the neighbourhood of a shadow boundary since the image intensity on each side of the shadow boundary is due to a different virtual point source. We now briefly examine the two types of shadow boundary.

In general, an *attached shadow* boundary lies between two nearby pixels (x_0, y_0) and (x_1, y_1) when for some point source, $\mathbf{L}_{(j)}$, it is the case that $\mathbf{N}(x_0, y_0) \cdot \mathbf{L}_{(j)} > 0$ and $\mathbf{N}(x_1, y_1) \cdot \mathbf{L}_{(j)} < 0$.

The image intensity is continuous across the attached shadow boundary even though V_j is discontinuous since $\mathbf{N} \cdot \mathbf{L}_{(j)} = 0$ at the boundary. However, the shading flow is *typically* not continuous at the attached shadow boundary. The shading flow due to the source $\mathbf{L}_{(j)}$ will be parallel to the shadow boundary on the side where $V_j = 1$, but it will be zero on the side where $V_j = 0$. The shading flow due to the rest of the light sources will be smooth across the boundary but this shading flow will not typically be parallel to the boundary. Hence the sum of the two shading flows will typically be discontinuous at the boundary.

A *cast shadow* boundary is produced between two nearby points (x_0, y_0) and (x_1, y_1) when for some point source, $\mathbf{L}_{(j)}$, it is the case that both $\mathbf{N}(x_0, y_0) \cdot \mathbf{L}_{(j)} > 0$ and $\mathbf{N}(x_1, y_1) \cdot \mathbf{L}_{(j)} > 0$, and either $V_j(x_0, y_0) = 0$ or $V_j(x_1, y_1) = 0$ (but not both).

Examining (2), we note that for cast shadows, the discontinuity in V_j results in a discontinuity in image intensity, since the $\mathbf{N} \cdot \mathbf{L}_{(j)} > 0$. Furthermore, there is *typically* a discontinuity in the shading flow since the virtual light source defined on the side of the boundary where $V_j = 0$ will usually produce a different shading flow across the boundary than is produced by $\mathbf{L}_{(j)}$ on the side of the boundary where $V_j = 1$.

In the special case of parabolic (e.g. cylindrical) surfaces, the shading flow remains continuous across both cast and attached shadow boundaries because the flow is parallel to the axis of the cylinder. Note however that the attached shadow is necessarily parallel to the the shading flow field. This case is illustrated in Fig. 6.

To summarize, the image intensity in the neighbourhood of a pixel (x_0, y_0) can be modelled using a single virtual point source as long as there are neither attached nor cast shadow boundaries in that neighbourhood. Attached shadow boundaries produce continuous image intensities, but discontinuities in the shading flow. Cast shadows produce both intensity discontinuities and shading flow discontinuities.

5.2 Geometric Discontinuities

There are two different ways that the geometry of the scene can produce discontinuities in image. There can be a discontinuity in N along a continuous surface, or there can be a discontinuity in the surface itself when one surface occludes another. In the latter case, even if there is no discontinuity in the virtual light source direction there will still typically be a discontinuity in N which will usually result in both discontinuities in the image intensity and in the shading flow.

5.3 Material Discontinuities

If there is a discontinuity in the albedo along a smooth surface, then there will be a discontinuity in luminance across this material boundary. However, the shading flow will not vary across the boundary in the sense that the magnitude of the luminance gradient will change but the direction will not.

5.4 Summary of Discontinuities

In summary, shading flow discontinuities which are not accompanied by intensity discontinuities usually indicate attached shadows on a smooth surface. Intensity discontinuities which are not accompanied by shading flow discontinuities usually indicate material changes on a smooth surface. The presence of both types of image discontinuities indicates that either there is a cast shadow on a smooth surface, or that there is a geometric discontinuity.

6 Conclusions

We have proposed a new solution to the shape from shading problem based on notions from modern differential geometry. It differs from the classical approach in that light source and surface material consistency are solved for concurrently with shape properties, rather than independently. This has important implications for understanding light source and surface interactions, e.g., shadows, both cast and attached, and an example illustrating a cast shadow is included.

The approach is based on the notion of scenel, or unit scene element. This is defined to abstract the local photometry of a scene configuration, in which a single (virtual) light source illuminates a patch of surface. Since the image irradiance equation typically admits many solutions, each patch of the image gives rise to a collection of scenels. These are organized into a fibre space at that point, and the collection of scenel fibres is called the scenel bundle. Algebraic and topological properties of the scenel bundle will be developed in a subsequent paper.

The solution of the shape from shading problem thus reduces to finding sections through the scenel bundle, and these sections are defined by material, light source, and surface shape consistency relationships. The framework thus provides a unification of these different aspects of photometry, and should be sufficently powerful to indicate the limitations of unification as well.

References

Biederman, I.: "Recognition-by-Components: A Theory of Human Image Understanding," *Psychological Review*, **94** (1987) 115–147

Hopfield, J.J.: "Neurons with Graded Response Have Collective Computational Properties like Those of Two-State Neurons," *Proc. Natl. Acad. Sci. USA*, **81** (1984) 3088–3092

Horn, B.K.P.: "Obtaining Shape from Shading Information," P.H. Winston. Ed. in **The Psychology of Computer Vision**, McGraw-Hill, New York (1975)

Horn, B.K.P. and Brooks, M.J.: "The Variational Approach to Shape from Shading," *Comput. Vision, Graph. Image Process.*, **33** (1986) 174–208

Hummel, A.R. and Zucker, S.W.: "On the Foundations of Relaxation Labeling Processes," *IEEE Trans. Pattern Anal. Machine Intell.*, **5** (1983) 267–287

Husemoller, D.: **Fibre Bundles**, Springer, New York (1966)

Ikeuchi, K. and Horn, B.K.P.: "Numerical Shape from Shading and Occluding Boundaries," *Artificial Intelligence*, **17** (1981) 141–184

Iverson, L. and Zucker, S.W.: "Logical/Linear Operators for Measuring Orientation and Curvature," *TR-CIM-90-06*, McGill University, Montreal, Canada (1990)

Kœnderink, J.J. and van Doorn, A.J.: "Photometric Invariants Related to Solid Shape," *Optica Acta*, **27** (1980) 981–996

Kœnderink, J.J. and van Doorn, A.J.: "The Shape of Smooth Objects and the Way Contours End," *Perception*, **11** (1982) 129–137

Kœnderink, J.J.: "What Does the Occluding Contour Tell us about Solid Shape?," *Perception*, **13** (1984) 321–330

Kœnderink, J.J.: **Solid Shape**, MIT Press, Cambridge, Mass. (1990) p. 320

Miller, D.A. and Zucker, S.W.: "Efficient Simplex-like Methods for Equilibria of Nonsymmetric Analog Networks," *TR–CIM–91–3* McGill University, Montreal, Canada (1991); *Neural Computation*, (in press)

Nayar, S.K., Ikeuchi, K. and Kanade, T.: "Shape From Interreflections," *International Journal of Computer Vision*, **6** (1991) 173–195

Oliensis, J.: "Shape from Shading as a Partially Well-Constrained Problem," *Comp. Vis. Graph. Im. Proc.*, **54** (1991) 163–183

Pentland, A.: "Finding the Illuminant Direction," *J. Opt. Soc. Amer*, **72** (1982) 448–455

Pentland, A.: "Linear Shape From Shading," *International Journal of Computer Vision*, **4** (1990) 153–162

Ratliff, F.: **Mach Bands: Quantitative Studies on Neural Networks in the Retina**, Holden-Day, San Francisco (1965)

Sander, P., and Zucker, S.W.: "Inferring Surface Trace and Differential Structure from 3-D Images," *IEEE Trans. Pattern Analysis and Machine Intelligence*, **9** (1990) 833–854

Spivak, M.: **A Comprehensive Introduction to Differential Geometry**, Publish or Perish, Berkeley (1979)

Zucker, S.W., Dobbins, A., and Iverson, L.: "Two Stages of Curve Detection Suggest Two Styles of Visual Computation," *Neural Computation*, **1** (1989) 68–81

DYNAMIC VISION

Alex P. Pentland

Vision and Modeling Group, The Media Laboratory,
Massachusetts Institute of Technology
Room E15-387, 20 Ames St., Cambridge MA 02139

ABSTRACT: Almost all computational theories of vision are *static*, that is, they think of visual analysis as happening independently at each instant in time. Even motion theories confine themselves to a brief instant in time. Yet it is clear that human vision is not static in this sense: we cannot perceive thirty (or ten, or five, or even one) random images per second as well as we perceive the continuous flow of imagery typical of the real world. Moreover, results from control theory *prove* that static analysis is insufficient for the real-time, interactive behavior exhibited by people: a being that interacts with a changing, dynamic environment must also perceive it using computational models based on *physical dynamics*. In robotics such a perceptual theory is known as an *optimal observer* because it provides optimal estimates in a changing environment. In this paper I show how standard theories of visual perception — which are inadequate to account for people's ability for real-time interaction — can be modified to perform optimally in changing environments, and describe simple neural networks that carry out the required computations.

1 Introduction

Often the hardest part of research is defining the goal. What *is* vision? What capabilities would a successful visual system have? What information would it output? What input would it require? These are some of the most important, yet most ill-defined and least discussed questions in vision research. The simple answer is that the goal is defined by the larger system in which the vision system is embedded. Unfortunately, this answer doesn't help much in building general theories of visual function.

Still, the question is important because it defines the context for all subsequent research effort. So, despite the ill-defined nature of the question, I will put forward a suggestion about the goal of a "general purpose" vision system. In this paper I will take the goal of a vision system to be:

- Providing the motor control system with the estimates *and predictions* of geometry, velocity, and momentum that are necessary for manipulation and locomotion.

- Providing cognitive faculties with canonical, stable descriptions *and predictions* of 2-D appearance and 3-D structure, so that we can perform a wide variety of recognition, learning, and comparison operations.

In general, I will assume that the range of cognitive and motor control operations we want to perform are roughly the same as those thought to be performed by all mammals.

The major difference between these goals and those adopted by others is the inclusion of the requirement of *prediction* as well as estimation. This change reflects the fundamental difference between the optimal observer of a non-stationary or *dynamic* system, and the optimal observer of a stationary or static system. In the case of a static system, one simply collects all of the observations and makes an estimate. In the case of a dynamic system, however, one must also take into account how the system is changing in order to make an optimal estimate.

The fundamental change, therefore, is to define "the vision problem" as observation of a dynamic system rather than as observation of a static situation. It is exactly the change in perspective that lead to modern control theory some thirty years ago, and which has lately come to vision research through the work of researchers such as Brodia and Chellappa [Chellappa(1986)], Faugeras [Faugeras(1986)], Ayache [Ayache89], and Matthies, Szeliski, and Kanade [Matthies(1987)], to name but a few of the pioneers.

Redefining vision as observation of a dynamic system requires rethinking not only the algorithms within our vision system, but also redesigning the representations used. The primary change is inclusion of not only whatever system variables are to be estimated (e.g., range) but also a model of the dynamics of the system, e.g., velocity, inertia, and covariance. Developing the ramifications and implications of including system dynamics in our visual representations is the focus of this paper.

2 Background: Optimal Observers, Kalman Filtering

How can we use sensor measurements to estimate the instantaneous value of a rapidly changing variable? This is the problem that motivated development of observer theory and Kalman filtering, and it is a central problem for any visual process that attempts to deal with an environment that changes over time.

Consider, for instance, the problem of estimating distance to a parked car using a sequence of noisy stereo disparity measurements. After the first measurement, your estimate of the distance will be the measured value. After the second measurement, your estimate will be one-half the first measurement plus one-half the second. After the nth draw, your estimate will be $(n-1)/n$ times your last estimate plus $1/n$ times the newest measurement. As your confidence in your estimate of the distance grows, you assign more and more weight to it and less to any new observation.

Now consider the case of estimating the distance to a moving car. In this case very old observations will be useless, because the car will have moved, whereas more recent observations will be more reliable. Consequently, the simple weighted average that worked in the case of distance to a static object will produce a poor estimate.

To obtain a good estimate of distance it is necessary to compensate for the car's motion. To accomplish this, one could compute a weighted average of the most recent measurement and a *prediction* of the car's distance based on your last estimate of the car's distance *and velocity*. Mathematically, this approach is equivalent to computing a weighted average of the newest observation and the last estimate, just as in the static measurement case, but where the weighting also takes into account the fact that the true distance is changing.

The formal version of this approach is the *Kalman filter*, also called the *optimal linear observer*. It is the standard technique for obtaining estimates of the state of dynamic models, and for predicting the state at some later time. Outputs from the Kalman filter are the

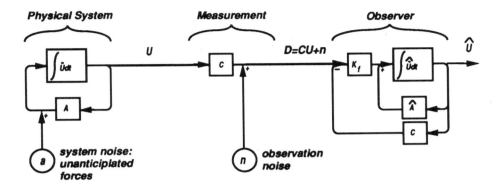

Figure 1: (a) Physical dynamic system in surrounding environment, (b) measurement data, (c) the observer: an internal dynamic system tracking the state of the exterior system.

optimal (weighted) least-squares estimate; for Gaussian noises this is also the maximum likelihood estimate.

In the following sections I will develop the ideas of an *observer* and the *Kalman filter*, and show how they are related to standard vision techniques such as Wiener filtering and regularization. This development will parallel the presentation of Friedland [Friedland(1986)]; this reference is recommended for readers seeking additional technical background.

2.1 Observer Theory

Control theory provides us with a formal definition of an *observer*, a notion which is critical to the control of dynamic systems. This definition, due to Luenberger [Luenberger(1963)], is as follows:

> *A dynamic system whose state variables are the estimates of the state variables of another system is called an* observer *of that system.*

Luenberger showed that it is possible to design observers such that their estimation error will go to zero as quickly as is desired.

The ideas of observer theory are illustrated in Figure 1. In this illustration the state of a non-stationary physical system, shown at the right, is being observed. Measurement data **D** are taken of the unknown state variables **U**. The measurement data are presumed to have been corrupted by a linear distortion **C** and noise n. form an estimate $\hat{U}$ the unknown state variables, we construct an *observer system*, shown at the right of Figure 1. This system mimics the behavior of the real, physical system, and tries to simultaneously "undo" the noise and distortion corrupting the measurement data.

The idea that this notion of observer is the basic framework of most of vision is perhaps the main point of this paper. In the remainder of this section I will develop the ideas of observer theory and of dynamic systems, and show how many of the standard, static image

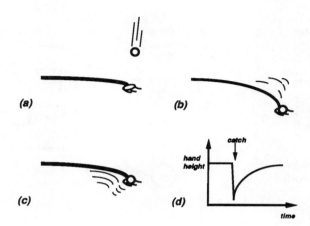

Figure 2: (a) A flexible robot arm, (b) the arm being bent by an outside force, (c) internal stresses cause the arm to straighten, (c) position of the arm's hand as a function of time.

analysis techniques fit into this framework. The remainder of the paper will show how some apparently difficult problems can be made easier by posing them as a dynamic observer system.

A simple example will help to make all of this more concrete. Imagine that we are interested in estimating the unknown heights **U** of a set of n points or nodes spaced along a flexible robot arm, as shown in Figure 2(a).

If the arm were to catch a ball the ball's inertia would cause the arm to bend, as shown in Figure 2(b). However internal stresses within the arm will produce a force causing the arm to straighten out, as shown in Figure 2(c). If the arm has a large amount of internal friction, so that its motion is critically damped, then it will return smoothly to nearly its original position, as shown in Figure 2(d), with velocity proportional to the amount of bending.

This *dynamic system* can be described using *state-space notation* by the following system of equations:

$$\dot{\mathbf{U}} = \mathbf{A}\mathbf{U} \qquad (1)$$

where **U** is an 1 x n vector of the unknown state variables of the system, and **A** is an n x n matrix that relates the state variables **U** to their time derivatives $\dot{\mathbf{U}}$. In our example, the state variables are the heights of points along the arm, and $\mathbf{A} = -\lambda \mathbf{I}$ where λ is a scalar constant and **I** is the n x n identity matrix.

Now, let us suppose that we cannot observe the unknowns **U** directly, but rather only a linear transformation of them. In our example this might occur if we were viewing the arm though a camera, so that our observations are distorted by perspective projection. The observer equations are:

$$\mathbf{D} = \mathbf{C}\mathbf{U} \qquad (2)$$

where **D** is an 1 x k vector of observation data, and **C** is an n x k matrix describing the observation tranformation. The case where $k < n$ can occur when, for instance, some of the points blur together or are occluded in the image.

To obtain estimates $\hat{\mathbf{U}}$ of the unknowns, we can set up the following *observer system*:

$$\dot{\hat{\mathbf{U}}} = \mathbf{A}\hat{\mathbf{U}} + \mathbf{K}(\mathbf{D} - \mathbf{C}\hat{\mathbf{U}}) \tag{3}$$

where $\mathbf{K}$ is a $k \times n$ matrix, and then integrate $\dot{\hat{\mathbf{U}}}$ to obtain an estimate $\hat{\mathbf{U}}^t$ at time t,

$$\hat{\mathbf{U}}^t = \int_0^t \dot{\hat{\mathbf{U}}} \tag{4}$$

Luenberger showed that for the correct choice of $\mathbf{K}$, one can always reduce the error of observation to within acceptable bounds.

2.2 Optimal Observers: The Kalman Filter

In the early 1960's Kalman and Bucy [Kalman(1961)] developed a state estimator that is *optimal* with respect to system and measurement noise under very general conditions. This estimator, now called the *Kalman filter*, fits the definition of an observer, and so the Kalman filter is now commonly referred to as the *optimal observer*.

As with the simple observer, we start by defining a dynamic system

$$\dot{\mathbf{U}} = \mathbf{A}\mathbf{U} + \mathbf{B}\mathbf{a} \tag{5}$$

and observations

$$\mathbf{D} = \mathbf{C}\mathbf{U} + \mathbf{n} \tag{6}$$

where $\mathbf{a}$ and $\mathbf{n}$ are white noise processes having known autocorrelation matrices. This dynamic system is the same as in the simple observer case, except that now we have added $\mathbf{a}$, the *system excitation noise*, and $\mathbf{n}$, the *observation noise*.

Kalman and Bucy showed that the optimal estimate $\dot{\hat{\mathbf{U}}}$ of $\dot{\mathbf{U}}$ is given by the following *Kalman filter*

$$\dot{\hat{\mathbf{U}}} = \mathbf{A}\hat{\mathbf{U}} + \mathbf{K}_f(\mathbf{D} - \mathbf{C}\hat{\mathbf{U}}) \tag{7}$$

given that the *Kalman gain matrix* $\mathbf{K}_f$ is chosen correctly. As in the simple observer case, $\dot{\hat{\mathbf{U}}}$ is then integrated to obtain the optimal estimate $\hat{\mathbf{U}}^t$ at time t,

$$\hat{\mathbf{U}}^t = \int_0^t \dot{\hat{\mathbf{U}}} dt \quad \approx \quad \hat{\mathbf{U}}^{t-1} + \dot{\hat{\mathbf{U}}}^{t-1} \tag{8}$$

The Kalman gain matrix $\mathbf{K}_f$ is chosen to minimize the covariance matrix $\mathbf{P}$ of the error $\mathbf{e} = \mathbf{U} - \hat{\mathbf{U}}$. Assuming that the cross-variance between the system excitation noise $\mathbf{a}$ and the observation noise $\mathbf{n}$ is zero, then

$$\mathbf{K}_f = \mathbf{P}\mathbf{C}^T \mathcal{N}^{-1} \tag{9}$$

where the observation noise autocorrelation matrix $\mathcal{N}$ must be nonsingular [Friedland(1986)]. Assuming that the noise characteristics are constant, then the optimizing covariance matrix $\mathbf{P}$ is obtained by solving the *Riccati equation*

$$0 = \dot{\mathbf{P}} = \mathbf{A}\mathbf{P} + \mathbf{P}\mathbf{A}^T - \mathbf{P}\mathbf{C}^T \mathcal{N}^{-1}\mathbf{C}\mathbf{P} + \mathbf{B}\mathcal{A}\mathbf{B}^T \tag{10}$$

where $\mathcal{A}$ is the autocorrelation matrix of the system noise **a**. In practice **P** is often recalculated in parallel with the Kalman filter estimates.

The simple filter of Equation 7 provides an efficient, robust method of estimating the state variables of non-stationary processes in the presence of noise with constant characteristics. Readers seeking information about yet more complex estimators are referred to reference [Friedland(1986)].

2.2.1 Important Example 1: A Kalman Filter with Velocities

Most of the Kalman filters that have appeared in the vision literature have state variables that include position and velocity, but not higher-order system dynamics. This example illustrates this order of filter, using the example of measuring the height u of the flexible arm's hand.

In state-space notation the behavior of the arm is described by the following scalar equation:

$$\dot{u} = -\lambda u + \mathbf{a} \tag{11}$$

where **a** is the system noise due to outside forces. The observed variable is the measured height d of the hand:

$$d = u + \mathbf{n} \tag{12}$$

where **n** is the observation noise. The Kalman filter is therefore

$$\dot{\hat{u}} = -\lambda \hat{u} + k_f(d - \hat{u}) \tag{13}$$

When we solve the Riccati equation we find that $p = n\sqrt{\lambda^2 + a/n} - n\lambda$, where a and n are the variance of the system and observation noise, respectively. Consequently the Kalman gain is $k_f = \sqrt{\lambda^2 + a/n} - \lambda$, and the Kalman filter is simply

$$\dot{\hat{u}} = -\lambda \hat{u} + k_f(d - \hat{u}) \tag{14}$$

so that the estimate of height at time $t + 1$ is

$$\begin{aligned}\hat{u}^{t+1} &= \hat{u}^t + \dot{\hat{u}}^t \\ &= (1 - \lambda)\hat{u}^t + k_f(d^t - \hat{u}^t)\end{aligned} \tag{15}$$

That is, the optimal estimate of hand height consists of two parts:

- the prediction $(1 - \lambda)\hat{u}^t$ made by combining the current estimate $\hat{u}^t$ and the velocity prediction $-\lambda\hat{u}^t$ obtained from the system dynamics,

- a correction factor that depends on the error $d^t - \hat{u}^t$ made in the last prediction, weighted by a factor k_f that depends on the signal-to-noise ratio $\sqrt{(a/n)}$.

The implications for representation are clear: one needs to represent not only the observed variable, as is standard, but also the signal-to-noise ratio of the observer system. The need to represent the signal-to-noise ratio can be quite a departure from current practice; for instance, if noise characteristics vary across an image then the signal-to-noise ratio must be represented at each pixel.

2.2.2 Important Example 2: A Kalman filter with Accelerations

From the point of view of modeling a real flexible arm, the above model is too simple to be realistic. This is because it does not consider the mass m of the arm, nor does it characterize changes in the system as being due to outside forces. To incorporate mass and force into the arm's dynamic model, we must model not only velocity $\dot{u}$ but also acceleration $\ddot{u}$, because force equals mass times acceleration. The system noise will be forces from outside the system. The force generated by catching a ball is a good example of such an external force, as illustrated in Figure 2.

The dynamic system describing the arm is therefore system of equations is

$$\begin{bmatrix} \dot{u} \\ \ddot{u} \end{bmatrix} = \begin{bmatrix} 0 & 1 \\ 0 & 0 \end{bmatrix} \begin{bmatrix} u \\ \dot{u} \end{bmatrix} + \begin{bmatrix} 0 \\ \frac{1}{m} \end{bmatrix} \mathbf{a} \tag{16}$$

where $\mathbf{a}$ is the unknown external force, and m is the mass of the arm. The observed variable will again be the hand height u

$$d = u + \mathbf{n} \tag{17}$$

where $\mathbf{n}$ is the observation noise.

The Kalman filter is therefore

$$\begin{bmatrix} \dot{\hat{u}} \\ \ddot{\hat{u}} \end{bmatrix} = \begin{bmatrix} \dot{\hat{u}} \\ 0 \end{bmatrix} + \begin{bmatrix} k_{f,1} \\ k_{f,2} \end{bmatrix} \left(d - \begin{bmatrix} 1 & 0 \end{bmatrix} \begin{bmatrix} \hat{u} \\ \dot{\hat{u}} \end{bmatrix} \right) \tag{18}$$

where $k_{f,1}$ and $k_{f,2}$ are the Kalman gains for velocity and acceleration, respectively.

Solving the Riccati equation for the Kalman gains, we find that $k_{f,1} = a/mn$ and $k_{f,2} = 2\sqrt{a/mn}$, where again a and n are the variance of the system and observation noise, respectively. The prediction of arm height for time $t+1$ is therefore

$$\begin{aligned} \hat{u}^{t+1} &= \hat{u}^t + \hat{\dot{u}}^t \Delta t + \hat{\ddot{u}}^t \tfrac{\Delta t^2}{2} \\ &= \hat{u}^t + \hat{\dot{u}}^t + \gamma \left(d - \hat{u}^t \right) \end{aligned} \tag{19}$$

where $\gamma = a/mn + 2\sqrt{a/mn}$.

As before, the optimal estimate of hand height consists of two parts:

- the prediction made by combining the current estimate u^t and the velocity prediction obtained from the system dynamics,

- a correction factor that depends on the error $d^t - \hat{u}^t$ made in the last prediction, weighted by a factor that depends on the signal-to-noise ratio $\sqrt{(a/n)}$.

The major difference between this example and the previous one is that now the prediction requires knowing both the current estimate of position and the current estimate of velocity. Consequently, to employ this more physically-correct model of system dynamics one needs to represent the observed variable, *its velocity*, and the signal-to-noise ratio of the observer system for that variable.

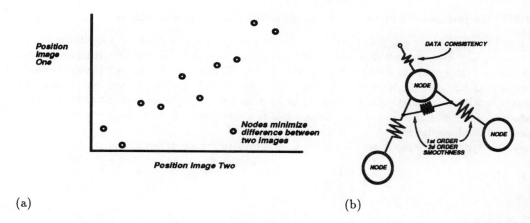

Figure 3: (a) Point-by-point stereo matches are often too noisy. (b) A spring mechanism to smooth stereo matches.

3 An Illustrative Example: Stereo Networks

Algorithms to compute stereo disparity have a long history in vision research, and so provide a good example for illustrating the Kalman filtering approach to visual processing. The central problem of stereopsis is usually taken to be finding corresponding points in two different views of a scene. By measuring the change in position between the two views, we can then calculate the distance to each of the matched points (assuming, of course, that we know the camera model for each image). A standard method for accomplishing this is to take corresponding epipolar lines (typically horizontal rows) from each of the two images, and compare the local pattern of intensity between the two lines. A correspondence or match between the two images occurs when a patch from one image very closely matches a patch from the other image. The difference in position between the centers of the two patches is called the stereo disparity of those points.

The results of such a match are illustrated by Figure 3(a). The axes are position within the epipolar line taken from each of the two images. Perfect matches for a flat surface would form a 45° line; unfortunately real matches are often more like the matches illustrated here.

3.1 Previous Approaches: The Spring Model

To obtain better stereo matches, "spring models" such as shown in Figure 3(b) have been suggested. This particular mechanism is a composite of stereo algorithms proposed by Julesz [Julesz(1968)], Sperling [Sperling(1970)], Marr and Poggio [Marr(1982)], Grimson [Grimson(1981)], Terzopoulos [Terzopoulos(1987)], and Blake and Zisserman [Blake(1987)]. Such "spring models" are the basis for most neural network models of stereopsis.

In this model, there are a series of active *nodes* placed at each pixel along one image's epipolar line. A "data consistency" spring attaches the the node to the best-matching disparity, and "smoothness" springs attach each node to its neighbors. Typically there are either first-order or second-order smoothness springs, but not both. The springs then relax, moving the entire system to its lowest energy state, and the resting position of the nodes provides an improved estimate of the point-by-point stereo disparity.

The energy ϵ_d within the data consistency spring is proportional to

$$\epsilon_d = (d_i - u_i)^2 \qquad (20)$$

where d_i is the disparity measured using simple patch-by-patch matching, and u_i is the position of the corresponding node. The energy ϵ_s within the smoothness springs is proportional to either the first derivative of u_i,

$$\epsilon_{s1} = (du_i)^2 = (u_i - u_{i-1})^2 \qquad (21)$$

if first-order springs are used, or the second derivative (curvature) of u_i,

$$\epsilon_{s2} = (d^2 u_i)^2 = (u_{i+1} - 2u_i + u_{i-1})^2 \qquad (22)$$

if second-order springs are employed.

If we collect all of the variables into vectors, we have that the system energy is proportional to

$$\epsilon = \epsilon_d + \epsilon_s = (\mathbf{U} - \mathbf{D})^T(\mathbf{U} - \mathbf{D}) + \lambda \mathbf{U}^T \mathbf{K} \mathbf{U} \qquad (23)$$

where the matrix $\mathbf{K}$ contains all of the spring coefficients, the vector $\mathbf{U}$ is the unknown nodal positions, the vector $\mathbf{D}$ the original disparity matches, and λ is the ratio between the spring constants for the smoothness and data springs. The minimum-energy state $\hat{\mathbf{U}}$, which will be the estimate of stereo disparity, occurs when

$$\lambda \mathbf{K} \hat{\mathbf{U}} + \hat{\mathbf{U}} - \mathbf{D} = 0 \qquad (24)$$

3.2 Spring Models are Special Case of Kalman Filtering

How good is the estimate of disparity provided by this spring model? To see the relationship between Kalman filtering and the spring model, let us again use the simple dynamic system for the flexible arm. Recall that the dynamic system equations are

$$\dot{\mathbf{U}} = -\lambda \mathbf{U} + \mathbf{a} \qquad (25)$$

with observation equation

$$\mathbf{D} = \mathbf{C}\mathbf{U} + \mathbf{n} \qquad (26)$$

and that the Kalman filter is

$$\dot{\hat{\mathbf{U}}} = -\lambda \hat{\mathbf{U}} + \mathbf{K}_f(\mathbf{D} - \mathbf{C}\hat{\mathbf{U}}) \qquad (27)$$

If we assume that constant loads are being applied to this dynamic system, and that we are observing the system after it has come to rest, then $\dot{\mathbf{U}} = 0$. The Kalman filter then reduces to

$$0 = -\lambda \hat{\mathbf{U}} + \mathbf{K}_f(\mathbf{D} - \mathbf{C}\hat{\mathbf{U}}) \qquad (28)$$

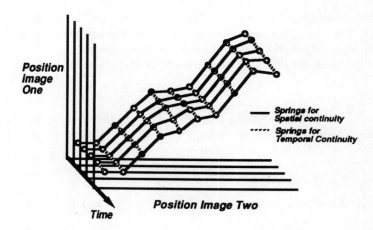

Figure 4: Illustration of a stereo network that implements a Kalman filtering approach.

It is now clear that the spring model's equilibrium point (Equation 24) is the same as the steady-state Kalman filter solution (Equation 28), given that we can observe all of the nodal positions, so that $\mathbf{C} = \mathbf{I}$, and that we have choosen the spring constants such that $\mathbf{K} = \mathbf{K}_f^{-1}$. This steady-state Kalman filter is also equivalent to a Wiener filter, and to the well-known interpolation technique called regularization [Poggio(1985)]. This equivalence is discussed in Appendix A.

Perhaps the main drawback of this solution is that it gives the wrong answer in the case of complete and noiseless data, as setting $\mathbf{U} = \mathbf{D}$ does not produce a solution to Equation 28 except when $\mathbf{U} = \mathbf{O}$. This is because there is always residual tension in the springs for any surface but a flat plane. One solution to this problem is not to make the rest state of the springs be some *a priori* model, but rather to be the data itself. This is discussed further in the next section.

3.3 The Time-Varying Spring Model: Full Kalman Filtering

We have seen that many of the stereo models previously proposed can be viewed as steady-state variants of the Kalman filter. Given the reality of active observers and a changing world, however, it is clear that we would like to remove the steady-state restriction to obtain a full-fledged Kalman filter stereo mechanism. Such a Kalman filter mechanism is illustrated by Figure 4; the central idea is to smooth both between adjacent nodes and also between successive instants in time. This allows the filter to take advantage of the surface's coherence in both space and time.

The Kalman filter approach to this type of problem was first explored by Matthies, Szeliski, and Kanade [Matthies(1987)]. Their approach was (very roughly) to simply remove the steady-state assumption of Equation 28, so that the spring model's outputs were treated as velocity increments that where smoothed over time by integration.

An alternative that seems more natural is to treat the spring output as *forces* acting on the nodes. This is the physically correct description of the spring's behavior, and is easy to implement using physical mechanisms. In this approach the spring forces act to accelerate the nodes, and the inertia of the nodes provides smoothing over time. This is exactly the approach we used to model the flexible arm, and so we can again use that Kalman filter:

$$\begin{bmatrix} \dot{\hat{U}} \\ \ddot{\hat{U}} \end{bmatrix} = \begin{bmatrix} \dot{\hat{U}} \\ 0 \end{bmatrix} + \begin{bmatrix} K_{f,1} \\ K_{f,2} \end{bmatrix} \left(D - \begin{bmatrix} C & 0 \end{bmatrix} \begin{bmatrix} \hat{U} \\ \dot{\hat{U}} \end{bmatrix} \right) \qquad (29)$$

and as before the prediction for time $t+1$ is

$$\hat{U}^{t+1} = \hat{U}^t + \dot{\hat{U}}^t + K \left(D^t - C\hat{U}^t \right) \qquad (30)$$

where $K = K_{f,1}\Delta t + 1/2 K_{f,2}\Delta t^2$. We will see further development of this Kalman filter in the next two sections.

Perhaps the most important difference between this model and the previous ones is that the rest state of the model is the data itself, rather than some combination of the data and the matrix K. This is because the spring coefficients K are now applied to the error between the node positions and disparity measurements, rather than to the node positions directly. In effect, the spring's resting length has been set to match the data, rather than to match a flat plane.

4 Modal Analysis

The most obvious drawback of the these methods is their large computational expense. These methods require roughly $O(nm_k)$ operations per time step, where n is the number of nodes and m_k the bandwidth[1] of the stiffness matrix K. Further, $O(n)$ time steps (iterations) are required to obtain an equilibrium solution, so that these methods are implausible as a biological model unless one supposes the existence of some "infinitely fast" analog neural network processing. Moreover, for a maximally accurate 3-D model, where the stiffness matrix is no longer sparse, the computational cost scales as $O(n^3)$, so that a neural network model must be fully connected and take $O(n)$ iterations.

Thus there is a need for a method which transforms our equations into a form which is not only less costly, but also allows closed-form solution. Since the number of operations is proportional to the bandwidth m_k of the stiffness matrix K, a reduction in m_k will greatly reduce the cost of step-by-step solution. Moreover, if we can actually diagonalize the system of equations, then the degrees of freedom will become uncoupled and we will be able to find closed-form solutions. To accomplish this goal, we utilize a generalized frequency analysis technique known as *Modal Analysis*. For further details the reader is referred to Bathe[Bathe(1982)].

4.1 The Modal Coordinate Transform

To diagonalize the system of equations, a linear coordinate transformation P of the nodal point displacements U can be used:

$$U = P\tilde{U} \qquad (31)$$

---

[1] See Bathe[Bathe(1982)] Appendix A.2.2 for complete discussion on bandwidth of a stiffness matrix.

where $\mathbf{P}$ is a square orthogonal transformation matrix and $\tilde{\mathbf{U}}$ is a vector of generalized displacements in the new coordinate system. The columns of $\mathbf{P}$ are the basis vectors of this new coordinate system.

Substituting Equation 31 into the equilibrium spring equation $\mathbf{KU} = \mathbf{R}$, and premultipling by $\mathbf{P}^T$ yields:

$$\tilde{\mathbf{K}}\tilde{\mathbf{U}} = \tilde{\mathbf{R}} \quad (32)$$

where $\tilde{\mathbf{K}} = \mathbf{P}^T\mathbf{K}\mathbf{P}$ and $\tilde{\mathbf{R}} = \mathbf{P}^T\mathbf{R}$. With this transformation of basis set a new system of equations is obtained which has a smaller bandwidth than the original system.

The optimal transformation matrix $\mathbf{\Phi}$ is derived from the the following eigenvalue problem

$$\mathbf{K}\phi_i = \omega_i^2 \phi_i \quad (33)$$

which has $3n$ solutions $(\omega_1^2, \phi_1), (\omega_2^2, \phi_2), \ldots, (\omega_n^2, \phi_{3n})$. For dynamic systems, these eigenvectors are called the *free vibration modes* of the system.

Using these modes we can define a transformation matrix $\mathbf{\Phi}$, which has for its columns the eigenvectors ϕ_i,

$$\mathbf{\Phi} = [\phi_1, \phi_2, \phi_3, \ldots, \phi_{3n}] \quad (34)$$

such that $\mathbf{\Phi}^T\mathbf{K}\mathbf{\Phi} = \tilde{\mathbf{K}}$ where $\tilde{\mathbf{K}}$ is a *diagonal* matrix with the eigenvalues ω_i^2 on its diagonal:

$$\tilde{\mathbf{K}} = \begin{bmatrix} \omega_1^2 & & & \\ & \omega_2^2 & & \\ & & \ddots & \\ & & & \omega_{3n}^2 \end{bmatrix} \quad (35)$$

The coordinate system defined by $\mathbf{\Phi}$ is known as the *modal coordinate system*. From the fact that $\mathbf{K}$ is diagonal in this coordinate system, it is obvious that it is the optimal coordinate system for solving this system of equations, and that $\mathbf{P}$ is the optimal transformation matrix.

Thus by setting $\tilde{\mathbf{U}} = \mathbf{\Phi}^{-1}\mathbf{U} = \mathbf{\Phi}^T\mathbf{U}$ premultipling by $\mathbf{\Phi}^T$ the system of equations is decoupled into $3n$ independent equations

$$\tilde{k}_i \tilde{u}_i = \tilde{r}_i \quad (36)$$

where $\tilde{u}_i$ and $\tilde{r}_i$ are the ith entries of the transformed vectors $\tilde{\mathbf{U}}$ and $\tilde{\mathbf{R}}$, and $\tilde{k}_i$ is the ith diagonal element of the transformed matrix $\tilde{\mathbf{K}}$. Because the equations are decoupled, the equilibrium solution is simply

$$\tilde{u}_i = \frac{\tilde{r}_i}{\tilde{k}_i} \quad (37)$$

Aside: Perhaps even more important than the ability to obtain closed-form solutions, is the fact that surface or object descriptions posed in this coordinate system are *canonical*. This permits the sort of canonical representations needed to perform recognition, learning, and comparison. Recognition examples using the modal coordinate s technique are presented in Turk and Pentland [Turk(1991)] (face recognition), Mase and Pentland [Mase(1990)] (lip reading), and Pentland and Sclaroff [Pentland(1991b)] (face recognition using range data). In each case accuracies in the high 90% range were achieved, with recognition times averaging a small fraction of a second.

4.2 Modes for 2-D Surface Interpolation Problems

2-D surface interpolation is a common and perhaps prototypical problem in vision and neural network research [Grossburg(1976), Carpenter(1987), Poggio(1985)]. However despite the use of sophisticated numerical approximation methods, at least $O(n)$ iterations are required to solve these problems, where n is the number of nodes used in the problem discretization. As a consequence of this scaling behavior, problem solution generally involves large computational expense. Further, the large number of iterations required make such algorithms implausible as biological models.

As we have seen in the previous section, the cost of finding such solutions can be greatly reduced by by transforming the problem to a modal coordinate system. However 2-D interpolation problems present a special difficulty to this approach, because their dynamics depend not only on the smoothness or stiffness matrix $\mathbf{K}$ but also on the observation or data sampling matrix $\mathbf{C}$. In the case of regularization (Equation 28) the optimal basis $\mathbf{\Phi}$ has columns that are the modes or eigenvectors of $(\mathbf{K} + \mathbf{C})$. As a consequence, the optimal basis will be different for each data sampling. Further, as discontinuities are introduced the basis vectors change, and must again be recomputed. As a consequence, the straightforward modal analysis approach is too expensive to be useful.

4.2.1 Wavelet "Modes"

Instead of finding the exact modes or eigenvectors of our system of equations, we would like to find a good set of "general purpose" basis vectors for solving 2-D surface interpolation problems. The idea is that if we can find a transform that is close enough to the true eigenvectors, then we can still get closed-form approximations that are within a few percent of the exact solution. Moreover, iterating this approximate solution technique on the residuals will be an efficient way to obtain arbitrarily accurate solutions.

This type of "approximate modes" transform is commonly described as a *decorrelating transform*, or as an approximation to the Karhunen-Loeve transform. Although not exact, such a transform can still both approximately decouple the system of equations and improve its condition number. This can lead to either closed-form approximations, or to very efficient iterative schemes.

For the class of physically-motivated systems, the ideal transform would very fast to compute, and would have basis vectors that are both spatially and spectrally localized. The desire for spectral localization stems from the fact that, in the absence of boundary conditions, fractures, etc., these sort of physical equilibrium problems can usually be solved in closed form in the frequency domain. In similar fashion, the transformation of the dynamic system equations into a spectrally-localized basis will tend to produce a banded stiffness matrix $\mathbf{K}$. The requirement for spatial localization stems from the need to account for local variations in $\mathbf{K}$'s band structure due to, for instance, boundary conditions, fracture, or material inhomogeneity.

4.2.2 Orthogonal Wavelet Bases

A class of bases that provide the desired properties are generated by functions known as *orthogonal wavelets* [Grossman(1984), Meyer(1986), Mallat(1987), Daubechies(1988), Simoncelli (1990)].

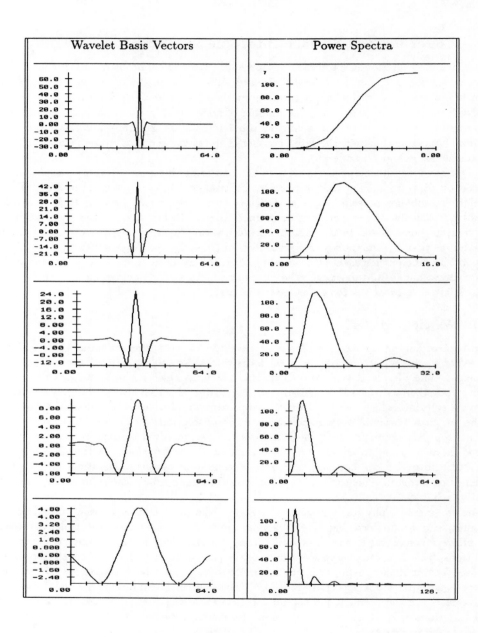

Figure 5: Five elements of the wavelet basis set "closest" to the Wilson-Gelb psychophysical model of human spatial vision. Left column: Wavelet filter family "closest" to Wilson-Gelb model. Right column: The power spectra of these filters on a linear scale.

Orthogonal wavelet functions and receptive fields are different from the wavelets previously used in biological and computational modeling because *all* of the functions or receptive fields within a family, rather than only the functions or receptive fields of one size, are orthogonal to one another. A family of orthogonal wavelets $h_{a,b}$ is constructed from a single function h by dilation of a and translation of b

$$h_{a,b} = |a|^{-1/2} h\left(\frac{x-b}{a}\right), \qquad a \neq 0 \tag{38}$$

Typically $a = 2^j$ and $b = 2^j$ for $j = 1, 2, 3...$. The critical properties of wavelet families that make them well suited to this application are that:

- For appropriate choice of h they can provide an orthonormal basis of $\mathbf{L}^2(\Re)$, i.e., *all* members of the family are orthogonal to one another.

- They can be simultaneously localized in both space and frequency.

- Digital transformations using wavelet bases can be *recursively* computed, and so require only $O(n)$ operations.

Such families of wavelets may be used to define a set of multiscale orthonormal basis vectors. I will call such a basis $\mathbf{\Phi}_w$, where the columns of the $n \times n$ matrix $\mathbf{\Phi}_w$ are the basis vectors. Because $\mathbf{\Phi}_w$ forms an orthonormal basis, $\mathbf{\Phi}_w^T \mathbf{\Phi}_w = \mathbf{\Phi}_w \mathbf{\Phi}_w^T = \mathbf{I}$. That is, like the Fourier transform, the wavelet transform is self-inverting.

The left-hand column of Figure 5 shows a subset of $\mathbf{\Phi}_w$; from top to bottom are the basis vectors corresponding to $a = 1, 2, 4, 8, 16$ and $b = n/2$. [2] The right-hand column shows the Fourier power spectrum of each of these bases; it can be seen that they display good joint spatial-spectral localization. All of the examples presented in this book will be based on the wavelet basis illustrated in this figure. For additional detail about the construction of wavelets, see references [Simoncelli(1990), Pentland(1991c)].

The shapes shown in the left-hand column of Figure 5 are the "receptive field profiles" that transform an input signal into, or out of, the wavelet coordinate system. I developed this particular set of wavelets to match as closely as possible the human psychophysical model of Wilson and Gelb [Wilson(1984)]; there is only a 7.5% MSE difference between this set of wavelet receptive fields and the Wilson-Gelb model. [3] *This set of wavelets, therefore, provides a good model of human spatial frequency sensitivity, and of human sensitivity to changes in spatial frequency.* Coefficients for this family of wavelets are given in reference [Pentland(1991c)]

On digital computers, transformation to the wavelet coordinate system is normally computed recursively using separable filters. A two-dimensional example is illustrated in Figure

---

[2] The literature is confused about what to call approximatly orthogonal wavelets such as the ones shown here. Mathematicians prefer to call these quadrature mirror filters, whereas most engineers seem to prefer grouping approximate wavelets along with exact wavelets. I will side with what appears to be engineering practice, and use the term wavelet to refer to both exact and approximate wavelets.

[3] Wavelet receptive fields from only five octaves are shown, although the Wilson-Gelb model has six channels. Wilson, in a personal communication, has advised us that the Wilson-Gelb "b" and "c" channels are sufficiently similar that it is reasonable to group them into a single channel.

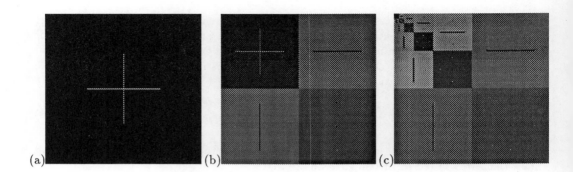

Figure 6: (a) A 128 x 128 node input image $\mathbf{D}$, (b) subbands comprising the first level of the wavelet transform, which are the product of the highest frequency members of $\mathbf{\Phi}_w^T$ and the input image $\mathbf{D}$, (c) the complete transform $\mathbf{\Phi}_w^T \mathbf{D}$. Total execution time: approximately 2.0 seconds on a Sun 4/330.

6. At the first level of recursion, the input image (a) is split into low-pass, horizontal high-pass, vertical high-pass, and diagonal high-pass sub-bands, as shown in (b). This is accomplished by convolving the input image with filters whose coefficients come from the highest-frequency wavelet basis, shown at the top left of Figure 5, and then subsampling by taking every other value. The high-pass subbands that result from these three convolutions are the product of the first $\frac{3}{4}n$ rows of $\mathbf{\Phi}_w^T$ (the highest-frequency basis vectors) and the input image $\mathbf{D}$.

At each successive level of recursion the low-pass image is split into four more bands, in the limit producing $\mathbf{\Phi}_w^T \mathbf{D}$, the complete wavelet transform of $\mathbf{D}$, shown in (c). Note that at each iteration the same filters are used to split the image; it is the process of recursive application that generates the entire family of filters shown in Figure 5. To go from the wavelet coordinate system to the original coordinate system, one simply reverses the process, recursively summing these same basis vectors by first inserting extra zeros between each transform value and then convolving with the same filters. For further information see reference [Simoncelli(1990), Pentland(1991c)].

5 2-D Kalman Filtering Using Wavelet Modes

In this paper I have argued that visual problems such as surface interpolation occur within the context of a dynamic, changing environment, and therefore should be considered within the framework of Kalman filtering. Support for this view is provided by the recent work of Matthies, Szeliski, and Kanade [Matthies(1987)], Heel [Heel(1990)], Singh [Singh(1990)], and others who have shown that Kalman filtering can provide very good solutions to surface interpolation and data integration problems even in rapidly changing environments.

However, as in static surface interpolation (e.g., regularization), the major computational cost of these Kalman filter formulations is the inversion of a large matrix, in this case one whose off-diagonal entries are the autocorrelation of the surface. This makes such methods

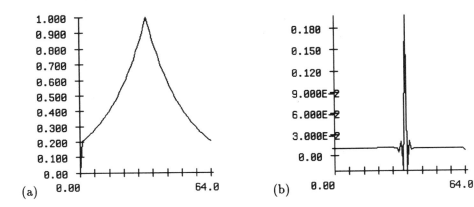

Figure 7: (a) Second-order Markov process autocorrelation; vertical axis is autocorrelation, horizontal axis is pixel separation. (b) After transformation to the wavelet basis set.

very expensive for computer implementation. Further, as a consequence of the large number of iterations required, such models are not biologically plausible.

In this section I will show that by using wavelet "modes", we can obtain closed-form Kalman filter solutions to surface interpolation and information integration problems. Consequently such techniques become very efficient for computer implementation, and biologically-plausible as a neural network.

5.1 Constructing A 2-D Kalman Filter: An Example

As an example I will show how the wavelet "modes" can be used to diagonalize the Kalman filter of Equation 29, for the case of an observable point-by-point quantity U (e.g., range) and its derivative $V = \dot{U}$ (e.g., velocity) at each point in an $\sqrt{n} \times \sqrt{n}$ image. Because each point is observable, the observation matrix $C = I$.

I will assume that n and a originate from independent second-order Gauss-Markov noise processes. Figure 7(a) illustrates what this autocorrelation function looks like for a typical image; as can be seen, there are significant correlations over a dozens of pixels of image separation. It is this large area of correlation that makes 2-D Kalman filters expensive, because all of the equations are linked together.

Figure 7 illustrates the performance of the wavelet transform at diagonalizing the second-order Gauss-Markov autocorrelation matrix, and thus at diagonalizing the Kalman filter system of equations. Figure Figure 7(a) shows one row of the autocorrelation matrix in the original nodal coordinate system. Figure 7(b) shows the autocorrelation function in the wavelet coordinate system. As can be seen, the autocorrelation matrix is nearly diagonalized. In other words, the wavelet basis functions are very nearly the modes of the Kalman filter equations.

Thus we have that

$$\mathcal{N} \approx \Phi_w \mathbf{N} \Phi_w^T \qquad \mathcal{A} \approx \Phi_w \mathbf{A} \Phi_w^T \qquad (39)$$

where **N** and **A** are *diagonal* matrices. Given this approximation to $\mathcal{N}$ and $\mathcal{A}$ we may then use Equations 10 and 9 to determine the Kalman gain matrices, which are

$$\mathbf{K}_1 = \mathbf{\Phi}_w (2\mathbf{A}\mathbf{N}^{-1})^{1/2} \mathbf{\Phi}_w^T \qquad \mathbf{K}_2 = \mathbf{\Phi}_w (\mathbf{A}\mathbf{N}^{-1}) \mathbf{\Phi}_w^T \tag{40}$$

Substituting this result into the Kalman filter surface interpolation mechanism of Equation 29 we obtain

$$\begin{bmatrix} \hat{\dot{\mathbf{U}}} \\ \hat{\mathbf{U}} \end{bmatrix} = \begin{bmatrix} \hat{\dot{\mathbf{U}}} + \mathbf{\Phi}_w (2\mathbf{A}\mathbf{N}^{-1})^{1/2} \mathbf{\Phi}_w^T \left(\mathbf{D} - \hat{\mathbf{U}} \right) \\ \mathbf{\Phi}_w (\mathbf{A}\mathbf{N}^{-1}) \mathbf{\Phi}_w^T \left(\mathbf{D} - \hat{\mathbf{U}} \right) \end{bmatrix} \tag{41}$$

Letting $\tilde{\mathbf{U}} = \mathbf{\Phi}_w^T \mathbf{U}$, and premultiplying by $\mathbf{\Phi}_w^T$, we obtain

$$\begin{bmatrix} \hat{\dot{\tilde{\mathbf{U}}}} \\ \hat{\tilde{\mathbf{U}}} \end{bmatrix} = \begin{bmatrix} \hat{\dot{\tilde{\mathbf{U}}}} + (2\mathbf{A}\mathbf{N}^{-1})^{1/2} \left(\tilde{\mathbf{D}} - \hat{\tilde{\mathbf{U}}} \right) \\ (\mathbf{A}\mathbf{N}^{-1}) \left(\tilde{\mathbf{D}} - \hat{\tilde{\mathbf{U}}} \right) \end{bmatrix} \tag{42}$$

as $\mathbf{\Phi}_w^T \mathbf{\Phi}_w = \mathbf{I}$.

That is, in the wavelet coordinate system defined by $\mathbf{\Phi}_w$, the Kalman filter equations are decoupled into n independent two-variable Kalman filters. The major consequence of this decoupling is that only $O(n)$ computations and $O(n)$ storage locations are required. Even in the variable-noise case (not discussed here due to space limitations) only $O(n)$ computations are required, as even space-varying Markov **N** are approximately diagonal in the wavelet coordinate system [Simoncelli(1990)].

Because the equations are decoupled, we may write the Kalman filter separately for each point in the wavelet coordinate system. We can now formulate the position prediction for time $t + \Delta t$:

$$\hat{\tilde{u}}_i^{t+\Delta t} = \hat{\tilde{u}}_i^t + \hat{\dot{\tilde{u}}}_i^t \Delta t + k_i \left(\tilde{d}_i - \hat{\tilde{u}}_i^t \right) \tag{43}$$

where $\hat{\tilde{u}}_i$ is the ith element of $\hat{\tilde{\mathbf{U}}}$, and $k_i = (a_i/n_i)\Delta t^2 + (2a_i/n_i)^{1/2} \Delta t$.

Discontinuities. The Kalman gain matrices **K** describe the connectivity between adjacent points on a continuous surface; thus whenever a discontinuity occurs **K** must be altered. Following Terzopoulos [Terzopoulos(1987)], we can accomplish this by disabling receptive fields that cross discontinuities. In a computer implementation, the simplest method is to locally halt the recursive construction the wavelet transform whenever one of the resulting bases would cross a discontinuity.

5.2 A Possible Neural Mechanism

Equation 43 is exactly the central-difference update rule for direct time integration of the second order differential equation

$$\tilde{m}_i \ddot{\tilde{u}}_i + \tilde{u}_i = \tilde{d}_i \tag{44}$$

which describes the time behavior of a spring with unit stiffness, mass $\tilde{m}_i$, and loads $\tilde{d}_i$. Thus the Kalman filter equations may be interpreted as a physical process of hysteresis,

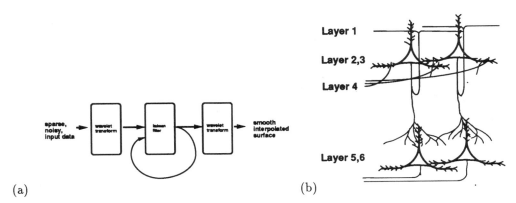

Figure 8: (a) A biological mechanism for Kalman filtering, (b) one possible neural implementation.

integrating data over time by giving a certain amount of mass or inertia to the previous estimates.

This suggests the neural mechanism illustrated in Figure 8(a). The input measurements $\mathbf{D}^t$ are passed through a layer of neurons with receptive fields such as are shown in Figure 5 at each location. This computes $\tilde{\mathbf{D}}^t = \mathbf{\Phi}_w^T \mathbf{D}^t$, the wavelet transform of $\mathbf{D}^t$. The activity of each neuron is then averaged with predictions based on the previous estimates, using a weighting that reflects the relative confidence of the old estimates and new measurements, thus computing new estimates $\hat{\tilde{\mathbf{U}}}^t$ in the wavelet coordinate system. Finally, each neuron's output is summed with a spatial distribution equal to its receptive field, thus computing the inverse wavelet transform and obtaining the estimated values in the original coordinate system $\hat{\mathbf{U}}^t = \mathbf{\Phi}_w \hat{\tilde{\mathbf{U}}}^t$.

Figure 8 (b) illustrates one way this computation can be mapped onto neurons. In this figure the input layer arborizes very locally, with the pyramidal cell's basal dendrites producing receptive fields shaped as in Figure 5. It is important to note that wavelet receptive fields can be produced from an initial unspecific center-surround receptive field structure by, for instance, Kohonen's or Linsker's "learning" mechanisms.

This transforms the input to the wavelet basis. Recurrent axons from these neurons form the core of the Kalman feedback loop, by allowing predictions based on the previous instant's activity to be averaged with new inputs. Such prediction might be accomplished by hysteresis, however the details of such a neural mechanism are far from clear. Finally, the output axons then arborize with the spatial distribution of Figure 5 among the apical dendrites of a second layer of neurons, producing receptive fields similar to those of the basal dendrites. This produces the inverse wavelet transform, so that the output of the second layer of neurons is the new estimated surface.

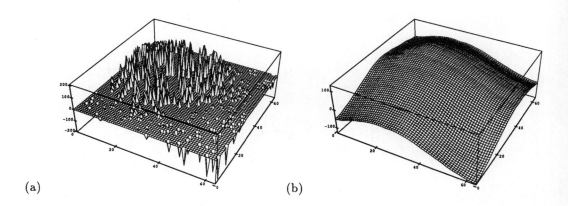

(a)   (b)

Figure 9: A typical surface interpolation problem. (a) Height data for a 64 x 64 node surface; these data were generated by a 10 % density random sampling of the function $z = 100[\sin(kx) + \sin(ky)]$, vertical axis is height. (b) Interpolated surface. After one iteration (approximately 1 second on a Sun 4/330) the algorithm converged to within 1% of the true equilibrium state.

5.3 Static Examples: Regularization

I will first illustrate the application of this Kalman filter formulation in the static case, where it becomes equivalent to the spring or regularization discussed in Sections 3.1 and 3.2 (see Equation 28). In this case the input data **D** is transformed to the wavelet coordinate system, weighted on a point-by-point (or neuron by neuron) basis, and then transformed back to the original nodal coordinate system. This regularization process is performed by the same mechanism shown in Figure 8, but without necessity of the feedback loop.

Figure 9(a) shows the height measurements input to a 64 x 64 node interpolation problem (zero-valued nodes have no data); the vertical axis is height. These data were generated using a sparse (10%) random sampling of the function $z = 100[\sin(kx) + \sin(ky)]$. Figure 9(b) shows the resulting interpolated surface. In this example Equation 28 converged to within 1% of its true equilibrium state with a single iteration. Execution time was approximately 1 second on a Sun 4/330.

Figure 10 illustrates a second example that uses real image data and incorporates discontinuities. Shown in the top row are two 128 x 128 images of the same scene taken with different apertures, thus varying the depth of field. By comparing the amount of blur present in these two images depth estimates can be extracted, as first described by Pentland [Pentland(1987)]. The resulting depth estimates are shown in Figure 10(c). Figure 10(d) shows the first iterations of the wavelet-based surface interpolation process. Point and line breaks were introduced into the estimated surface based on examination of the local strain energy, and the estimation process repeated for two additional iterations. The final estimated surface (after a total of three iterations) is shown in Figure 10(f). Execution time is approximately 15 seconds on a Sun 4/330 computer. For additional detail see references [Pentland(1990a), Pentland(1991c)].

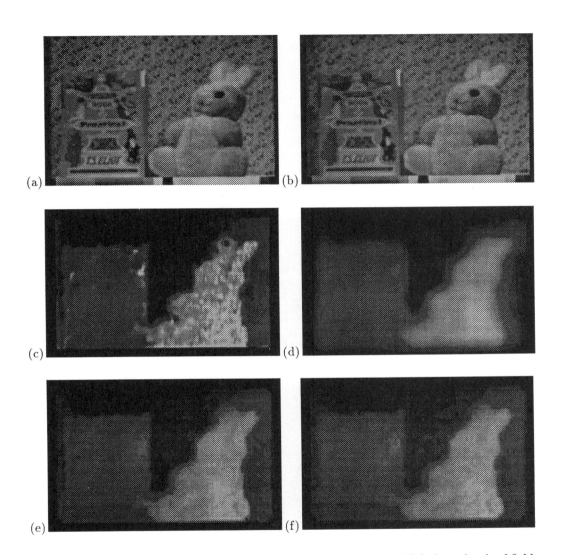

Figure 10: Two 128 x 128 images of a scene taken with (a) long and (b) short depth of field, (c) Range extracted by comparing the amount of blurring in the two images, as described in Pentland [Pentland(1987)], (d) first iteration of estimation, (e) second iteration, and (f) third iteration of estimation with breaking. Total time: approximately 15 seconds on a Sun 4/330 computer.

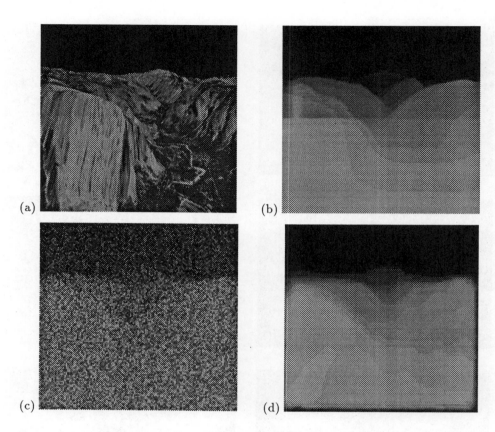

Figure 11: (a) The sixth frame of a fly-though of Yosemite valley, (b) the true range map associated with frame 6, (c) range plus additive correlated noise with a signal-to-noise ration of 1:1, (d) Kalman filter range estimates at frame 6.

5.3.1 A Dynamic Example: Full Kalman Filtering

Figure 11 shows an example using the Kalman filter of Equation 42 on an image sequence. Illustration (a) of this figure is the sixth frame from a synthetic image sequence of a fly-through of Yosemite Valley from the vantage point of a small plane flying down the center of the valley floor. A sequence of corresponding range images, the sixth of which is shown in Figure 11(b), were generated by reprojecting a digital terrain map of the area from the same set of viewpoints. In these range images brighter points are closer to the camera, and darker points are further away. These 128 x 128 range images were then corrupted by the addition of uniformly distributed *correlated* noise ($P(\omega) = \omega^{-0.2}$, where ω is spatial frequency) resulting in a sequence of range images with a signal-to-noise ratio of 1:1, as is illustrated by Figure 11(c).

Accurate estimates of pixel-by-pixel range cannot be obtained by averaging successive frames of the noisy range images, as the range values vary as a nonlinear function of both space and time. This nonlinear variation is caused by the curved camera path, and by the resulting perspective distortion. Similarly, regularization of individual frames does not produce as accurate estimates of range because of the large amount of correlated noise. Therefore a Kalman filter was constructed as described above, and the equations solved in the wavelet coordinate system. Both surface position and velocity were estimated, with breaks introduced into the surface based on strain energy, as reference [Pentland(1990a), Pentland(1991c)]. The computation time was approximately 5 seconds per frame on a Sun 4/330.

The estimated surface range after frame 6 is shown in Figure 11(d). By comparing the estimated surface range shown in (d) to the true range shown in (b), it can be seen that a good estimate of surface shape is obtained. At frame 6 the mean per-pixel error in the position estimate was 8.5% of the initial error, an improvement of 21 db. The mean per-pixel error in the velocity estimate was 5.5% of the initial error, an improvement of 25 db.

6 Summary

The problem of analyzing image sequences is quite different than that of analyzing static images. The major change is that continuity in time is fundamentally different from the sort of continuity that exists in the spatial coordinates. Whereas spatial autocorrelations are at least plausibly modeled using a Gauss-Markov process, temporal relationships are not.

To model continuity over time, I have argued that we must incorporate the physical relationships that cause that continuity: mass, inertia, force, and acceleration. Incuding the dynamic models in this way leads naturally to the theory of optimal observers and Kalman filtering as developed in control theory. This theory allows optimal estimates to be made even in changing environments, by combining new measurements with extrapolations made using the dynamic model.

I have also shown that physically-based methods for simulating nonrigid dynamics provide an elegant way of implementing Kalman filters that include both velocity and acceleration. So, for instance, by associating state variables such as stereo disparity with the position of points on a nonrigid thin plate with mass, one can obtain an optimal estimate of disparity even in a changing environment. By incorporating accelerations into our dynamic model, we can remove the estimation bias (due to residual plate stresses) that exists in previous algorithms.

The ability to implement optimal observers using physically-based methods also provides an elegant method of sensor and data fusion. This is because it is easy to formulate shape constraints in terms of energy potentials, and so the energy-minimizing behavior of a physical model provides an elegant method of integrating information from many different vision modules.

Finally, I have shown that by transforming the dynamics equations (used in both the Kalman filter and physical simulation/prediction) to a coordinate system defined by the *modes* of the system, all of the equations become decoupled and ordered by frequency. These modes derive from the system's characteristic function, e.g., from the system's eigenvectors.

Decoupling the parameters of dynamic system allows simple, closed-form solutions. The linear transformation to the modal coordinate system not only makes solution very efficient, but also makes it simple enough that it can be done by simple mechanisms, e.g., by a single analog circuit or a single neuron.

References

[Ayache89)] Ayache, N., Faugeras, O., (1989) Maintaining representations of the environment of a mobile robot, *IEEE Trans. Robotics Automation,* 5(6): 804–819.

[Bathe(1982)] Bathe, K-J., (1982)   *Finite Element Procedures in Engineering Analysis.* Prentice-Hall

[Blake(1987)] Blake, A., and Zisserman, A., (1987) *Visual Reconstruction* Cambridge, MA.: M.I.T. Press

[Carpenter(1987)] Carpenter, G. A., and Grossburg, S. (1987) *A massively parallel architecture for a self-organizing neural pattern recognition machine*, Computer Vision, Graphics, Image Proc., 37, pp. 54-115.

[Chellappa(1986)] Broida, T., and Chellappa, R., (1986) Estimation of Object Motion Parameters from Noisy Images, *IEEE Trans. Pattern Analysis and Machine Intelligence,* 8(1):90–99.

[Daubechies(1988)] Daubechies, I., (1988)   Orthonormal Bases of Compactly Supported Wavelets. *Communications on Pure and Applied Mathematics*, XLI:909-996.

[Faugeras(1986)] Faugeras, O., Ayache, N., and Faverjon B., (1986) Building Visual Maps by Combining Noisy Stereo Measurements, *Proc. IEEE Conf. on Robotics and Automation,* San Francisco, CA.

[Friedland(1986)] Friedland, B., (1986) *Control System Design.* McGraw-Hill

[Grossburg(1976)] Grossburg, S. (1976) *Adaptive Pattern Recognition and universal recoding, Biol. Cyber.*, 23, pp. 187-202.

[Grimson(1981)] Grimson, W., (1981) *From Images to Surfaces: A Computational Study of the Human Early Visual System*, Cambridge, MA.: M.I.T. Press

[Grossman(1984)] Grossmann, A. and Morlet, J., (1984) Decomposition of Hardy functions into square integrable wavelets of constant shape, *SIAM J. Math*, 15:723–736

[Heel(1990)] Heel, J., (1990) Temporally-Integrated Surface Reconstruction *IEEE Int'l Conf. on Computer Vision*, Osaka, Japan, pp. 292-295.

[Julesz(1968)] Julesz, B. (1968) *Cyclopean Vision*, Cambridge, MA.: M.I.T. Press.

[Kalman(1961)] Kalman, R., and Bucy, R., (1961) New Results in Linear Filtering and Prediction Theory. *Transaction ASME (Journal of Basic Engineering)*, 83D(1):95–108.

[Kohonen(1982)] Kohonen, T., (1982) Self-organized formation of topologically correct feature maps, *Biol. Cyber.*, 43, pp. 59-69.

[Linsker(1986)] Linsker, R. (1986) *From basic network principles to neural architecture*, Proc. Nat. Acad. Sci, U.S.A., 83, pp. 7508-7512, 8390-8394, 8779-8783.

[Luenberger(1963)] Luenberger, D., (1979) *An Introduction to Dynamic Systems* New York: Wiley

[Mallat(1987)] Mallat, S. G., (1989) *A theory for multiresolution signal decomposition: the wavelet representation. IEEE Trans. Pattern Analysis and Machine Intelligence*, 11(7):674–693.

[Matthies(1987)] Matthies, L., Kanade, T., and Szeliski, R. (1989) Kalman-Filter Based Algorithms for Estimating Depth from Image Sequences *Int'l Journal of Computer Vision*, 3:209-236.

[Marr(1982)] D. Marr, (1982) *Vision*, San Francisco,: W. H. Freeman and Co.

[Mase(1990)] Mase, K. and Pentland, A., (1990) Automatic Lipreading by Computer, *Trans. Inst. Elec. Info. and Comm. Eng.*, vol. J73-D-II, No. 6, pp. 796-803.

[Meyer(1986)] Meyer, Y., (1986) Principe d'incertitude, bases hilbertiennes et algebres d'operateurs. *Bourbaki Seminar*, No. 662, 1985-1986.

[Pentland(1987)] Pentland, A. P. (1987) A New Sense for Depth of Field. *IEEE Trans. Pattern Analysis and Machine Intelligence*, 9(4):523–531.

[Pentland(1990a)] Pentland, A., (1990) Physically-Based Dynamical Models for Image Processing and Recognition, *Mustererkennung 1990, Informatik-Fachberiche 254*, pp. 171-193, R. E. Grosskopf, Ed., Springer-Verlag.

[Pentland(1991b)] A. Pentland, and S. Sclaroff, (1991) Closed-Form Solutions to Physically-Based Modeling and Recognition. *IEEE Trans. Pattern Analysis and Machine Intelligene*, July.

[Pentland(1991c)] Pentland, A., (1991) Spatial and Temporal Surface Interpolation using Wavelet Bases, *Computer Vision III*, SPIE Conf. No. 1570, San Diego, CA.

[Poggio(1985)] Poggio, T., Torre, V., and Koch, C., (1985) Computational vision and regularization theory. *Nature*, 317:314–319, Sept. 26.

[Singh(1990)] Singh, A. (1990) An Estimation-Theoretic Framework for Image-Flow Computation, *IEEE Int'l Conf. on Computer Vision*, Osaka, Japan, pp. 168-177.

[Simoncelli(1990)] Simoncelli, E., and Adelson, E., (1990) Non-Separable Extensions of Quadrature Mirror Filters to Multiple Dimensions, *Proceedings of the IEEE*, 78(4):652–664.

[Sperling(1970)] Sperling, G., (1970) Binocular vision: a physical and a neural theory, *Amer. J. of Psychology*, 83, 461-534.

[Terzopoulos(1987)] Terzopoulis, D., (1987) Regularization of inverse visual problems involving discontinuities, *IEEE Pattern Analysis and Machine Intelligence*, Vol. 8, No. 6, pp. 413-424.

[Turk(1991)] Turk, M., and Pentland, A., (1991) Eigenfaces for Recognition, *Journal of Cognitive Neuroscience*, 3(1):71-87.

[Wilson(1984)] Wilson, H., and Gelb, G., (1984) Modified line-element theory for spatial-frequency and width discrimination, *J. Opt. Soc. Am. A* 1(1):124-131.

Appendix: Spring Models, Regularization, Wiener and Kalman Filtering

The steady-state Kalman filter solution produced by the spring model of Sections 3.1 and 3.2 is equivalent to Wiener filtering. This means that we can derive $\mathbf{K}_f$, and thus $\mathbf{K}$, in terms of the power spectrum of the noise and disparity measurements. To accomplish this, we first define $\mathbf{\Phi}_c$ to be the discrete cosine transform basis matrix (i.e., a matrix whose columns are are sampled cosine functions of all frequencies, the cosine transform basis vectors). Then the discrete cosine transforms of $\mathbf{U}$ and $\mathbf{D}$ are

$$\tilde{\mathbf{U}}(\omega) = \mathbf{\Phi}_c \mathbf{U} \qquad \tilde{\mathbf{D}}(\omega) = \mathbf{\Phi}_c \mathbf{D} \tag{45}$$

Similarly, the cosine domain representation of $\mathbf{K}$ is

$$\tilde{\mathbf{K}}(\omega) = \mathbf{\Phi}_c^T \mathbf{K} \mathbf{\Phi}_c \tag{46}$$

where $\tilde{\mathbf{K}}(\omega)$ is a diagonal matrix with real-valued entries (i.e., $\mathbf{KU}$ can be represented as a convolution).

Using this change of basis, then Equation 24 becomes

$$\lambda \tilde{k}_\omega \tilde{u}_\omega + \tilde{u}_\omega - \tilde{d}_\omega = 0 \tag{47}$$

or

$$\tilde{u}_\omega = \left[\frac{1}{1 + \lambda \tilde{k}_\omega}\right] \tilde{d}_\omega \tag{48}$$

where $\tilde{u}_\omega$ and $\tilde{d}_\omega$ are the ω-frequency cosine components of $\mathbf{U}$ and $\mathbf{D}$, and $\tilde{k}_\omega$ are the diagonal components of $\tilde{\mathbf{K}}(\omega)$.

If we choose $\lambda \tilde{k}_\omega = \tilde{n}_\omega / \tilde{u}_\omega$, then Equation 48 is exactly the Wiener filter. The first-order smoothness springs, also called the membrane model, have $\tilde{k}_\omega = \omega$. The second-order smoothness springs, also called the thin-plate model, have $\tilde{k}_\omega = \omega^2$.

The technique of regularization, as standardly employed in the vision literature, is also identical to the steady-state Kalman filter of Equation 28 for the correct choice of $\mathbf{K}_f^{-1}$. Thus we may view regularization as a special case of the Kalman filtering approach.

In practice regularization algorithms have employed either the membrane or thin-plate models, i.e., either the first- or second-order smoothness springs, in order to define $\mathbf{K}$. When data is present for all nodes, they are the Wiener filters (and thus optimal) in the case that the ratio of signal-to-noise is ω (first-order or membrane model) or ω^{-2} (second-order or thin plate model). For other signal-to-noise characteristics, or when the data's sampling distribution is known, more accurate estimates of the surface shape can be obtained by choosing $\mathbf{K}$ using either the Riccati equation or the Wiener filter criterion (Equation 48).

FIGURE-GROUND SEPARATION OF CONNECTED SCENIC FIGURES:

BOUNDARIES, FILLING-IN, AND OPPONENT PROCESSING

Stephen Grossberg† and Lonce Wyse‡
Center for Adaptive Systems
and
Department of Cognitive and Neural Systems
Boston University
111 Cummington Street
Boston, MA 02215

Abstract

A neural network model is described for automatic parallel separation of connected scenic figures from one another and from their backgrounds. The model is part of a self-organizing architecture for invariant pattern recognition in a cluttered environment. The figure-ground separation process iterates operations from a Feature Contour System (FCS) and a Boundary Contour System (BCS) in the order FBF. The FCS discounts the illuminant and fills-in surface properties, such as brightness and color, using the discounted signals. A key idea of the FBF network is to use filling-in for figure-ground separation. The BCS generates boundary segmentations that define the regions in which filling-in occurs. The BCS is modelled by a feedforward network, call the CORT-X 2 filter, that combines oriented receptive fields with rectifying, competitive, and cooperative interactions to detect, regularize, and complete boundaries in up to 50% analog noise. This filter combines complementary properties of large receptive fields and small receptive fields, and of on-cells and off-cells, to generate positionally more accurate and less noisy boundaries. Double opponent interactions of on-cells and off-cells facilitate separation of figures with incomplete CORT-X boundaries. The results clarify why an FBF network can rapidly separate figures that humans cannot separate during visual search tasks.

1. Theoretical Background: Figure-Ground Separation by Humans and Machines

This chapter contributes to the development of a self-organizing neural network architecture for invariant pattern recognition in a cluttered environment. Carpenter and Grossberg (1988) described a version of this architecture. A related scheme is considered herein (Figure 1), whose primary functional stages are:

---

† Supported in part by the Air Force Office of Scientific Research (AFOSR 90-0175), the Army Research Office (ARO DAAL-03-88-K0088), DARPA (AFOSR 90-0083), and Hughes Research Laboratories (S1-804481-D and S1-903136).

‡ Supported in part by the American Society for Engineering Education and Hughes Research Laboratories (S1-804481-D).

Acknowledgements: The authors wish to thank Cynthia E. Bradford, Carol Yanakakis Jefferson, and Diana Meyers for their valuable assistance in the preparation of the manuscript.

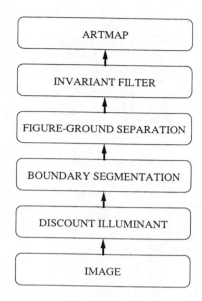

Figure 1. Stages of a neural network architecture for real-time automatic target recognition.

Stage 1. Discount the illuminant.

Stage 2. Detect, regularize, and complete figure boundary. Suppress interior and exterior image noise.

Stage 3. Detach figure from ground.

Stage 4. Filter to give invariance under translation, rotation, and contraction.

Stage 5. Let invariant spectra of the boundary-enhanced, noise-suppressed, detached figures be the input patterns to an ART or ARTMAP architecture for stable self-organization of recognition categories (Carpenter and Grossberg, 1987a, 1987b, 1988; Carpenter, Grossberg, and Reynolds, 1991; Carpenter, Grossberg, Markuzon, Reynolds, and Rosen, 1991, and this volume).

In Carpenter, Grossberg, and Mehanian (1989), a neural network preprocessor for the second stage—the boundary segmentation stage—of such an architecture was described. This boundary preprocessor, called the CORT-X filter, detects, regularizes, and completes sharp (even one pixel wide) boundaries, while simultaneously suppressing the noise. The CORT-X filter is based upon the biologically derived Boundary Contour System of Grossberg and Mingolla (1985a, 1985b, 1987). The Boundary Contour System uses nonlinear feedback interactions to select and complete sharp boundaries over long gaps in image contours. The CORT-X model uses only feedforward interactions that are faster to simulate and easier to implement in hardware. Their ability to complete boundaries is also more limited, but is adequate for many applications. The CORT-X filter is modified herein to deal with at least

50 percent analog noise. This modified filter is called CORT-X 2 to distinguish it from the original model.

We use the CORT-X 2 model to develop a new approach to designing the first three stages of the architecture. We show how Stages 1 and 2 can be designed to facilitate figure-ground separation by Stage 3. Figure-ground separation is the process whereby a figure, or object, in a scene is separated from other figures and background clutter. Whereas knowledge about a figure may facilitate its separation, such knowledge is not necessary for biological vision systems to carry out figure-ground separation. Experiences abound of unfamiliar figures that "pop out" from their backgrounds before they ever enter our corpus of learned knowledge about the world. The fact that figure-ground separation can occur even for unfamiliar figures contributes to the general-purpose nature of biological vision, which can process both unfamiliar and familiar scenes, and does not require prior instruction about an environment in order to operate effectively.

In this chapter, a new type of system is described that is capable of automatic figure-ground separation (Figure 2). This process separates scenic figures whose emergent boundary segmentations (defined below) surround a connected region. As a result of this property, such a system can automatically distinguish between connected and disconnected spirals (Figure 3), a benchmark that gained fame through its emphasis in the book by Minsky and Papert (1969, 1988) on perceptrons. Why the present biologically-motivated algorithm can distinguish interleaved spirals in a way that humans cannot is described below in Section 7. This analysis also clarifies why humans cannot rapidly detect conjunctions of some visual features, such as shape and color, but can rapidly detect conjunctions of other visual features, such as disparity and color, or motion and color (Nakayama and Silverman, 1986; Treisman and Gelade, 1980), that distinguish target figures from surrounding distractors.

Figure-ground separation is an essential step in pattern recognition whenever the objects to be recognized may vary in their position, orientation, and size in a scene. Once a scenic figure has been separated from other figures and background clutter, as in Figure 2, it can be input to an invariant filter at Stage 4 of the architecture (Figure 1). The output of the filter is invariant under translations, rotations, and contractions of the figure. If the figure is not first detached from the scenic background, then output of the filter is not invariant under translations, rotations, and contractions of the figure with respect to the fixed background.

There exist at least three different approaches to automatic figure-ground separation. Section 2 reviews a method for figure-ground separation that uses combinations of laser radars, or related artificial detectors. Sections 3–5 review a method that arises in a neural network model of biological vision that has been called FACADE Theory (Grossberg, 1987, 1990; Grossberg, Mingolla, and Todorović, 1989). Section 6 begins the exposition of the model, which we call an FBF model for reasons that are noted below. The FBF model is capable of separating connected figures from their backgrounds in response to either monochromatic images, such as a grey-scale high-altitude photograph, or from images derived from multiple detectors. Its mechanisms are based upon those described in FACADE Theory, which are adapted for use in a setting where only a single detector may be available.

Figure-ground separation is accomplished in the model by iterating operations adapted from the Feature Contour System (FCS) and the Boundary Contour System (BCS) of FACADE Theory in the order FCS–BCS–FCS; hence the name FBF model. The FCS operations include the use of nonlinear shunting networks to compensate for variable illumi-

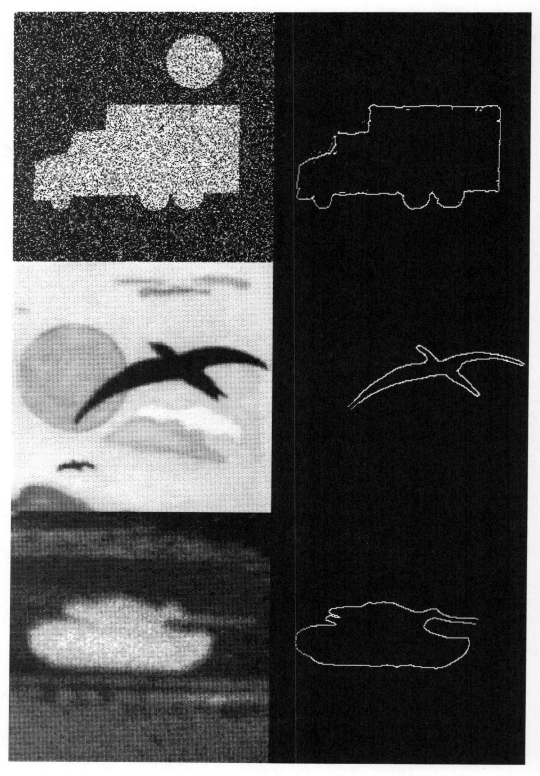

Figure 2. Three examples of figure-ground separation by Stages 1–3 of the architecture.

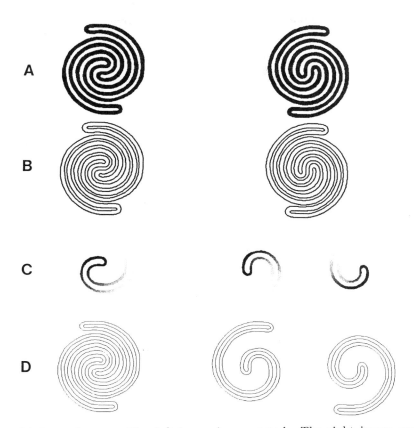

Figure 3. (a) Input images. The left image is connected. The right image consists of two disconnected components. (b) Boundary segmentations. (c) Beginning of filling-in. (d) Separated outputs. The left boundary lies on a single Stage 3 slab. Each connected component of the right boundary lies on a different slab.

nation ("discount the illuminant") and nonlinear diffusion networks to control filling-in. A key new feature of an FBF network is the use of filling-in for figure-ground separation. The BCS operations include oriented filters joined to competitive and cooperative interactions designed to detect, regularize, and complete boundaries in up to 50 percent noise, while suppressing the noise. The new CORT-X 2 filter achieves this competence by using both on-cells and off-cells to generate a boundary segmentation from a noisy image. On-cells and off-cells exhibit two different types of complementary responses that are useful for processing noisy images: they respond in a complementary way to noise pixels (light on dark, vs. dark on light) and to convex and concave image contours. These complementary reactions are joined together to form a boundary segmentation that overcomes the weaknesses of either detector taken separately (Section 6). As a result, the output of such a segmentation process is capable of joining together image features derived from opposite directions-of-contrast. Another new idea is the use of a double-opponent network, defined in Section 6, to facilitate figure-ground separation of regions with incomplete boundary segmentations.

2. Figure-Ground Separation by Artificial Detectors: Laser Radar Arrays.

A conceptually simple technique for figure-ground separation is to utilize a detector which is itself capable of automatically separating figure from ground. Such an approach motivated the original architecture of Carpenter and Grossberg (1988). There it was assumed that the detector consists of pairs of laser radars whose outputs are and-gated to separate figure from ground. For example, a range detector focussed at the distance of the figure can extract the figure and a contiguous piece of the ground. The figure can be detached from the ground by spatially intersecting the range pattern with a pattern from another detector that is capable of differentiating figure from ground. A doppler image can be intersected with the range image when the figure is moving. The intensity of laser return can be intersected with the range image when the figure is stationary (Gschwendtner, Harney, and Hull, 1983; Harney, 1980, 1981; Harney and Hull, 1980; Hull and Marcus, 1980; Kolodzy, 1987; Sullivan, 1980, 1981; Sullivan, Harney, and Martin, 1979).

More generally, arrays of laser radar detectors may be used to separate image figures into distinct network levels, or slabs. This can be accomplished by intersecting the output signals from multiple detectors that simultaneously inspect the image. For example, a series of range detectors can register all objects at a regular series $D, 2D, 3D, \ldots, ND$ of distances, within some tolerance ΔD; a series of doppler detectors can register all objects at a regular series $S, 2S, 3S, \ldots, MS$ of speeds, within some tolerance ΔS; and an $N \times M$ matrix of intersection images can be generated which extract the figure at each combination of distance iD and speed jS within this tolerance. These $N \times M$ intersection images can be processed simultaneously in parallel with CORT-X and invariant filters before they activate a parallel array of ART or ARTMAP pattern recognition architectures.

3. Figure-Ground Separation in FACADE Theory

During biological vision, the retinal detectors do not, in themselves, separate figure from ground. One task of neural network research is to suggest how subsequent network processes which are activated by the retinal detectors may generate this competence. Grossberg (1987; reprinted in Grossberg, 1988) has, for example, introduced a neural theory of binocular vision in which image figures are separated from one another into distinct network levels, or slabs. A macrocircuit of this theory is shown in Figure 4, where the vertically hatched boxes form part of the Boundary Contour System (BCS) and the dotted boxes form part of the Feature Contour System (FCS). The theory describes how parallel and hierarchical interactions between the BCS and FCS generate a multiplexed, multiple-scale representation, called a FACADE representation, of the scene's Form-And-Color-And-DEpth. Within this representation, figural components which encode distinctive combinations of features, such as prescribed combinations of color, depth, and size, are segregated from one another into different network levels. These levels, in turn, activate subsequent stages of network processing that are designed for visual object recognition (Figure 5).

For present purposes, the main insight that may be derived from FACADE theory is that a properly designed sequence of FCS-BCS-FCS operations can separate figure from ground. Henceforth all networks that use this strategy, including the present model, will be called *FBF networks*. To indicate how such a sequence of operations can separate figure

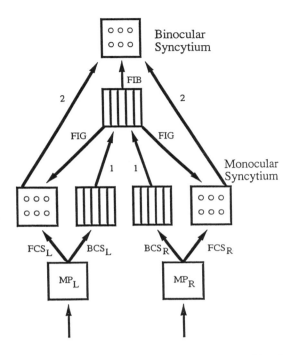

Figure 4. Macrocircuit of monocular and binocular interactions within the Boundary Contour System (BCS) and the Feature Contour System (FCS): Left and right monocular preprocessing stages (MP_L and MP_R) send parallel monocular inputs to the BCS (boxes with vertical lines) and the FCS (boxes with three pairs of circles). The monocular BCS_L and BCS_R interact via bottom-up pathways labelled 1 to generate a coherent binocular boundary segmentation. This segmentation generates output signals called filling-in generators (FIGs) and filling-in barriers (FIBs). The FIGs input to the monocular filling-in domains, or syncytia, of the FCS. The FIBs input to the binocular filling-in domains, or syncytia, of the FCS. Inputs from the MP stages interact with FIGs at the monocular syncytia where they select those monocular FC signals that are binocularly consistent. The selected FC signals are carried by the pathways labelled 2 to the binocular syncytia, where they interact with FIB signals from the BCS to generate a multiple scale representation of form-and-color-and-depth within the binocular syncytia.

from ground, we review two different competences of FACADE theory: discounting variable illumination and multidimensional fusion.

4. Discounting Variable Illumination and Filling-In

The theory provides an explanation of how variable illumination conditions are automatically discounted and used to trigger a filling-in process that completes a surface representation over image regions which are suppressed by the discounting process. A monocular version of this process was modelled by Cohen and Grossberg (1984) and Grossberg and Todorović (1988) to explain data about monocular brightness perception. This monocular

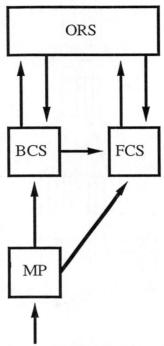

Figure 5. A macrocircuit of processing stages: Monocular preprocessed signals (MP) are sent independently to both the Boundary Contour System (BCS) and the Feature Contour System (FCS). The BCS preattentively generates coherent boundary segmentations from these MP signals. These structures send outputs to both the FCS and the Object Recognition System (ORS). The ORS, in turn, rapidly sends top-down learned expectation signals to the BCS. These expectations can modify the preattentively completed boundary structures using learned, attentive information. The BCS passes these modifications along to the FCS. The signals from the BCS organize the FCS into perceptual regions wherein filling-in of visible brightnesses and colors can occur. This filling-in process is activated by signals from the MP stage. The completed FCS representation, in turn, also interacts with the ORS. Fusion of BCS and FCS representations occurs in the ORS.

model is schematized in Figure 6.

In this model, variable illumination conditions are discounted by a shunting on-center off-surround network (Level 2), which constitutes the first FCS stage. (See the Appendix for all equations.) Image regions of high relative contrast are amplified and regions of low relative contrast are attenuated as a consequence of the discounting process. The shunting network, in turn, topographically activates a filling-in network (Level 6) which constitutes the second FCS state. This filling-in network uses a nonlinear diffusion process to complete a brightness representation over both the amplified and attenuated image regions.

Filling-in is restricted to compartments whose boundaries are defined by topographic signals from the Boundary Contour System, or BCS (Levels 2–5). The BCS converts signals

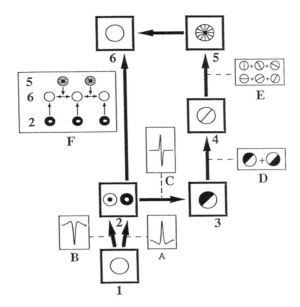

Figure 6. Grossberg-Todorović monocular model of how the FCS discounts variable illuminants and regulates featural filling-in: The thick-bordered rectangles numbered from 1 to 6 correspond to the levels of the system. The symbols inside the rectangles are graphical mnemonics for the types of computational units residing at the corresponding model level. The arrows depict the interconnections between the levels. The thin-bordered rectangles coded by letters A through E represent the type of processing between pairs of levels. Inset F illustrates how the activity at Level 6 is modulated by outputs from Level 2 and Level 5. This simplified model directly extracts boundaries from image contrasts, rather than generating emergent segmentations from image contrasts. The model's key elements concern how the Level 2 network of shunting on-center off-surround interactions discounts variable illuminants while extracting Feature Contour signals, and how Level 5 fills-in these signals via a nonlinear diffusion process within the compartments defined by BCS output signals.

from the first FCS stage (Level 2) into a boundary segmentation one of whose functions is to trigger a BCS-FCS interaction that contains the filling-in process at the second FCS stage. The result of this FBF interaction is a surface representation of featural quality, such as brightness or color, that is relatively uncontaminated by illumination conditions.

Figures 7 and 8 summarize computer simulations of Grossberg and Todorović (1988) that illustrate how the illuminant is discounted in FCS Level 2, and how the subsequent BCS-FCS interaction at Level 6 controls the filling-in process that completes the brightness representation. The image schematized in Figure 7a is called a McCann-Mondrian (Land, 1977). It is a patchwork of rectangular regions each with a different luminance. The image is uniformly illuminated. In Figure 7a, each circle's radius is proportional to the luminance registered by a network node located at the center of the circle. Figure 7b represents the activation pattern of the shunting on-center off-surround network at Level 2, Figure 7c represents the boundary representation at Level 5. Only on-cells were used to generate this

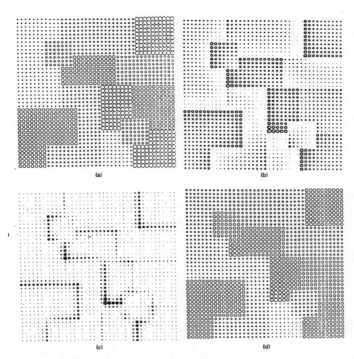

Figure 7. The evenly illuminated Mondrian. (a) The stimulus distribution consists of 13 homogeneous polygons with 4 luminance levels. Note that the square in the upper left portion of the stimulus has the same luminance as the square in the lower right portion. However, the average luminance of the regions surrounding the lower square is higher than the corresponding average luminance for the upper square. (b) The on-cell distribution. The amount of on-cell activity within the upper square is higher than within the lower square. (c) The Boundary Contour output. (d) The filled-in syncytium. The upper square is correctly predicted to look brighter than the lower square.

boundary, which consequently has uneven strength at concave and convex boundary shapes. Figure 7d represents the filled-in representation at Level 6. The diffusion spatially averages the activation patterns of Figure 7b within the compartments defined in Figure 7c.

In Figure 8a, the same image depicted in Figure 7a is illuminated from the lower right corner. Because the shunting on-center off-surround network at Level 2 effectively discounts the illuminant, the Level 2 activation patterns in Figures 8b and 7b are essentially identical. Hence the subsequent boundary patterns (Figures 8c and 7c) and filled-in patterns (Figures 8d and 7d) are also essentially identical.

In Figure 8d, the brightness, or activation level, of the square region in the upper left corner is larger than that of the square region in the lower right corner. In Figure 8a, the luminance, or activation level, of the upper left corner is smaller than that of the square region in the lower right corner. This luminance-to-brightness reversal compensates for the larger intensities of illumination in the lower right region.

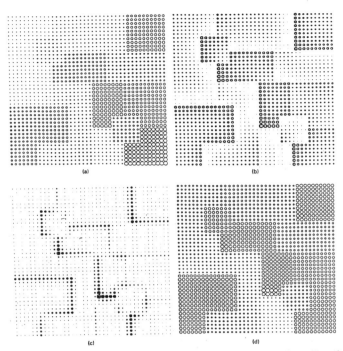

Figure 8. The unevenly illuminated Mondrian. (a) The stimulus distribution simulates the transformation of Figure 7a caused by the presence of a light source whose intensity decreases linearly from the lower right corner toward the upper left corner of the stimulus. The lower square is now more luminant than the upper square. (b) The on-cell distribution. (c) The Boundary Contour output. (d) The filled-in-syncytium. Figures 8b, 8c, and 8d are very similar to the corresponding figures for the evenly illuminated Mondrian (Figure 7). This illustrates the model's discounting of the illuminant. In addition, the upper square is still predicted to appear brighter than the lower square.

5. Multidimensional Fusion

An FBF interaction may be used to represent scenic form, notably scenic surface properties, as it separates figure from ground. This is achieved by embedding an FBF interaction into a binocular version of the theory. In the binocular theory, form, color, and depth are multiplexed together in the final representation; hence the mnemonic Form-And-Color-And-DEpth, or FACADE, in the theory's name.

The binocular version of the theory suggests how monocular image data from both eyes can be selectively processed so that only the binocularly consistent monocular data from each eye is allowed to influence the FACADE representation. Figure 4 schematizes the network that is used. In it, monocular BCS signals are derived from the monocular FCS patterns that discount the illuminant for each eye. These monocular BCS signals interact topographically to form the binocular boundary segmentation along the pathways labelled 1. This boundary segmentation regularizes and completes all the boundary data, across multiple spatial scales, that are capable of being binocularly fused. Binocularly discordant, or rivalrous, data are suppressed by competitive interactions. The binocular boundary segmentation sends topographic signals, called *filling-in barriers* (FIBS), to the monocular filling-in networks within the FCS that correspond to the left eye (FCS_L) and the right eye (FCS_R). This BCS-FCS interaction allows only those monocular featural data from (FCS_L) and (FCS_R) that are consistent with the binocular boundary segmentation to generate topographic output signals, labelled 2, to the binocular FCS stage. This BCS-FCS interaction carries out a type of figure-ground separation, since an FCS region can generate output signals only if it is surrounded by binocular FIGS from the BCS.

The binocular FCS stage is called the binocular syncytium. In the binocular syncytium, the selected monocular FCS signals from both eyes interact once again with the binocular BCS signals. Here the surviving monocular FCS signals activate a filling-in process within the compartments that are defined by the binocular BCS signals. These BCS signals are called *filling-in barriers* (FIBS) because they contain the filling-in process within their boundaries. Both FIGs and FIBs obey the same equations. Their different effects are due to their action at different locations in the network hierarchy. The FACADE representation that is generated within the binocular syncytium completes the preattentive figure-ground separation process by grouping distinctive combinations of features into figures within separate network levels, or slabs. These slabs thereupon send adaptively filtered signals to subsequent processing levels for purposes of visual object recognition.

Within such a biological theory of vision, the process of separating figures into different slabs exploits the fact that the retina contains photodetectors with different spectral sensitivities; for example, three types of retinal cones and one type of retinal rod. The theory suggests how figures may be spatially parsed into separate slabs based, in part, upon the distinctive colors that are derived from these detectors. In addition, there exist multiple spatial scales and multiple binocular disparity computations within the theory that further parse figural components into separate slabs based upon different size-disparity correlations (Grossberg, 1987; Grossberg and Marshall, 1989). Thus, although FACADE theory uses an FBF network—actually an FBFBF network—to achieve figure-ground separation, this network exploits the existence of multiple detectors and multiple-scale reactions to these detectors to carry out the separation. For moving images, relative motion provides another measure that is used for figure-ground separation.

The present FBF model shows how figural components may be separated into separate slabs even if only a single detector is used that, in itself, cannot separate figure from ground, and if subsequent processing stages cannot use multiple scales, binocular interactions, or relative motion as cues for separation. For example, how can individual figures be separated from the cluttered ground of a picture taken with a camera that uses monochromatic film? We now show how a suitably designed FBF network can accomplish this task for at least certain classes of images. Section 6 provides an intuitive description of network stages and their effects. The Appendix describes network equations and parameters. These sections can be read in either order.

6. Figure-Ground Separation by a Monochromatic FBF Network: The Dye-Injected FBF

In the FBF model, a key assumption is that the filling-in process is activated by *internally*

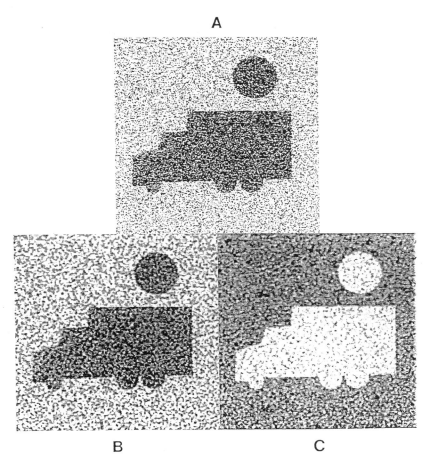

Figure 9. (a) The original figure in 50% noise (half the pixels are random). (b) The result of the ON-C shunting filter. (c) The result of the OFF-C shunting filter.

generated input sources. In particular, the network "paints" each connected figure of the image by using an internally generated "dye" that triggers the filling-in of that figure. This heuristic is realized by using the following procedure.

Step 1 (Discount the Illuminant). At the first FCS stage, variable illumination conditions are discounted both by a shunting on-center/off-surround network ("ON-C") and an off-center/on-surround network ("OFF-C"), operating in parallel. The ON-C network has a zero baseline activity. Hence, a cell's activity decays to zero if there is no signal within its entire receptive field. In contrast, the OFF-C network has a positive baseline activity. Because the OFF-C filter has a positive baseline activity and is inhibited by positive signal values, the network performs an image inversion. Both the ON-C network and the OFF-C network compensate for variable illumination and normalize the image using their shunting interactions (Figure 9).

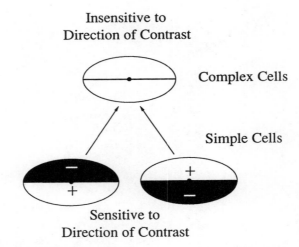

Figure 10. The simple cell and complex cell layers that process each of the two shunting network images in parallel. A horizontally oriented set of cells is shown. In Figures 11 and 12, only the outputs of horizontally oriented cells are displayed.

The ON-C and OFF-C networks operate in a complementary fashion. Along a straight boundary between a region of strong signal and one of no signal, both types of networks respond similarly by enhancing the contrast. At a concave corner of a high signal region, the ON-C network responds more strongly than the OFF-C network, while at a convex corner the converse is true (Grossberg and Todorović, 1988, Figure 6). This ON-OFF Complementarity Property also plays an important role in noise suppression when it interacts with the CORT-X 2 filter, as the next paragraphs explain.

Step 2 (CORT-X 2 Filter). The ON-C and OFF-C shunted images are transformed by a CORT-X 2 filter into a boundary representation. Each processing layer of this filter has the same number of cells as pixels in the image. The architecture is completely feedforward. Cells at a given layer have input fields ("IFs") that integrate over an area in the previous layer around its position in the field. Two separate scales (input field sizes) are used in parallel in the early stages of processing and are subsequently combined to take advantage of the best of their respective processing capabilities. The term *input field*, or in-field, is used instead of *receptive field* because the latter term from neurophysiology typically refers to the region at the first processing layer that influences the activity of a cell at any subsequent layer. Our layer-by-layer analysis of scale sizes requires a more microscopic analysis of network geometry.

The model's first stage, called the *simple cell layer*, is an oriented contrast detector that is sensitive to the orientation, amount, direction, and spatial scale of image contrast at a given image location. The orientation sensitivity is the result of an elliptically shaped IF (Figure 10). In-fields placed at equally spaced orientations operate in parallel at each position. The sensitivity to amount and direction of contrast is produced by exciting the cell with the inputs to one half of its IF, inhibiting it with the inputs to the other half, and thresholding the result to derive an output signal. The result is a half-wave rectification of

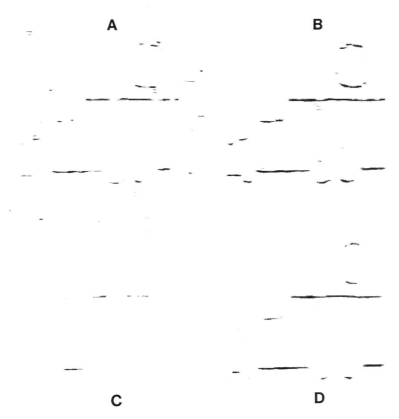

Figure 11. (a) Small scale horizontal complex cell output from the ON-C image of Figure 9b. (b) Small scale horizontal complex cell output from the OFF-C image of Figure 9c. (c) Large scale horizontal complex cell output from the ON-C image of Figure 9b. (d) Large scale horizontal complex cell output from the OFF-C image of Figure 9c.

the input.

The ON-C shunting network and the OFF-C shunting network each input to separate networks of simple cells at each receptive field size. Thus the ON-C and OFF-C networks together activate four networks of simple cells.

The outputs of these parallel simple cell networks are combined at each position to activate the second stage of the CORT-X 2 filter, called the *complex cell layer*. Complex cells are sensitive to the orientation, amount, and spatial scale of the contrast of the image at a given point, but not to the direction-of-contrast. This latter property is achieved by summing the outputs of all like-oriented simple cells at each position, including cells that are sensitive to opposite direction-of-contrast and that receive inputs from either the ON-C or OFF-C shunting networks. Two networks of complex cells, each sensitive to a different scale size, are generated in this way. Adding the half-wave rectified outputs from pairs of simple cells that are sensitive to opposite direction-of-contrast has the same net effect as full-wave

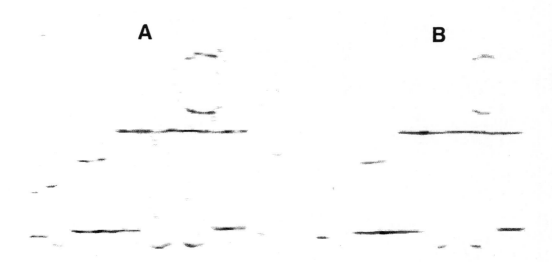

Figure 12. (a) ON-OFF small scale horizontal complex cell response derived by adding Figures 11a and 11b. (b) ON-OFF large scale horizontal complex cell response derived by adding Figures 11c and 11d.

rectification.

The complementary processing properties of the ON-C and OFF-C cells help to suppress noise when their outputs are further processed by simple cells and complex cells. To a first approximation, such a detector responds to the ratio of the inputs to each half of its oriented receptive field. A small amount of noise signal against a background of no signal affect them more than the same amount of "drop out" noise against a background of strong signal. The inversion of the image performed by the OFF-C filter changes the direction-of-contrast between any noise and its signal background. Thus, the noise will be disruptive in only one of the two parallel networks of simple cells and complex cells, while actual region boundaries will be strongly detected in both (Figure 11).

The output from both networks of contrast detectors are then summed, an operation which takes advantage of the ON-OFF Complementarity Property. First it yields roughly equal responses to both concave and convex curvatures. Second, since noise in a given region is suppressed in either the ON-C or the OFF-C network, while boundaries one are strong in both, the summation strengthens the boundary signals relative to the noise (Figure 12).

Figure 12 demonstrates that the two scales also exhibit another type of complementary processing capabilities (Carpenter, Grossberg, and Mehanian, 1989). The smaller scale filter does a better job of boundary localization than the larger scale filter, especially at positions of high boundary curvature, whereas the larger filter does a better job of noise suppression and boundary completion (Figure 13). In particular, the large scale filter achieves good noise suppression far from the image boundaries, but not within the radius of large scale IF's near these boundaries. The small scale filter is relatively poor at noise suppression anywhere. Canny (1986) has suggested how a single spatial scale can trade off between these virtues,

	SMALL SCALE	LARGE SCALE	SMALL + LARGE SCALE
BOUNDARY LOCALIZATION	YES	NOT AT HIGH CURVATURE BOUNDARIES	YES
NOISE SUPPRESSION	NO	YES	YES
BOUNDARY COMPLETION	NOT AT SEGMENTS MISSING DUE TO NOISE	YES	YES

Figure 13. Complementary processing properties of large and small oriented receptive fields. Reprinted from Carpenter, Grossberg, and Mehanian (1989) with permission.

NOISE SUPPRESSION NEAR BOUNDARY

ORIENTED SPATIAL COMPETITION:
Complex cells $C_s(x, k)$ output to an oriented spatial competition which inputs to target cells $D_s(x, k)$. Target cells:

at a boundary are activated;
near a boundary are suppressed;
far from a boundary may be activated by noise.

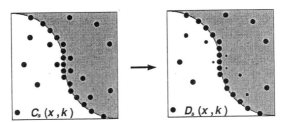

Figure 14. Oriented spatial competition inhibits noise pixels near the boundary. Reprinted from Carpenter, Grossberg, and Mehanian (1989) with permission.

but notes that "we cannot improve both simultaneously" (p. 684). The CORT-X family of models suggests a strategy whereby two or more scales can be combined to realize the best features of each, using the following operations.

The next stage of filtering is designed to control noise near boundaries. It converts complex cells into hypercomplex cells. Each complex cell excites the hypercomplex cell at the next level that corresponds to its position and orientation, while inhibiting hypercomplex cells at nearby positions that are not colinear with its axis of symmetry (Figure 14). This

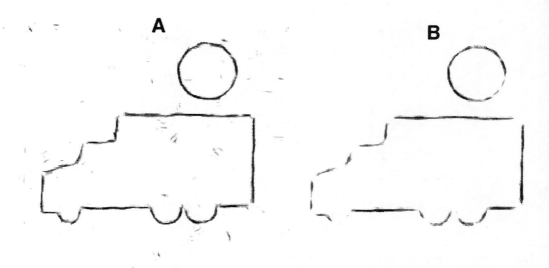

Figure 15. Output from the second competitive stage due to competition between orientations at each location. (a) Small scale. (b) Large scale.

interaction is called the first competitive stage. The level after that, is called the second competitive stage. It sharpens the activation pattern across orientations at each position and scale. In particular, at each position and scale, that cell is chosen whose orientation receives the maximal input from the first competitive stage (Figure 15).

The final operations include cooperative interactions between both filter sizes that select their desirable properties and eliminate their undesirable ones. The small scale's ability to localize boundaries and the large scale's ability to suppress noise and complete gaps in the boundaries are maintained by these cooperative interactions (Figure 16a).

Gaps in the boundary become more likely as the noise level of the original image increases. To overcome this problem, the cooperative level also includes oriented long-range interactions that are activated by the selected maximal responses in Figure 15. These long-range interactions activate an inactive cell if enough like-oriented cells at the previous level are active on both sides of the cell (Figure 16b). The cooperative cells play the role of the bipole cells in the Boundary Contour System (Grossberg and Mingolla, 1985a, 1985b). The final output of the CORT-X 2 filter is the sum of the combined-scales image and the completed-gap image (Figure 17).

Step 3 (Filling-In). The output from the CORT-X 2 filter is topographically mapped into M filling-in networks F_m, $i = 1, 2, \ldots, M$. In the Grossberg and Todorović (1988) article, the signals that trigger filling-in are generated by the image (Section 4). In the present application, they are generated by input sources that lie within the network. Moreover, each internally generated input is delivered to its filling-in network at a different position.

Figure 16. (a) Unoriented cooperation between both scales. (b) Oriented cooperation within the large scale.

Imagine, for definiteness, an $n \times n$ grid of $M = n^2$ nodes laid out over the boundary image generated by the CORT-X 2 filter. Each filling-in network F_m is associated with a different grid point where it will receive a featural "dye" injection into its copy of the boundary image. The injection then spreads unimpeded where there is no boundary signal, but does not spread through points where a boundary signal exists. Thus, each injection fills-in the connected figure that surrounds the injection point (Figure 18). Injection points do not need to occur at each pixel position. Rather, they should be dense enough to be enclosed, with high probability, by the smallest connected boundaries of regions that the CORT-X 2 filter can detect.

If the grid of injection points is dense enough, then all connected figural components will receive an injection within its boundary in at least one filling-in network F_m. This process is easy to replicate in large numbers because all the networks are identical except for the different, but regular, locations of the injected inputs, and all can operate independently and asynchronously in parallel.

Step 4 (Figure-Ground Separation by a Double Opponent Network). Each filling-in network feeds its activation pattern in parallel to another pair of shunting networks, one on-center/off-surround (ON-C) and one off-center/on surround (OFF-C). This completes the second "F" operation in the FBF model. Because of their contrast enhancing and ratio-processing properties, these filters amplify the filled-in activity near figural boundaries while tending to suppress the low-contrast regions generated by the spreading of activation across the interiors of figures and background regions. In order to achieve figure-ground separation of a connected region whose boundary segmentation contains holes, the OFF-C output is subtracted from the ON-C output.

Such an operation generates cells that are called *double opponent* cells in neurobiology. The "double" opponency describes the two successive operations of inhibition, one in each

Figure 17. Final CORT-X 2 filter output.

shunting network and one between shunting networks, that defines the net output. Double opponent cells have traditionally been described as color processing cells *in vivo*. Grossberg (1987) noted that double opponency also helps to separate figures onto separate slabs in FACADE Theory. The FBF model adapts this observation to the single-detector single-scale case.

In order to understand how a double opponent network helps, consider first the case where there are no holes in a boundary at which the spreading activation could leak through. If the filling-in process has been given enough time to reach all of the enclosing boundary signals, then the ON-C filter produces an output only inside the enclosing boundaries, due to the injected activity (Figure 19E).

The OFF-C filter produces signal only outside the enclosing boundaries, due to the spontaneous baseline activity which has not been quenched by the input injection (Figure 19H). In this hole-free case, the ON-C filtered images effectively separate each connected region from all others in the original image. However, because of the Gaussian shapes of the kernels used in these filters, the edges in the ON-C patterns also exhibit a Gaussian spatial spread. Also, smooth gradients, such as those generated by activation spreading across regions where there are no boundary signals, may not be quenched to zero if the center and surround kernels are of unequal area. A better boundary can be generated by multiplicatively gating each ON-C pattern with the original CORT-X 2 output. The resulting M images, one on each slab, then contain among them all the separated figural boundaries of the original image (Figure 19J–L). This method can be run in real-time until the output from each copy generates a recognition event at its ART network, or the input image is removed.

If the original image contains so much noise that the CORT-X 2 filter is unable to produce boundaries without weak points or small gaps, then significant activity could leak out of figural components during the filling-in process (Figure 19A). On the other hand, the ART recognition process, operating in real-time, could recognize the figure before the

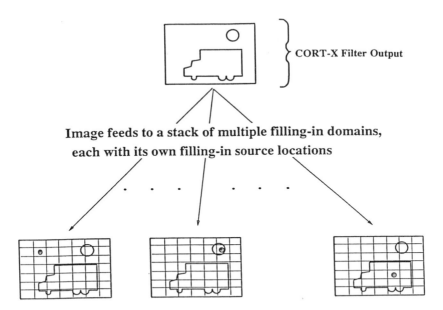

Figure 18. Copies of the CORT-X filter output are sent to M filling-in networks (three of which are pictured here). Activity is injected into a different place in each (gray disks) and begins to spread.

equilibrium state of equal activation on both sides of the boundary is reached.

Leakage of diffusing activity causes no problem if there are no other nearby object boundaries in the original scene. The spread of activation would produce a smooth gradient through the gap, and the shunting operation would not detect any contrast at the point of leakage or outside the object until well after the recognition event occurred (Figure 19J).

If, however, another object's boundary were near a boundary gap, then leaking activation followed by the ON-C shunting operation could detect the spurious boundary if the injection site were closer to the spurious boundary than to other boundaries of the figure.

In this case, the OFF-C filtered image is helpful (Figure 20). The ON-C signal at boundary regions exterior to the desired object is not as strong as the OFF-C filter signal at these points (Figure 20A), unless the boundary and the injection site are both proximal to the boundary gap. This property is due to the fact that very little injected activity would have spread there to excite the ON-C field, and the OFF-C field is tonically active. Subtracting the OFF-C image from the ON-C image to generate a double opponent image (Figure 20B), followed by CORT-X 2 gating (Figure 20C), therefore helps to suppress leakage from an incomplete boundary. The OFF-C subtraction does not distort the desired boundary signal when the ON-C signal is stronger than the OFF-C signal. In Figure 21, for example, a hole was made in the boundary of the truck near the boundary of another figure before filling-in occurs. The ON-C network clearly detects the outer boundary of the moon (Figure 21A). The OFF-C filtered signal is, however, stronger at these points (Figure 21B). Without taking the OFF-C network information into account, the output could cause difficulties for a

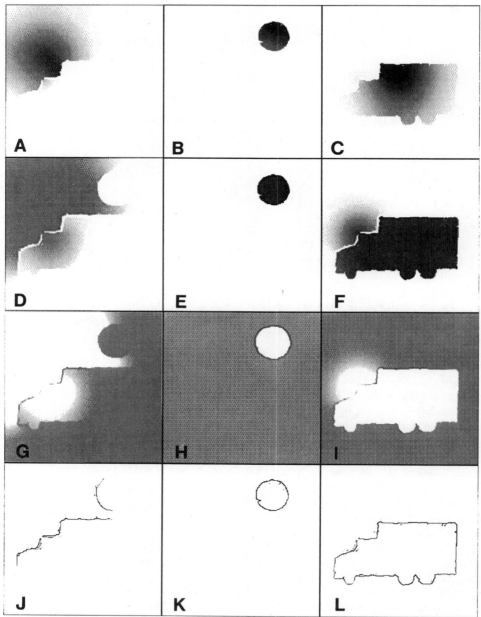

Figure 19. Row 1: Filling-in of activity is initiated at three different places in parallel filling-in networks. Due to the quantization of the gray scale, small filled-in activations do not print, even though they are detected by the shunting networks, as noted in rows 2 through 4. Row 2: The ON-C filter output of the respective filled-in regions. Row 3: The OFF-C filter output of the respective filled-in regions. Row 4: The final "separated figure" outputs to be passed along to a pattern recognizer such as an ART network.

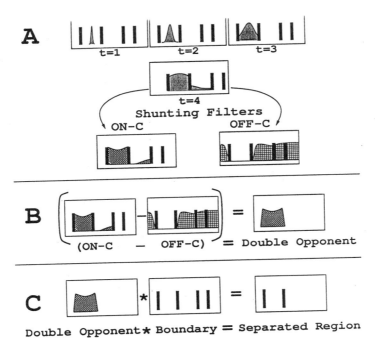

Figure 20. Each box represents a one-dimensional cross-section taken vertically through the truck/moon image of Figure 21. The 2 left-most black bars are from the truck boundary, the two rightmost, from the moon. (A) The injected activity at time=1 spreads, impeded only at locations where a boundary signal exists. At time t=4, we see leakage due to a nearby hole in the truck boundary which has flowed as far as the moon boundary. The ON-C and OFF-C filters of the activation contour are shown, with the boundary signals superimposed as black vertical bars. (B) The double opponency interaction suppresses the relatively weak leakage signal in the ON-C filter output. (C) The final gating operation with the original boundary from the CORT-X2 filter produces the desired representation of the separated truck figure.

pattern recognizer since the objects would not be separated (Figure 21C). Combining the two shunting network filters using a double opponent interaction produces the desired separation (Figure 21D).

A single figure is typically large enough to enclose several injection sites across the set of filling-in networks F_m. Thus, even if the injection site is close to a boundary gap and to a nearby spurious boundary in one network F_m, the injection sites of other filling-in networks will be further from the boundary gap. In some of these networks, double opponent processing can compensate for the gap, and trigger a correct recognition of the figure from the corresponding ART network. If the spurious boundaries are strong enough, as in Figure 20J, then the corresponding ART network will remain silent because the combination of partial figure and background is not similar enough to a previously learned recognition

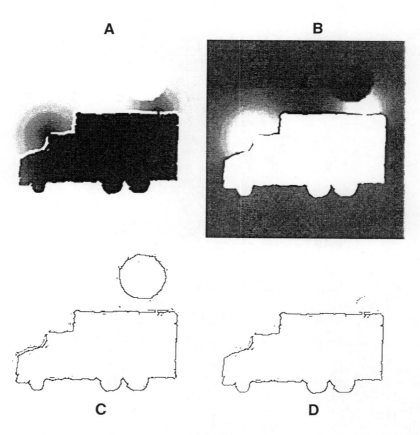

Figure 21. (a) The ON-C filter output of the filled in region with a hole in the boundary proximal to another figure. (b) The OFF-C filter of the filled in region. (c) The output as it would look without using the complementary OFF-C image. (d) The final image which uses the OFF-C filter to control the effects of the leakage leaving only the desired boundary signal.

code to be recognized. Parallel separation of multiple connected figures from other figures and background can thus be performed by an FBF architecture under noisy conditions by exploiting the parallel recognition by ART networks of boundary segmentations that are generated by sampling multiple filling-in perspectives.

In summary the FBF network separates figure-from-ground by using regular arrays of feedforward networks to discount the illuminant and to generate boundary segmentations, nearest-neighbor feedback signals for filling-in, and a proliferation of these circuits in parallel copies that input to parallel arrays of ART pattern recognition networks. The FBF networks thus seem to be appropriate designs for implementation in parallel hardware capable of operating at high rates in real time.

7. Recognition of Conjunctive Features

Why can the biologically-motivated FBF network automatically distinguish between the pair of connected and disconnected Minsky-Papert figures in Figure 3 that humans cannot distinguish? The main reason is that FBF networks use internally generated "dye injections". These inputs are topographically distributed across the entire perceptual space such that each dye injection is delivered to a different filling-in domain, or slab. In human perception, by contrast, the same Feature Contour signals that initiate Boundary Contour formation also act as input sources that trigger featural filling-in, as illustrated in Figures 6–8. The regions used by Minsky and Papert (1969) were all of the same color, of similar over-all shape, and occupied essentially the same region of their respective images. They would therefore tend to fill-in the same slab, or set of slabs, when they are being perceived by humans. Hence they could not be rapidly distinguished by filling-in a different set of slabs as a function of their connectivity.

This observation clarifies a recent controversy about human perception; namely whether target figures that differ from distractor figures by more than one type of feature can be separated from them by rapid parallel processing that does not require serial search. Treisman and her colleagues (Treisman and Gelade, 1980; Treisman and Souther, 1985) have suggested that such parallel processing can occur only if the target is distinguished from distractors along a single stimulus dimension, whereas if a target is defined by the conjunction of two or more stimulus dimensions, then it can only be separated from the distractors by a serial search process.

An exception to this rule was discovered by Nakayama and Silverman (1986), who showed that targets which differ from distractors by a combination of disparity and color, or of disparity and motion, can be rapidly separated without serial search. This result is consistent with the fact that FACADE representations of different disparity-and-color combinations activate different combinations of slabs (Grossberg, 1987). They are *structurally* separated in the representation, and hence can be rapidly detected.

In summary, the difficulty of distinguishing the connected and disconnected Minsky-Papert displays can now be explained by the same mechanisms that explain rapid search of Nakayama-Silverman displays, and that provide the heuristics for designing an FBF network for automatic figure-ground separation.

APPENDIX: FBF NETWORK EQUATIONS

Input Images

In the FBF model computer simulations, the images are 256 × 256 arrays with signal values in the interval [0, 1]. The simulations pictured herein represent maximum signal strength by black and minimum signal strength by white. Noise was generated by randomly choosing a percentage of pixels and setting their values to a random number, or gray level, in the interval [0, 1]. For the simulation pictured herein, 50% of the pixels were randomized. The input pattern $\{I_{ij}\}$ is thus represented as gray levels on a set of square pixels $\{P_{ij}\}$. Pixel P_{ij} attains the value I_{ij} at the set of image points $\{(u,v) : i \leq u < i+1, j \leq v < j+1\}$.

Step 1 (Discount the Illuminant).

ON-C Network

Each node v_{ij} is placed, for notational convenience, at the center of the corresponding pixel P_{ij} where it receives input I_{ij}. The activity x_{ij} of the node v_{ij} at lattice position (i,j) obeys the shunting on-center off-surround equation:

$$\frac{d}{dt}x_{ij} = -Ax_{ij} + (B - x_{ij})C_{ij} - (x_{ij} + D)E_{ij} \tag{1}$$

where C_{ij} is the on-center interaction and E_{ij} is the off-surround interaction. Each C_{ij} and E_{ij} is a discrete convolution of the input pattern $\{I_{ij}\}$ with a Gaussian kernel (Figure 22A). Thus

$$C_{ij} = \sum_{p,q} I_{pq} C_{pqij} \tag{2}$$

and

$$E_{ij} = \sum_{p,q} I_{pq} E_{pqij} \tag{3}$$

where

$$C_{pqij} = C \exp\left\{-\alpha^{-2} \log 2[(p-i)^2 + (q-j)^2]\right\} \tag{4}$$

and

$$E_{pqij} = E \exp\left\{-\beta^{-2} \log 2[(p-i)^2 + (q-j)^2]\right\}. \tag{5}$$

In our simulations, $A = 134$, $B = 1$, $C = 7$, $D = .5$, $E = 3.333$, $\alpha = 1.3$, and $\beta = 1.875$. For this choice of parameters, the ON-C and OFF-C Gaussians are of equal area. At equilibrium $(dx_{ij}/dt = 0)$, (1) yields:

$$x_{ij} = \frac{\sum_{(p,q)}(BC_{pqij} - DE_{pqij})I_{pq}}{A + \sum_{(p,q)}(C_{pqij} + E_{pqij})I_{pq}} \tag{6}$$

OFF-C Network

The activity $\overline{x}_{ij}$ of the node v_{ij} at lattice position (i,j) obeys the shunting off-center on-surround equation:

$$\frac{d}{dt}\overline{x}_{ij} = -A(\overline{x}_{ij} - S) + (\overline{B} - \overline{x}_{ij})\overline{C}_{ij} - (\overline{x}_{ij} + \overline{D})\overline{E}_{ij} \tag{7}$$

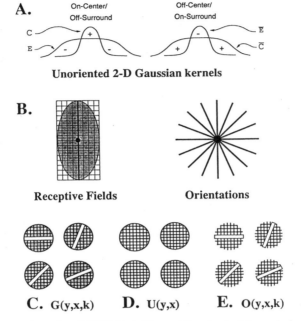

Figure 22. Kernels used in the CORT-X 2 filter. All oriented kernals have orientations at every $\pi/8$ radians.

where the on-center kernel of $\overline{x}_{ij}$ is the off-surround kernel of x_{ij}, and the off-surround kernel of $\overline{x}_{ij}$ is the on-center kernel of x_{ij}. In particular,

$$\overline{B} = D, \qquad (8)$$

$$\overline{C}_{ij} = E_{ij}, \qquad (9)$$

$$\overline{D} = B, \qquad (10)$$

and

$$\overline{E}_{ij} = C_{ij}. \qquad (11)$$

By Equations (8)–(11), (7) may be written as

$$\frac{d}{dt}\overline{x}_{ij} = -A(\overline{x}_{ij} - S) + (D - \overline{x}_{ij})E_{ij} - (\overline{x}_{ij} + B)C_{ij}. \qquad (12)$$

At equilibrium,

$$\overline{x}_{ij} = \frac{AS + \sum_{(p,q)}(DE_{pqij} - BC_{pqij})I_{pq}}{A + \sum_{(p,q)}(C_{pqij} + E_{pqij})I_{pq}} \qquad (13)$$

It follows by summing (6) and (13) that

$$x_{ij} + \overline{x}_{ij} = \frac{AS}{A + \sum_{(p,q)}(C_{pqij} + E_{pqij})I_{pq}} \qquad (14)$$

which shows that for images $\{I_{pq}\}$ with constant Gaussianly filtered total activity

$$\sum_{p,q}(C_{pqij} + E_{pqij})I_{pq}, \tag{15}$$

the sum $x_{ij} + \overline{x}_{ij}$ is conserved and maintained at a positive value that increases with the tonic activity level S of $\overline{x}_{ij}$ in the dark. In our simulations, $S = .2$. The value for the parameter A (given the other parameter values) was chosen so that the activation of cell $\overline{x}_{ij}$ takes on values between S and 0 under spatially uniform illumination between 0 and 1 within its receptive field. Under such spatially uniform illumination conditions, $I_{pq} = I$ for all (p,q). Since for our choice of parameters the two Gaussian kernels are of equal area, let $\Phi = \sum_{(p,q)} E_{pqij} = \sum_{(p,q)} C_{pqij}$. Then Φ factors out so that the equilibrium equation (13) becomes

$$\overline{x}_{ij} = \frac{AS + I(D-B)\Phi}{A + 2I\Phi} \tag{16}$$

We derive the value for A by

$$A = \frac{(B-D)\Phi}{S} \tag{17}$$

so that when $I = 1$, the numerator vanishes. When $I = 0$, $\overline{x}_{ij} = S$ independent of A.

Step 2 (CORT-X 2 Filter).

All simple cell input fields are elliptical. Two sizes of input fields were used, indexed by the subscript s. The smaller scale, $s = 1$, had a major axis of 12 pixels and a minor axis of 6 pixels. The larger scale, $s = 2$, had a major axis of 20 pixels and a minor axis of 10 pixels. Orientations were chosen around the clock spaced by $\pi/8$ degrees. They are indexed below by the subscript k.

Simple Cells

A simple cell with index (i,j) is centered at the lower left hand corner of pixel P_{ij}. By this convention, the nodes v_{pq} of the shunting variables x_{pq} do not lie on the oriented axes that separate the excitatory and inhibitory halves of vertically and horizontally oriented receptive fields. The output of the pair of simple cells of scale s centered at index (i,j) with receptive field orientation k is defined by

$$S_{sL}(i,j,k) = \max[L_s(i,j,k) - \alpha_s R_s(i,j,k) - \beta_s, 0] \tag{18}$$

and

$$S_{sR}(i,j,k) = \max[R_s(i,j,k) - \alpha_s L_s(i,j,k) - \beta_s, 0] \tag{19}$$

where $L_s(i,j,k)$ and $R_s(i,j,k)$ are the contributions of the
left-half $l_s(i,j,k)$ and right-half $r_s(i,j,k)$, respectively, of the oriented input field; that is,

$$L_s(i,j,k) = \frac{\sum_{(p,q)\in l_s(i,j,k)} x_{pq} w_{pq}}{\sum_{(p,q)\in l_s(i,j,k)} w_{pq}} \tag{20}$$

and

$$R_s(i,j,k) = \frac{\sum_{(p,q)\in r_s(i,j,k)} x_{pq} w_{pq}}{\sum_{(p,q)\in r_s(i,j,k)} w_{pq}}, \tag{21}$$

where w_{pq} is a weighting factor equal to the proportion of the area of pixel P_{pq} (taken to be one square unit) covered by the receptive field. An activity x_{pq} was included in L_s or R_s if its pixel had a non-zero intersection with the corresponding half of the receptive field. Parameters α_s are threshold contrast parameters and parameters β_s are threshold noise parameters. We chose $\alpha_1 = 1.4$, $\alpha_2 = 2.0$, and $\beta_1 = \beta_2 = \beta = .012$. Each simple cell in (18) and (19) is sensitive to the opposite direction-of-contrast than its companion, as indicated by the indices L and R in S_{sL} and S_{sR}, respectively.

The ON-C and OFF-C networks each input to a different network of simple cells. The simple cells that receive signals from the ON-C network are denoted by S_{sL}^+ and S_{sR}^+. The simple cells that receive signals from the OFF-C network are denoted by S_{sL}^- and S_{sR}^-.

The complex cells pool inputs from all simple cells of like orientation and scale that are centered at the same location, as described below.

Complex Cells

The complex cell output $C_s(i,j,k)$ is defined by

$$C_s(i,j,k) = F\left[S_{sL}^+(i,j,k) + S_{sR}^+(i,j,k) + S_{sL}^-(i,j,k) + S_{sR}^-(i,j,k)\right]. \qquad (22)$$

Such a cell is sensitive to spatial scale s and amount-of-contrast centered at cell x with orientation k, but it is insensitive to direction-of-contrast. In our simulations, $F = .5$.

Hypercomplex Cells (First Competitive Stage)

The hypercomplex cells $D_s(i,j,k)$ at the first competitive stage receive input from the spatial competition among the complex cells; that is,

$$D_s(i,j,k) = \max\left[\frac{C_s(i,j,k)}{\epsilon + \mu \sum_m \sum_{p,q} C_s(p,q,m) G_s(p,q,i,j,k)} - \tau, 0\right] \qquad (23)$$

where $\epsilon = .1$, $\mu = 5$, $\tau = .01$. The oriented competition kernel $G_s(y,x,k)$ is normalized so that

$$\sum_{p,q} G_s(p,q,i,j,k) = 1. \qquad (24)$$

As in Figure 22C, they are circular. Complex cells at the kernel periphery are weighted by the proportion of their area (taken to be one square unit) that are covered by the kernel. The grey areas in Figure 22C are inhibitory. Any cells whose defining pixel location lies within the one unit wide band through the middle of the kernel do not contribute to the inhibition. In our simulations, the small scale kernel is 8 units in diameter, and the large scale is 16 units in diameter.

Hypercomplex Cells (Second Competitive Stage)

The hypercomplex cells $D_s(i,j)$ at the second competitive stage realize a competition among the oriented activities $D_s(i,j,k)$ at each position x. For simplicity, the process is modelled as a winner-take-all competition; namely,

$$D_s(i,j) = D_2(i,j,K) = \max_k D_s(i,j,k), \qquad (25)$$

where index K denotes the orientation of the maximally activated cell.

Multiple Scale Interaction: Boundary Localization and Noise Suppression

The interaction between scales is defined by the equation

$$B_{12}(i,j) = D_1(i,j) \sum_{p,q} D_2(p,q) U(p,q,i,j). \tag{26}$$

The unoriented excitatory kernel $U(p,q,i,j)$ is circular (Figure 22D), and normalized so that

$$\sum_{p,q} U(p,q,i,j) = 1 \tag{27}$$

and had a diameter of 8 units. All cells covered by the kernel contribute to the excitation to the extent that their area (taken to be one square unit) is covered by the kernel. The small scale hypercomplex cell $D_1(i,j)$ accurately localizes boundary segments and suppresses noise near the boundary. The large scale hypercomplex cell $D_2(p,q)$ supresses noise far from the boundary. The product $D_1(i,j)D_2(p,q)$ would simultaneously realize both constraints except that, due to the poor spatial localization of D_2, this term may be zero at boundary points of high curvature, thereby cancelling the good localization properties of $D_1(i,j)$. The effect of $D_2(p,q)$ in the equation is made more spatially diffuse by the kernel $U(i,j,p,q)$. The size of $U(i,j,p,q)$ is chosen to scale with that of $D_2(p,q)$ in order to compensate for the positional uncertainty of $D_2(p,q)$; a larger choice of $D_2(p,q)$ would necessitate a larger choice of $U(i,j,p,q)$. Although term $\sum_{p,q} D_2(p,q) U(p,q,i,j)$ in (26) localizes the boundary even less accurately than $D_2(p,q)$ does, the *product* of $D_1(i,j)$ with $\sum_{p,q} D_2(p,q) U(p,q,i,j)$ in (26) restores this loss of boundary localization. Moreover, kernel $U(p,q,i,j)$ causes no harm at locations p,q that are far from the boundary, since $D_2(p,q) = 0$ there (Carpenter, Grossberg and Mehanian, 1989).

Long-Range Cooperation: Boundary Completion

The function $B_{12}(i,j)$ represents the image boundary well except where boundary pixels are missing due to noise. More and larger boundary gaps are generated as the noise level increases.

The large detectors $D_2(i,j)$ can be used to partially overcome this problem. Because of the spatial uncertainty of the large detectors $D_2(i,j)$, they are capable of responding at locations where pixel signal strength has been reduced by noise. Such boundary signals may, however, be poorly localized. To overcome this tradeoff between boundary completion and localization, cooperative interactions among the large scale cells are defined by

$$B_2(i,j) = D_2(i,j) \max\left[\sum_{p,q} D_2(p,q,K) O(p,q,i,j,K) - \delta, 0\right]. \tag{28}$$

The oriented kernel $O(p,q,i,j,k)$ is defined by the one-unit-wide white strips in Figure 25E. Any cells with centers that lie within the one unit wide band contributes to the cooperative process. The kernel is normalized so that

$$\sum_{(p,q) \text{ in kernel}} O(p,q,i,j,k) = 1. \tag{29}$$

In our simulations, the length of the kernel was 12 units, and $\delta = .001$.

CORT-X 2 Output

The final output of the CORT-X 2 filter is the rectified sum of the multiple scale interaction and the cooperative interaction:

$$B(i,j) = 1[B_{12}(i,j) + B_2(i,j)] \tag{30}$$

where

$$1[w] = \begin{cases} 1 & \text{if } w > 0 \\ 0 & \text{otherwise} \end{cases} \tag{31}$$

is the Heaviside function.

Step 3: Filling-In.

In each filling-in network F_m, an input is injected into a different area, with the shape of either a narrow Gaussian or a single node. For example, let $I = I(m)$ and $J = J(m)$ define the injection indices (I, J) of F_m. Then the injected input pattern to F_m was chosen to be

$$X_{ij}^{(m)} = X \exp\{-\gamma^{-2} \log 2 [(I-i)^2 + (J-j)^2]\}. \tag{32}$$

This input pattern triggers filling in within F_m via the nonlinear diffusion equation (Grossberg and Todorović, 1988):

$$\frac{d}{dt} S_{ij}^{(m)} = -M S_{ij}^{(m)} + \sum_{p,q \in N_{ij}} (S_{pq}^{(m)} - S_{ij}^{(m)}) P_{pqij} + X_{ij}^{(m)}, \tag{33}$$

where $S_{ij}^{(m)}$ is the activity of the (i,j) node of F_m, and the index set N_{ij} of the summation contains the nearest neighbors of (i,j). The permeability coefficient P_{pqij} is defined by

$$P_{pqij} = \frac{\delta}{1 + \epsilon(B(p,q) + B(i,j))} \tag{34}$$

where $B(p,q)$ and $B(i,j)$ are the outputs (30) from the CORT-X filter at the positions (p,q) and (i,j) respectively. Thus, activity spreads poorly, if at all, between cells where boundary signals are large, and easily where boundary signals do not exist.

In the simulations, the parameters $M = .0001$, $X = 50$, $\gamma = .5$, $\epsilon = 100000$, $\delta = 10$.

Step 4: Figure-Ground Separation.

Each filling-in network F_m inputs its filled-in image to shunting ON-C and OFF-C networks using the same equations (1) and (7) as above, with parameters $A = 1, B_{ij} = \overline{D}_{ij} = 1, C = 18, D_{ij} = \overline{B}_{ij} = .5, E = 3.333, \alpha = 2.96, \beta = 7$, and $S = .2$. Here the on-activations $x_{ij}^{(m)}$ and off-activations $\overline{x}_{ij}^{(m)}$ are parameterized by the filling-in network F_m from which they are derived. The boundary representation at position (i,j) of a figure derived from network F_m is defined by

$$R_{ij}^{(m)} = 1[(x_{ij}^{(m)} - \overline{x}_{ij}^{(m)}) B(i,j)] \tag{35}$$

where $1[w]$ is the Heaviside function. In other words, the figural boundary that is separated by network F_m is computed from the double opponent filter $(x_{ij}^{(m)} - \overline{x}_{ij}^{(m)})$ of filled-in F_m activation, gated by the CORT-X boundary segmentation $B(i,j)$.

Image Rendering

Each image has been scaled so that the maximum signals strength is mapped to black, and the minimum signal strength is mapped to white. Intermediate values map linearly onto a grey scale. The maximum signal strengths for the images are: Figure 9: (a) 1.0, (b) .125, (c) .197. Figure 11: (a) .041, (b) .067, (c) .018, (d) .058. Figure 12: (a) .054, (b) .035. Figure 14: (a) .261, (b) .219. Figure 15: (a) .013, (b) .038. Figure 16: (a) 1.0. Figure 18: Row 1 (col 1) 143.2, (col 2) 753.3, (col 3) 195.3; Row 2 (col 1) .324, (col 2) .330, (col 3) .325; Row 3 (col 1) .287, (col 2) .396, (col 3) .333; Row 4 (col 1) 1.0, (col 2) 1.0, (col 3) 1.0. Figure 19: (a) .325 (b) .332 (c) 1.0 (d) 1.0.

REFERENCES

Canny, J. (1986). A computational approach to edge detection. *IEEE Transactions on Pattern Analysis and Machine Intelligence*, **8**, 679–698.

Carpenter, G.A. and Grossberg, S. (1987a). A massively parallel architecture for a self-organizing neural pattern recognition machine. *Computer Vision, Graphics, and Image Processing*, **37**, 54–115.

Carpenter, G.A. and Grossberg, S. (1987b). ART 2: Stable self-organization of pattern recognition codes for analog input patterns. *Applied Optics*, **26**, 4919–4930.

Carpenter, G.A. and Grossberg, S. (1988). The ART of adaptive pattern recognition by a self-organizing neural network. *Computer*, **21**, 77–88.

Carpenter, G.A., Grossberg, S., and Mehanian, C. (1989). Invariant recognition of cluttered scenes by a self-organizing ART architecture: CORT-X boundary segmentation. *Neural Networks*, **2**, 169–181.

Carpenter, G.A., Grossberg, S., and Reynolds, J.H. (1991). ARTMAP: Supervised real-time learning and classification of nonstationary data by a self-organizing neural network. *Neural Networks*, **4**, 565–588.

Carpenter, G.A., Grossberg, S., Markuzon, N., Reynolds, J.H., and Rosen, D.B. (1991). Fuzzy ARTMAP: A neural network architecture for incremental supervised learning of analog multidimensional maps. Boston University Technical Report CAS/CNS-TR-91-016.

Cohen, M.A. and Grossberg, S. (1984). Neural dynamics of brightness perception: Features, boundaries, diffusion, and resonance. *Perception and Psychophysics*, **36**, 428–456.

Grossberg, S. (1987). Cortical dynamics of three-dimensional form, color, and brightness perception, II: Binocular theory. *Perception and Psychophysics*, **41**, 117–158.

Grossberg, S. (1988). **Neural networks and natural intelligence**. Cambridge, MA: MIT Press.

Grossberg, S. (1990). Neural FACADES: Visual representations of static and moving Form-And-Color-And-DEpth. *Mind and Language*, **5**, 411–456.

Grossberg, S. and Marshall, J. (1989). Stereo boundary fusion by cortical complex cells: A system of maps, filters, and feedback networks for multiplexing distributed data. *Neural Networks*, **2**, 29–51.

Grossberg, S. and Mingolla, E. (1985a). Neural dynamics of form perception: Boundary completion, illusory figures, and neon color spreading. *Psychological Review*, **92**, 173–211.

Grossberg, S. and Mingolla, E. (1985b). Neural dynamics of perceptual grouping: Textures, boundaries, and emergent segmentations. *Perception and Psychophysics*, **38**, 141–171.

Grossberg, S. and Mingolla, E. (1987). Neural dynamics of surface perception: Boundary webs, illuminants, and shape-from-shading, *Computer Vision, Graphics, and Image Processing*, **37**, 116–165.

Grossberg, S., Mingolla, E., and Todorović, D. (1989). A neural network architecture for preattentive vision. *IEEE Transactions on Biomedical Engineering*, **36**, 65–84.

Grossberg, S. and Todorović, D. (1988). Neural dynamics of 1-D and 2-D brightness perception: A unified model of classical and recent phenomena. *Perception and Psychophysics*, **43**, 241–277.

Gschwendtner, A.B., Harney, R.C., and Hull, R.J. (1983). Coherent IR radar technology. In D.K. Killinger and A. Mooradian (Eds.), **Optical and laser remote sensing**. New York: Springer-Verlag.

Harney, R.C. (1980). Infrared airborne radar. **Proceedings of the IEEE 1980 electronic and aerospace systems conference**, 462–471.

Harney, R.C. (1981). Military applications of coherent infrared radar. In **Physics and Technology of Coherent Infrared Radar**. Proceedings of the SPIE, **300**, 2-11.

Harney, R.C. and Hull, R.J. (1980). Compact infrared radar technology. In CO_2 **laser devices and applications**. Proceedings SPIE, Vol. 227, 162–170.

Hull, R.J. and Marcus, S. (1980). A tactical 10.6 micrometer imaging radar. **Proceedings of the IEEE**. National Aerospace and Electronics Conference, Vol. 2, 662–668.

Kolodzy, P. (1987). Multidimensional machine vision using neural networks. In M. Caudill and C. Butler (Eds.), **Proceedings of the IEEE international conference on neural networks, II**, 747–758.

Land, E.H. (1977). The retinex theory of color vision. *Scientific American*, **237**, 108–128.

Minsky, M.L. and Papert, S.A. (1969). **Perceptrons**. Cambridge, MA: MIT Press.

Minsky, M.L. and Papert, S.A. (1988). **Perceptrons: Expanded edition**. Cambridge, MA: MIT Press.

Nakayama, K. and Silverman, G.H. (1986). Serial and parallel processing of visual feature conjunctions. *Nature*, **320**, 264–265.

Sullivan (1980). Infrared coherent radar. In CO_2 **laser devices and applications**. Proceedings SPIE, Vol. 227, 148–161.

Sullivan, L.J. (1981). Firepond laser radar. *Electro/81 Conference Record*, Session 34.

Sullivan, D.R., Harney, R.C., and Martin, J.B. (1979). Real-time quasi-3-diminsional display of infrared radar images. **Real-time signal processing II**. Proceedings of the SPIE, Vol. 180, 57–65.

Treisman, A. and Gelade, G. (1980). A feature integration theory of attention. *Cognitive Psychology*, **12**, 97–136.

Treisman, A. and Souther, J. (1985). Search asymmetry: A diagnostic for preattentive processing of separable features. *Journal of Experimental Psychology: General*, **114**, 285–310.

Toward a unified theory of spatiotemporal processing in the retina

Paolo Gaudiano

Boston University, Department of Cognitive & Neural Systems
111 Cummington Street, Boston, MA 02215 USA

Abstract

Why do stabilized images fade? How can X cells in the retina respond linearly to a broad range of spatiotemporal stimulation functions in spite of possible nonlinear preprocessing? Why have models of *spatial* vision that utilize *static* images and assume *linearity* of preprocessing led to reasonable results in spite of the two questions above? This chapter introduces the push-pull shunting network, a model of spatiotemporal visual processing that resolves these and other controversial findings. Development of the model is based on an analysis of the spatial and temporal response characteristics of networks of neurons that obey membrane equations. The resulting architecture is structurally similar to the mammalian retina, requiring a mechanism for temporal adaptation analogous to photoreceptors, followed by cells of opposite polarity analogous to on and off bipolar cells, and finally a layer of ganglion cells that summate bipolar cell inputs. The model predicts that X and Y cells consist of the same neural mechanism acting in different parametric regimes. In agreement with morphological and physiological data, analyses and numerical simulations show that an increase in receptive field center size changes the model's response from X-like to Y-like. This functional duality results from mathematical properties of the push-pull shunting network, which can selectively enhance sustained or transient response components on the basis of RF morphology. The model also explains how it is possible for X cells to respond linearly to a broad range of spatial and temporal modulation functions in spite of arbitrary nonlinear preprocessing. Finally, in general agreement with biological data, the model predicts that stationary images lead to loss of instantaneous contrast at equilibrium, i.e., that stabilized images fade, as an unavoidable side effect of photoreceptor adaptation.

1 Introduction

The retina, owing to its accessibility and to its fundamental role in the initial transduction of light into neural signals, is arguably the most extensively studied neural structure in vision. The anatomical structure of the retina has been known for almost exactly one century, beginning with the work of Ramón y Cajal in the late nineteenth century. It was not until approximately 50 years later that the technology became available for recording of individual cell activity, allowing researchers to gain insights on the information carried by neurons in the retina. Since that time detailed descriptions have been given for the physiological, pharmacological, and functional properties of most classes of retinal cells. In spite of these advances, a unified understanding of the functional role of the processing carried out in the retina is still lacking.

In this chapter I introduce the *push-pull shunting network*, which was designed as a theoretical model of spatiotemporal information processing by networks of neurons that obey membrane, or *shunting*, equations (Grossberg, 1970; Sperling, 1970; Sperling and Sondhi, 1968). I will show that owing to the nonnegative nature of luminance signals, feed-forward shunting neural network models are subject to an inescapable trade-off between accurate processing in the spatial and temporal domains. Accurate processing in both domains can be achieved by extending the model to include a total of three layers of feed-forward processing. The resulting model's architecture is structurally analogous to the feed-forward retinal circuit connecting photoreceptors to retinal ganglion cells (RGCs) through pairs of push-pull bipolar cells. This analogy is confirmed through computer simulations showing that the model's output layer exhibits many properties observed in cat RGCs.

In addition to qualitatively reproducing experimental data on retinal cells, the model is able to resolve a number of controversial experimental and theoretical findings, and offers a unified functional explanation for the retina's role as the first stage of spatiotemporal visual processing in biological systems.

This chapter begins with a brief description of the anatomical and functional organization of the mammalian retina in Section 2, with a brief summary of the properties of two functional classes of cat retinal ganglion cells known as X and Y cells in Section 3. This is followed by two sections outlining a number of unresolved and controversial issues on retinal processing. Section 6 introduces the shunting network on which this model is based, while Section 7 motivates the addition of a push-pull mechanism for accurate spatiotemporal processing. Sections 8–11 present the push-pull shunting equation and a necessary model of temporal adaptation. Following a number of simulation results in Sections 12–15, the chapter concludes by summarizing the properties of the model, discussing the controversial experimental findings presented in earlier sections, and outlining a number of predictions and future applications of the model.

2 Anatomy and physiology of the retina

The retina of all vertebrates is organized into three cellular (or *nuclear*) layers and two synaptic (or *plexiform*) layers (Dowling, 1987). Information flows from the photoreceptors, through a layer of bipolar cells, and finally through the retinal ganglion cells (RGCs), whose axons project via the optic nerve to subcortical and cortical areas. In addition to the feed-forward information processing carried out by these three cell types, there exist two classes of cells, horizontal cells and amacrine cells, that carry signals laterally through the retina for additional processing.

Although there is general agreement that retinal cells can be subdivided into these broad classes, morphological and anatomical studies have shown the existence of a great number of cell types within each class, including dozens of RGC and amacrine cell types (e.g., Kolb et al., 1981).

In addition to morphological and anatomical classification schemes, physiological studies have shown the existence of distinct, broad functional classes. Kuffler (1953) and Barlow (1953) first reported the existence of two RGC classes: on-center (ON) and off-center (OFF) cells. Both types have a concentrically-organized receptive field consisting of a central region and a surrounding annulus of opposite polarity. Presentation of a light stimulus over the receptive field center causes increased activation for ON cells, and decreased activation for OFF cells. Conversely, stimulation of the receptive field surround decreases activation in the ON cells, and increases activation in OFF cells. The distinct response characteristics of ON and OFF RGCs appear to originate in the bipolar cell layer (Werblin and Dowling, 1969).

An additional, independent functional classification scheme was proposed by Enroth-Cugell and Robson (1966), who reported the existence of two RGC classes on the basis of spatiotemporal response characteristics: X cells, which respond to inputs in a sustained fashion, and appear to linearly summate luminance signals throughout their receptive field; and Y cells, which respond to inputs in a transient fashion, and exhibit a more complicated, nonlinear spatial summation of luminance signals throughout their receptive field. This classification has been supported and

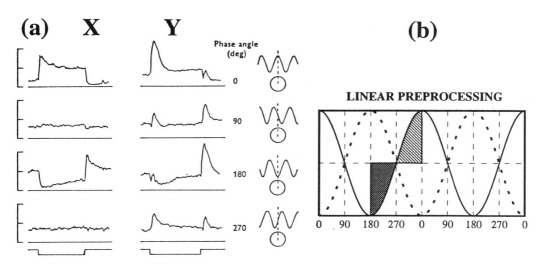

Figure 1: (a): The null test. Left column represents response of an X cell to introduction and withdrawal of a sinusoidal grating. Right column represents response of a Y cell to same stimulus conditions. See text for details. Reprinted with permission from Enroth-Cugell and Robson (1966). (b): Linear preprocessing of a temporally square-wave modulated sinusoidal grating leads to exact cancellation of input increments (hatched area) and decrements (stippled area) for cells located at ±90° spatial phase relative to the sinusoidal grating.

extended through a number of physiological and anatomical studies, and a relationship has been established between the morphological classes of *alpha* and *beta* ganglion cells, respectively, and the functional classes of Y and X cells (Boycott and Wässle, 1974; Cleland and Levick, 1974; Fukuda et al., 1984; Hochstein and Shapley, 1976a,b; Saito, 1983).

3 An overview of X and Y retinal ganglion cells

The functional distinctions between X and Y cells reported by Enroth-Cugell and Robson (1966) were revealed through stimulation with temporally modulated spatial patterns. In these experiments, the activity of ganglion cells in response to such patterns was recorded in the optic nerve. The patterns were projected on a screen located in front of paralyzed cats, in a way that allowed accurate control of the location of the stimulus pattern relative to the receptive field center of the cell being recorded.

A primary distinction between X and Y responses was found through what is referred to as the *null test*, which is carried out in the following manner: the response of a ganglion cell is recorded while the projected image of a sinusoidal grating is turned on and off, i.e., is modulated in time by a square wave of fixed temporal frequency. The sinusoidal spatial modulation is such that the average luminance of the grating superimposed on the background is equal to the luminance of the background alone. Measurement of the cell's response to such spatiotemporal input modulation is repeated as the grating's spatial phase relative to the cell's receptive field center is systematically varied across trials.

As shown in Fig. 1a, two characteristics distinguish X and Y cells. First, as shown in the top and third rows, the response of X cells (left column) to introduction and withdrawal of a grating at zero or 180 degrees spatial phase consists of a primarily sustained component, whereas that of Y cells (right column) consists of primarily transient components. The other noticeable difference between X and Y cells is that for an X cell it is generally possible to find two distinct relative spatial phases, usually at ±90°, at which no response is elicited by grating onset or offset (left column, second and fourth rows from the top of Fig. 1a). On the other hand, at a spatial phase of ±90° the Y cell exhibits an *on-off* or *frequency doubling* response, that is, it exhibits a positive going response to both onset and offset of the grating (right column, second and fourth rows).

The existence of null responses at ±90° in X cells indicates that the sudden decrease in input to one side of the cell's receptive field exactly cancels the concomitant increase in input to the other side of the receptive field, as illustrated in Fig. 1b. As discussed below, these results therefore suggest that X cells perform linear spatial summation over the extent of their receptive field, while Y cells perform a more complex form of nonlinear spatial summation.

Hochstein and Shapley (1976a,b) confirmed and extended the X/Y classification on the basis of the null test. They found that the linear/nonlinear classification holds for a wide range of temporal frequency of modulation and for various types of spatial and temporal waveform, which led them to the hypothesis that the distinct behavior of X and Y cells implies different underlying retinal circuits. Specifically, these authors proposed that the receptive field of both X and Y cells includes linear center and surround mechanisms of the type proposed by Rodieck (1965), but that the Y cell receptive field further includes small nonlinear subunits that overlap both the receptive field center and surround.

4 Some unresolved experimental findings on X and Y cells

The findings outlined in the previous section have been the basis of much experimental and theoretical work. However, there are a number of issues that remain unclear and have not been satisfactorily explained. This includes not only issues on retinal processing, but also on the importance of the retina as the first processing stage in the visual system.

Of great importance to many vision researchers is the issue of linearity: a number of vision models, both retinal and cortical, assume that retinal elements behave approximately linearly. In partial support of this assumption, Enroth-Cugell and Robson (1966) noted that the existence of null phases in X cells "... implies either that the signals from photoreceptors are linearly related to their illumination or that they are non-linearly related in a way which is symmetrical about the mean illuminance." These authors concluded that "the latter alternative can be ruled out by the observation that the null positions for grating patterns are not changed when the mean luminance is changed" (p. 524-525). In other words, the seemingly linear behavior of X cells regardless of average luminance and contrast forced these authors to conclude that the response of photoreceptors and bipolar cells must be linearly related to input intensity. The necessity of this conclusion is illustrated by Fig. 2, in which hypothetical nonlinear preprocessing mechanisms are shown to affect the ability of subsequent stages (such as X cells) to carry out linear spatial summation.

However, it is known that photoreceptor response is *not* linearly related to luminance; photoreceptors exhibit *temporal adaptation* that leads to a static nonlinear compression of the input range (e.g., Dowling, 1987; Rushton, 1958). From a functional point of view, it is important that the photoreceptors be able to compress the input into a finite range, since any information that is lost at the photoreceptors cannot be recovered at later stages. Such a static nonlinearity could prevent

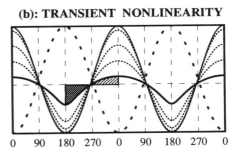

(a): STATIC NONLINEARITY

(b): TRANSIENT NONLINEARITY

Figure 2: The effect of hypothetical static and transient nonlinear preprocessing on the X cell's ability to exhibit null responses at ±90° when stimulated by a square-wave modulated sinusoidal grating. Each plot illustrates the input reaching various parts of the X cell's receptive field as a spatial sinusoidal grating is contrast-reversed. The dashed and solid lines, respectively, represent the grating just before and just after contrast reversal. (a): Static nonlinear preprocessing may distort the sinusoidal distribution depending on average input and contrast, so that input increments and decrements at ±90° may not cancel out, as indicated by hatched and stippled areas. (b) A transient nonlinearity may rapidly equilibrate to approximate linearity, but the initial nonlinear distortion (thick solid line) should prevent cancellation at ±90° immediately following contrast reversal.

linear spatial summation in the X cell, as diagrammed in Fig. 2a.

In addition to this static, compressive nonlinearity it is known that photoreceptors and bipolar cells exhibit transient overshoots and undershoots to sudden input changes. Such transients are the result of each photoreceptor trying to adjust its dynamic range in response to local changes in input intensity. The purpose of these adjustments, in principle, is to maintain each photoreceptor within the linear portion of its dynamic range. However, such adjustments cannot be *instantaneous* (see also Section 15), so that rapid input fluctuations may temporarily drive the photoreceptor into the nonlinear portion of its dynamic range, until the adaptation mechanism can drive it back into the linear range. Hence photoreceptors should exhibit *transient* nonlinearities in response to rapid temporal fluctuations. This type of transient nonlinearity is diagrammed in Fig. 2b, which portrays the responses of hypothetical photoreceptors to a counterphase-flickered sinusoidal grating. As shown in this figure, the photoreceptors' responses immediately following contrast reversal can lead to a nonlinear distortion of the *spatial* sinusoidal distribution, which may quickly converge to a more linear response. However, such a transient nonlinearity should prevent exact cancellation in the X cell response to square-wave temporal modulation of a sinusoidal grating. In other words, one would not expect null responses at ±90° spatial phase when the grating is square-wave modulated, as in the classical experiments of Enroth-Cugell and Robson (1966). How can X cells exhibit seemingly linear behavior in spite of nonlinear preprocessing?

Equally puzzling is the experimental finding that whereas X cells behave linearly even in response to fast *temporal* modulation, it has been noted that rapid changes in the *spatial* modulation function can result in X cell nonlinear behavior. For example, X cells sometimes exhibit on-off responses to temporal modulation of a bipartite field, or contrast edge. Is this behavior the result of a mechanism similar to the one generating on-off responses in Y cells? If so, why does this mechanism only affect X cell responses to a particular type of *spatial* modulation?

5 Some general issues in spatiotemporal processing

Resolution of the controversial findings outlined in the last section requires simultaneous analysis of the spatial and temporal response characteristics of retinal ganglion cells. However, even though data have shown an inextricable tie between RGC responses in the space and time domains, a majority of existing models of *spatial* vision assume that the incoming image is *stabilized*, i.e. it is constant in time; conversely, most models of RGC *temporal* response characteristics only consider individual cells responding to carefully defined spatial stimuli.

The use of stabilized stimuli in spatial vision models is perplexing for two reasons: first, it has been known for four decades that biological visual systems are unable to perceive stabilized images. Ditchburn and Ginsborg (1952) first reported that images stabilized with respect to the retina rapidly fade, giving rise in a matter of seconds to perception of a uniform gray field (loss of contrast), a finding that appears to hold for most biological visual systems. Second, most physiological and psychophysical experiments make use of stimuli that are modulated both in space and in time.[1] Therefore models that preclude the possibility of temporal modulation cannot be applied to the majority of physiological and psychophysical data.

This chapter introduces a model of spatiotemporal information processing that was developed to address these and related questions. The next section provides an overview of prior work on which the model is based.

6 The feed-forward shunting network

The present work is based on the feed-forward *shunting* network, which consists of continuous-time neurons obeying membrane equations. Membrane conductance to different ionic species is modulated by the input through a center-surround receptive field structure. Historically, the term *shunt* as applied to neural networks refers to the presence of multiplicative saturation terms in the equation describing membrane potential (Furman, 1965; Grossberg, 1970, 1973; Hodgkin, 1964; Sperling, 1970; Sperling and Sondhi, 1968). The present usage of the term *shunt*, based on Grossberg's (1970) original formulation, is more general than in its physiological definition, in the sense that the shunting network can exhibit either *divisive* (multiplicative) or *subtractive* (additive) effects depending on the choice of parameters and the input intensity.

The simplest differential equation describing an on-center, off-surround shunting network (Fig. 3a) can be written as follows:

$$\frac{dv_i}{dt} = -Av_i + (B - v_i)I_i - (D + v_i)\sum_{k \neq i} I_k. \qquad (1)$$

Here v_i represents the activation or potential (which are assumed to be proportional in this model) of the ith cell in the network; A is the rate of passive decay toward the resting potential (which is assumed here to be zero); B and D represent the excitatory and inhibitory saturation points, respectively; I_i represents the excitatory contribution from the ith component of the input pattern to the ith cell (on-center); and $\sum_{k \neq i} I_k$ represents the inhibitory contribution from all other components of the input pattern to the ith cell (off-surround). All parameters are assumed to be nonnegative. In the absence of inputs, activation v_i decays to zero at a rate $-Av_i$. Otherwise, (1) guarantees

---

[1] Note that because the human eye is constantly jittering, even those psychophysical experiments that involve fixation of a steady image result in a temporally modulated image at the retinal level.

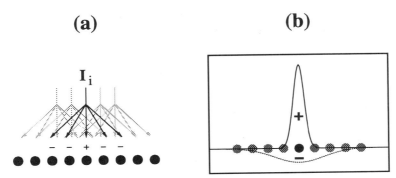

Figure 3: (a) A 1-D feed-forward, distance-dependent shunting network. Inputs I_i impinge upon the network in an on-center, off-surround fashion. (b) Typical Difference-Of-Gaussians (DOG) RF profile. Each cell is excited by inputs falling within a small central area, and is inhibited by inputs falling within a broader surround.

that activation is always bounded between the values B and $-D$, regardless of input intensity. Through a simple change of variables, (1) can be interpreted as a passive membrane equation (e.g., Grossberg, 1988, p. 35):

$$C\frac{dv}{dt} = (v^p - v)g^p + (v^+ - v)g^+ + (v^- - v)g^-. \tag{2}$$

In this case the inputs modulate membrane conductance (g^+ and g^-) of two ionic species which have opposite effects on membrane polarization v. The constant terms v^p, v^+, and v^-, respectively, represent passive, excitatory, and inhibitory equilibrium potentials.

It should be pointed out that although eqs. (1) and (2) are *linear* differential equations, the behavior of the system is *nonlinear*, due to the multiplicative relationship between the dependent variable v and the temporally modulated inputs. This is known as a *time-varying* system, and its nonlinear behavior can be observed for example by noting that an identical increment in excitatory and inhibitory inputs (viz., terms I_i and $\sum_{k \neq i} I_k$ above) will not cancel out unless the quantities $(B - v_i)$ and $(D + v_i)$ are equal, which depends on the cell's activation level.

The behavior of network (1) has been analyzed through various mathematical techniques. Details on the properties of shunting networks can be found elsewhere (Grossberg, 1973, 1983, 1988; Sperling, 1970). Many properties of this type of network are derived by analysis of eq. (1) at steady-state, since the system converges. The equilibrium potential of the ith cell in the network is found by setting $dv_i/dt = 0$ in eq. (1)

$$v_i = \frac{BI_i - D\sum_{k \neq i} I_k}{A + \sum_k I_k}. \tag{3}$$

In the area of static visual processing, shunting network properties include faithful transmission of relative luminance levels in a complex scene, Weber-law sensitivity, and the ability to suppress uniform backgrounds. Equation (3) can be used for example to show that the shunting network is able to represent the relative luminance at different positions in an image (local contrast processing) regardless of fluctuations in the overall input intensity. Denoting the total input by $I = \sum_k I_k$, and the relative input intensity at position i by $\theta_i = I_i/I$, equation (3) can be rewritten as

$$v_i = \frac{(B+D)I}{A+I}\left(\theta_i - \frac{D}{B+D}\right). \tag{4}$$

As long as the passive decay is small ($A \ll I$), equation (4) factorizes information about relative input size θ_i from overall intensity I, because activation v_i is approximately proportional to $\theta_i - \frac{D}{B+D}$ regardless of overall input intensity I. Grossberg (1980) has also shown that the region of maximal sensitivity of a cell obeying (1) shifts without compression as the background intensity is parametrically increased. This is known as the *shift property*, which has been demonstrated experimentally for certain classes of retinal cells (Werblin, 1971).

Similar results ensue when the center-surround mechanism consists of distinct, overlapping center and surround components, as shown in Fig. 3b. Equation (1) then becomes

$$\frac{dv_i}{dt} = -Av_i + (B-v_i)\sum_k \mathbf{C}(k-i)I_k - (D+v_i)\sum_k \mathbf{S}(k-i)I_k. \tag{5}$$

The distance-dependent terms $\mathbf{C}(k-i)$ and $\mathbf{S}(k-i)$ represent the center and surround mechanisms of the receptive field profile of each cell in the network. Most of the general results reported here are insensitive to the exact shape of the center and surround mechanisms. Where explicit closed-form results are required these terms are described as Gaussians:

$$\mathbf{C}(k-i) = C\exp\left[-\frac{(k-i)^2}{2\sigma_C^2}\right]; \quad \mathbf{S}(k-i) = S\exp\left[-\frac{(k-i)^2}{2\sigma_S^2}\right]. \tag{6}$$

Each Gaussian is described by its peak amplitude (C, S) and standard deviation (σ_C, σ_S). The overall receptive field shape is composed of the antagonistic excitatory center and inhibitory surround Gaussians, and is thus referred to as a Difference-Of-Gaussians profile. Throughout this chapter it is assumed that $\sigma_C < \sigma_S$ and $C > S$, which is representative of an on-center, off-surround anatomy, though all results generalize to off-center on-surround anatomies as well.

7 Spatiotemporal analysis

An analysis of the shunting network's response to spatiotemporal modulation revealed a tradeoff in the network's ability to process information in space and time simultaneously. Details and simulation results can be found elsewhere (Gaudiano, 1991a,b). Briefly, the conclusion stems from the realization that light, as a physical quantity, must be nonnegative. As a result the shunting network exhibits an asymmetry in its response to light increments and decrements. This can be seen from eq. (1), which shows that when inputs are *increased* the response of each neuron in the network changes at a rate that depends both on the neuron's activation level v_i and on the size of the inputs I_j. On the other hand reducing all inputs to zero leaves the neuron's activation to decay passively at a rate that only depends on the neuron's activation level v_i and the decay constant A.

The space-time tradeoff is revealed in eq. (4) by noting that the shunting network's ability to process relative luminance requires that A is much less than I, resulting in a slow passive decay in response to input decrements. Increasing the size of A relative to the other dynamic parameters (viz., B and D in eq. [1]) can partly compensate for this asymmetry, but only at the cost of reducing that portion of the neuron's dynamic range in which relative luminance is accurately processed.

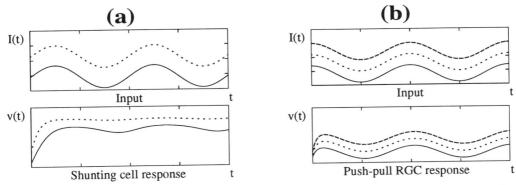

Figure 4: (a): The response of a shunting cell to whole-field sinusoidal modulation. (b): The response of a push-pull shunting cell to similar whole-field sinusoidal modulation. See text for details.

Fig. 4a illustrates the response of a single neuron whose activation obeys eq. (5) to whole-field temporal modulation of a 1-D luminance profile. When the input is first turned on (upper plot, solid line), the neuron's activation (lower plot) rapidly rises, but subsequently shows little response to the deep temporal modulation. The passive decay rate A in this example was chosen an order of magnitude larger than the positive saturation constant B. Such a large passive decay rate hampers the shunting neuron of its reflectance processing ability, as illustrated in Fig. 4a by the dashed lines, which shows that a small increase in background intensity results in decreased response modulation.

Although the above results are based on the shunting formalism as it applies to luminance stimuli, nonnegative inputs must have similar asymmetrical effects on *any* model that prescribes time-dependent interactions between input and activation. For example, a similar problem can occur in the control of limb movement: Bullock & Grossberg (1988; see also Gaudiano, 1991a) used a set of continuous-time equations to describe the behavior of a neural network for the generation and control of arm movement trajectories. It was found that if the neural command signal that causes muscle contraction is assumed to be nonnegative, then the muscle will contract at a variable rate that depends on the size of the signal, but will only be able to relax at a rate determined by a passive decay constant. The solution in that case was to postulate the existence of agonist-antagonist muscle pairs coupled in a push-pull fashion, so that *passive relaxation* of the agonist muscle is accompanied by an *active contraction* of the antagonist muscle, and *vice versa*. Such agonist-antagonist interactions are of course ubiquitous in skeleto-muscular systems.

In analogy with this motor control problem, Gaudiano (1991a) first proposed that a similar arrangement could take place in the visual system. Specifically, the asymmetry arising from the nonnegative nature of light can be corrected by means of two complementary input pathways to the network, one signaling luminance increments, the other signaling luminance decrements. Together, these pathways allow the shunting network to track temporal modulation faithfully (Fig. 4b).

As mentioned in Section 2, such complementary pathways are known to exist in the parallel ON and OFF retinal circuits (Barlow, 1953; Kuffler, 1953); thus one possibility is that signals from the ON and OFF pathways interact recurrently to generate a response to increasing as well as decreasing inputs. Although some evidence exists for feedback signals within the retina, this does

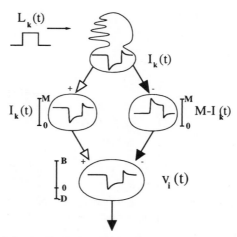

Figure 5: (a): Schematic of the push-pull mechanism. Light signals impinging upon each photoreceptor cause equal and opposite changes in membrane potential of two bipolar cells. These cells differentially activate excitatory and inhibitory membrane mechanisms in a single RGC.

not seem like a tenable solution, particularly since the ON and OFF pathways have been shown to be largely functionally independent (Schiller, 1982; Slaughter and Miller, 1981).

Alternatively, a mechanism that acts feed-forward *within* each pathway can force the shunting cells to decay actively at input signal cessation. In addition to having the desired functional properties, I will show that a model based on such a mechanism can fit anatomical, morphological, and physiological data on the circuit connecting cones, bipolar cells, and X and Y retinal ganglion cells.

Fig. 5 schematically illustrates such a mechanism. Each photoreceptor gives rise to equal and opposite signals through two bipolar cells: one whose activation is maximal when photoreceptor activation is maximal, and one whose activation is maximal when photoreceptor activation is minimal. Some evidence for such parallel, push-pull pathways was found in the cat retina. In particular, McGuire et al. (1986; see also Sterling, 1990) showed that a single cone may contact (among other cells) two pairs of bipolars: one pair (a_1, a_2) sends axons to the dendrites of a single OFF-beta RGC in layer a of the Inner Plexiform Layer (IPL), while the other pair (b_1, b_2) converges on a single ON-beta RGC in layer b of the IPL. Cone activation has opposite effects on each bipolar within a pair, and there is evidence that the bipolars within each pair in turn have opposite effects on the RGC they contact (Sterling, 1990). These converging push-pull bipolar pathways coexist with and overlap the well-known diverging bipolar pathways that give rise to ON-center and OFF-center ganglion cell classes. Similar push-pull *bipolar→RGC* connections may also exist for the ON- and OFF-alpha RGC classes (Freed and Sterling, 1988).

The push-pull mechanism has been interpreted as a means of increasing the cone's effective dynamic range (McGuire et al., 1986; Sterling, 1990) by ensuring that excitation is accompanied by disinhibition, and *vice versa*. The present analysis instead suggests that the push-pull mechanism is necessary for accurate, simultaneous processing of spatial and temporal visual information.

It should be pointed out that the evidence for such push-pull connectivity *in vivo* has been recently questioned by the authors that first reported it (Sterling, personal communication). Nonetheless, the conclusions drawn in this section are based on general constraints of spatiotemporal processing of nonnegative signals, so that the present work predicts that such a push-pull mechanism should be found in the retina (see also the closing section in this chapter).

8 The push-pull shunting network

I next introduce a simple extension of eq. (5) that embodies the proposed push-pull mechanism, and analyze its spatiotemporal response. Detailed derivation of all results presented here can be found elsewhere (Gaudiano, 1991a,b). This section may be skipped on a first reading.

With reference to Fig. 5, let the input impinging upon each element of the shunting network be bounded between zero and a maximum level M. This constraint is valid on assumption that inputs to the shunting network originate from layers of preprocessing elements that possess a finite dynamic range, as may result from photoreceptor adaptation. Each photoreceptor then gives rise to two signals, one that affects the excitatory mechanism of the target RGC through a depolarizing bipolar cell, and one that affects its inhibitory mechanism through a hyperpolarizing bipolar cell. It is assumed that the same push-pull mechanism exists for cells comprising both the center and the antagonistic surround mechanisms. This does not preclude the possibility that center and surround mechanisms be mediated by different cell classes: for example, amacrine cells may mediate the surround mechanism, in which case it is assumed that these cells also receive parallel opponent inputs from push-pull bipolar cells.

If the bipolars act on a fast time scale relative to the photoreceptors, they can be lumped into equation (5) by addition of two terms to the input. Specifically, the excitatory input through the receptive field center $[\sum C(k-i)I_k]$ is complemented by an opponent inhibitory input $[\sum C(k-i)(M-I_k)]$, also through the receptive field center; likewise, the inhibitory surround input $[\sum S(k-i)I_k]$ is complemented by an opponent excitatory term $[\sum S(k-i)(M-I_k)]$. The resulting push-pull shunting equation is

$$\frac{dv_i}{dt} = -Av_i + (B-v_i)\left[\sum C(k-i)I_k + \sum S(k-i)(M-I_k)\right]$$
$$-(D+v_i)\left[\sum S(k-i)I_k + \sum C(k-i)(M-I_k)\right], \quad (7)$$

where $I_k \leq M$ for all k. The equation shows that the center and surround each act on *both* excitatory and inhibitory channels, but in an opposing fashion.

Equation (7) can be rewritten as an integro-differential equation in space and time:

$$\frac{\partial v(x,t)}{\partial t} = -Av(x,t) + [B-v(x,t)]\left[\int_{-\infty}^{\infty}\mathbf{C}(x-\xi)I(\xi,t)d\xi + \int_{-\infty}^{\infty}\mathbf{S}(x-\xi)(M-I(\xi,t))d\xi\right]$$
$$- [D+v(x,t)]\left[\int_{-\infty}^{\infty}\mathbf{S}(x-\xi)I(\xi,t)d\xi + \int_{-\infty}^{\infty}\mathbf{C}(x-\xi)(M-I(\xi,t))d\xi\right] \quad (8)$$

Equation (8) can be simplified when the input function is *space-time separable*, i.e., when it consists of the product of two functions, one depending only on time, and the other depending only on space:

$$I(x,t) = I(x)\mathrm{m}(t). \quad (9)$$

Many of the stimuli used for experimental measurements can be expressed as space-time separable functions, including for example sinusoidally or square-wave modulated spatial gratings, and drifting sinusoidal gratings. Other functions sometime lead to closed-form solutions of equation (8).

More generally, the response of equation (8) to an arbitrary function can be numerically integrated. All analytical results presented here are also simulated on the computer through numerical integration of the discrete equation (7) above.

Substitution of equation (9) into (8) and simplification leads to the following equation in standard form

$$\frac{dv}{dt} + v(A + M_C + M_S) = \mathrm{m}(t)(B + D)(I_C - I_S) + BM_S - DM_C, \qquad (10)$$

where I_C and I_S represent the convolution of an arbitrary spatial input distribution with the receptive field center and surround mechanisms

$$I_C = \int_{-\infty}^{\infty} \overset{\circ}{\mathrm{C}}(x - \xi)I(\xi)d\xi \quad \text{and} \quad I_S = \int_{-\infty}^{\infty} \overset{\circ}{\mathrm{S}}(x - \xi)I(\xi)d\xi; \qquad (11)$$

M_C and M_S represent the area (or volume for a 2-D receptive field) of the receptive field center and surround mechanisms, each weighted by the maximum input level M:

$$M_C = M \cdot V_C = M \int_{-\infty}^{\infty} \overset{\circ}{\mathrm{C}}(x)dx \quad \text{and} \quad M_S = M \cdot V_S = M \int_{-\infty}^{\infty} \overset{\circ}{\mathrm{S}}(x)dx. \qquad (12)$$

Note that whereas the quantities I_C and I_S depend on the spatial input distribution, the terms M_C, M_S, V_C, and V_S are fixed for a given choice of receptive field dimension and shape.

Equation (10) shows that addition of the push-pull mechanism has caused cancellation of all time-varying terms multiplying the dependent variable v, thus reducing the network to a *linear time-invariant* (LTI) system. An LTI system such as (10) is generally simple to analyze, as I will now show, and possesses certain properties that will lead to important predictions in the next section.

For an arbitrary initial condition $v(x,0) \equiv v_0$, a general solution is found:

$$v(x,t) = \frac{1}{a}(BM_S - DM_C)\left[1 - e^{-at}\right] + e^{-at}\left[(B+D)(I_C - I_S)\int_0^t e^{a\tau}\mathrm{m}(\tau)d\tau + v_0\right], \qquad (13)$$

where

$$a = A + M_C + M_S = A + M(V_C + V_S). \qquad (14)$$

Equation (13) is easily solved for many different choices of temporal modulation function $\mathrm{m}(t)$. Note that (13) holds regardless of the dimension of the underlying network. The receptive field profiles, and thus the spatial input function I, can be one- or two-dimensional without affecting the general solution. Similarly, (13) is not dependent on the specific choice of receptive field profile: the quantities A, M_S and M_C are constant for any given choice of receptive field profile and maximum bipolar activation M.

9 Push-pull response to modulated sinusoidal gratings

The previous section shows that addition of a push-pull mechanism to restore response symmetry causes the shunting network to behave as a *linear, time-invariant* system. The symmetry in the network's response to increments and decrements regardless of average stimulus intensity or contrast is due to a general property of LTI systems. Namely, the response of an LTI system to an arbitrary input consists of two terms: a sustained component whose magnitude only depends on the sustained component of the input, and a transient or modulated component that only depends on modulated input components. This is illustrated in the following simple example.

Let the input to the push-pull network consist of a steady background $I(x)$ upon which is superimposed a spatial modulation function $J(x)$ multiplied by a temporal modulation function $m(t)$. Mathematically, this input distribution is represented as a sum of two terms:

$$I(x,t) = I(x) + J(x)m(t). \tag{15}$$

Based on this definition, the modulated component of the spatial input (the grating J) and the temporal modulation function need not be restricted to nonnegative values: as long as the maximum overall negative excursion of the modulated grating $J(x)m(t)$ is smaller in amplitude than the steady background $I(x)$, the net input distribution remains nonnegative.

The input function (15) only affects the right-hand side of (10), and direct substitution into (13) leads to the general solution

$$\begin{aligned} v(x,t) = &\frac{1}{a}[(B+D)(I_C - I_S) + BM_S - DM_C](1 - e^{-at}) + \\ &\left[(B+D)(J_C - J_S)\int_0^t e^{a\tau}m(\tau)d\tau + v_0\right]e^{-at}, \end{aligned} \tag{16}$$

with I_C, I_S, M_C, M_S, and a defined as before in (11), (12), (14), and

$$J_C = \int_{-\infty}^{\infty} C(x-\xi)J(\xi)d\xi \quad \text{and} \quad J_S = \int_{-\infty}^{\infty} S(x-\xi)J(\xi)d\xi, \tag{17}$$

for an arbitrary spatial modulation function $J(x)$.

The solution (16) is almost identical to (13), with the exception that the two components of the input distribution give rise to two distinct response components: a temporally modulated component depending only $J(x)$ and an asymptotically steady component depending only on $I(x)$.

Consider now the network response to stimuli that are sinusoidally modulated in both space and time against a steady uniform background, i.e., let $I(x) \equiv I$ (constant); the temporal modulation function is

$$m(t) = \sin \Omega t, \tag{18}$$

and the spatial modulation is

$$J(x) = \alpha \cos \omega x. \tag{19}$$

Substituting (18) and (19) into (16) leads to the following solution:

$$v(x,t) = v_0 e^{-at} + \frac{B+D}{a}\left[I(V_C - V_S) - \frac{DV_C - BV_S}{B+D}\right](1 - e^{-at}) +$$
$$\frac{B+D}{a^2 + \Omega^2}\left[V_C \exp\left(-\frac{\omega^2 \sigma_C^2}{2}\right) - V_S \exp\left(-\frac{\omega^2 \sigma_S^2}{2}\right)\right]\alpha \cos \omega x \left(\Omega e^{-at} + a \sin \Omega t - \Omega \cos \Omega t\right). \tag{20}$$

Here the Difference-Of-Gaussians receptive field profile of equation (6) was used to arrive at an explicit solution. It is possible to make some general remarks about the form of this solution, which consists of three main components: a transient component, an asymptotically constant component, and a spatiotemporally modulated component.

The first, transient component represents the network's dependence on its initial state v_0, and decays at a rate e^{-at}.

The second, asymptotically constant component reaches steady-state as rapidly as the transient component vanishes, and its equilibrium value depends jointly on the steady background I, the center and surround RF volumes V_C and V_S, and the physiological parameters A, B, D and M. However, this term is not affected by the superimposed, temporally modulated grating. The mathematical form of this component shows that interactions between input parameters (I), morphological parameters (V_C, V_S), and physiological parameters (A, B, D, M) can jointly modify sensitivity to steady input components, resting level, and integration rate of the cell.

The third, spatiotemporally modulated component exhibits a complex dependence on both the spatial and temporal modulation functions. However, this term is unaffected by the steady background I. Thus this term represents the network's ability to respond to spatiotemporal information in the modulated component of the input. Interaction of the modulated input with morphological and physiological parameters simultaneously affects spatial frequency selectivity, temporal frequency selectivity, and integration rate.

As claimed above, the addition of a push-pull mechanism allows the shunting network to behave as an LTI system, thus segregating sustained and modulated input components into mutually independent response components. This property is confirmed by numerical integration, as shown in Fig. 4b, which illustrates the network's ability to segregate a steady background from modulated input components.

It should be emphasized that the push-pull shunting network *behaves as a linear time-invariant system regardless of the nature of input preprocessing*. The example above assumes that the input is a perfect sinusoid in space and time only for clarity. In reality this input must be preprocessed by a layer of cells that can (nonlinearly) compress the luminance signal into a finite range. In such a case the response of the push-pull network exhibits a sustained component that depends exclusively on the average output level of the preprocessing element, and a transient component that depends on the modulated response component of the preprocessing element. As a result, the push-pull shunting network may appear to behave linearly in spite of nonlinear preprocessing, as will be shown later.

10 A unified mechanism for X and Y cells

Equation (20) can be used to demonstrate further properties of the push-pull shunting network. Specifically, I will show how the interaction between morphological, physiological, and input parameters can selectively enhance transient or sustained response components. In particular, a uniform increase in a (which is approximately equal to the total area or volume of the RF) decreases the amplitude of sustained response components through the factor $1/a$, and simultaneously increases the network's integration rate through the factor e^{-at}. The transition from more sustained to more transient response with an increase in receptive field size is suggestive of the functional and morphological distinction between the two main classes (X and Y) of cat retinal ganglion cells as first reported by Enroth-Cugell and Robson (1966).

Although the original work of Enroth-Cugell and Robson (1966) sparked a wave of qualitative and quantitative experiments on RGC response, and although a number of qualitative and quantitative RGC receptive field models have been proposed, the qualitative nonlinear receptive field model of Hochstein and Shapley (1976b) is still considered to be the authoritative account of the

Y-cell's receptive field. This model embodies the belief that different neural mechanisms must comprise the X and Y receptive fields.

The push-pull shunting network of eq. (7) instead provides a framework for a unified quantitative explanation of the behavior of both X and Y cells in response to spatiotemporally modulated inputs. Specifically, the interactions of the push-pull bipolar mechanism with the receptive field center and surround components may give rise to X-like response characteristics for small receptive field profiles, or Y-like response characteristics for larger receptive field profiles, in agreement with experimental data linking X cells with beta cells, and Y cells with alpha cells (Cleland and Levick, 1974; Fukuda et al., 1984; Saito, 1983).

In order to assess the validity of this claim it is first necessary to specify a mechanism capable of temporal adaptation, that is, one that can adjust its dynamic range in response to input fluctuations so as to compress a broad range of luminance signals into a bounded range of neural signals. Note that in the absence of such a mechanism the push-pull shunting network is unable to simulate X and Y cell data for at least two reasons: first, the model as it stands does not exhibit the overshoots and undershoots typically seen in X and Y cell response to sudden input fluctuations (e.g., Fig. 1). Second, if the input to the network were simply a linear copy of the luminance profile (such as the sinusoidal grating used above), then the network's response should be purely linear, and nonlinear phenomena such as Y cell frequency doubling cannot be generated.

I will argue in the remainder of the chapter that the transient (overshoots and undershoots) and nonlinear components of X and Y cell responses arise through preprocessing in the photoreceptor and bipolar cell layers, and that the push-pull mechanism causes the ganglion cells to behave as close to linear as possible in spite of nonlinearities in the preprocessing layers.

11 The photoreceptor model

The temporal adaptation mechanism used here is based on a model of signal transmission through chemical gating introduced by Grossberg (1968), which was later extended by Carpenter and Grossberg (1981) to simulate many properties of photoreceptor behavior. For present purposes, I motivate the model by proposing that the overshoots and undershoots arise from interactions between slow and fast components of a transduction mechanism whose purpose is to transmit input signals without bias through the photoreceptor by means of an intracellular transmitter substance. All derivations in this section follow those of earlier articles on which the temporal adaptation model is based (e.g., Carpenter and Grossberg, 1981), but are duplicated here for completeness. This section may be skipped on a first reading.

The simplest law describing the transduction of light into neural signals is a multiplicative, or *mass action* law

$$I(t) = L(t) \cdot z(t), \qquad (21)$$

where $L(t)$ represents the incoming luminance signal, $I(t)$ represents the photoreceptor activation, and $z(t)$ represents the amount of available intracellular transmitter substance.

If the amount of available transmitter is limited, and if transmitter is activated at a rate proportional to the input signal, then the signal through the photoreceptor adapts to the average input intensity on a time scale that depends on the rates of transmitter substance activation and regeneration, resulting in a simple form of temporal adaptation. Alternatively, the mechanism presented here can be interpreted as a transduction step involving a second messenger system, as appears to be the case in some photoreceptors (see below).

The amount of available transmitter can be described by an ordinary differential equation,

$$\frac{dz(t)}{dt} = F\left[G - z(t)\right] - HL(t)z(t) \tag{22}$$

which indicates that transmitter is activated at a rate proportional to the net signal through the photoreceptor (indicated by the term $HL(t)z(t)$), and is replenished at a rate which is proportional to the amount of used transmitter (indicated by the term $F\left[G - z(t)\right]$). A heuristically derived law of form (22) has been used to describe the amount of free sites to which cyclic GMP can bind for transduction of light signals into neural activity in the vertebrate rod (e.g., Pugh and Lamb, 1990).

The net signal through the photoreceptor under the influence of a steady input signal is a monotone *increasing* function of the input signal amplitude, even though the amount of available transmitter is a monotone *decreasing* function of the input signal amplitude. This can be seen by solving (22) at steady-state:

$$z_{ss} = \frac{FG}{F + HL_{ss}}, \tag{23}$$

where L_{ss} and z_{ss}, respectively, represent the steady-state input signal and available transmitter. This equation confirms that the steady-state amount of available transmitter Z_{ss} is a monotone decreasing function of input signal L_{ss}. Substituting (23) into (21) leads to

$$I_{ss} = \frac{L_{ss}FG}{F + HL_{ss}}, \tag{24}$$

where I_{ss} is the steady-state photoreceptor response, which is a monotone increasing function of the input intensity, as claimed.

The ability to adapt to time changes in average input intensity leads to transient overshoots and undershoots in response to sudden input fluctuations. To show the existence of such overshoots and undershoots, consider what happens if a photoreceptor which has adapted to a steady input level L_0 is suddenly switched to a new input level L_1 at time $t = 0$. Immediately before the input is switched (time $t = 0^-$), the equilibrium signal through the photoreceptor is

$$I(t = 0^-) \equiv I_0 = \frac{L_0 FG}{F + HL_0}. \tag{25}$$

If the regeneration rate is slow compared to the depletion rate—and hence to rapid fluctuations in the input signal—then the amount of transmitter available at time $t = 0^+$ (immediately after the input jumps to a new value L_1) remains approximately constant, and the net signal is

$$I(t = 0^+) \equiv I_0^+ \approx \frac{L_1 FG}{F + HL_0}. \tag{26}$$

This equation shows that the model exhibits a form of Weber's law: the system's response to a new input level L_1 is calibrated against the prior (equilibrium) input level L_0, so that increment in response is of the form

$$I_0^+ - I_0 = \Delta I \propto \frac{\Delta L}{L}. \tag{27}$$

After the photoreceptor has equilibrated to the new input L_1, the steady-state signal through the photoreceptor becomes

$$I(t \to \infty) \equiv I_1 = \frac{L_1 FG}{F + HL_1}. \tag{28}$$

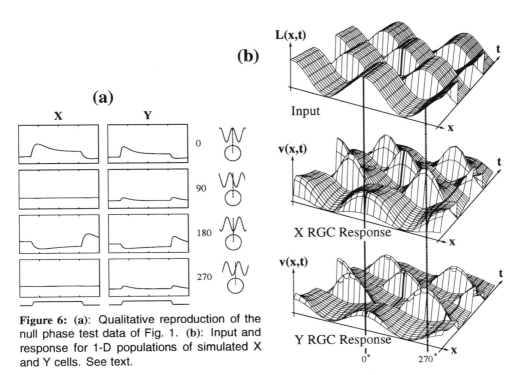

Figure 6: (a): Qualitative reproduction of the null phase test data of Fig. 1. (b): Input and response for 1-D populations of simulated X and Y cells. See text.

If $L_1 > L_0$, (26) and (28) jointly imply that $I_0^+ > I_1$, which is indicative of a transient overshoot. A similar analysis shows that an undershoot will occur if $L_1 < L_0$.

It is possible through a similar analysis to show that the overshoot attained in switching from a lower steady-state input signal L_1 to a higher signal L_2 is always larger than the corresponding undershoot attained in switching from the higher steady-state signal L_2 to the lower signal L_1. This property of photoreceptor adaptation has been used (Gaudiano, 1991a,c) to show mathematically that the push-pull network can sometimes exhibit nonlinear behavior (such as frequency doubling) in response to temporal modulation of spatial stimuli, depending on the input and model parameters.

12 The null test for X and Y cells

I will now present a number of illustrative simulation results to support the claims of previous sections. Fig. 6 shows a qualitative fit of the experimental data in Fig. 1. These results are based on numerical integration of (7) when the input—consisting of a square-wave modulated sinusoidal grating—is preprocessed by a layer of the simulated photoreceptors. To emphasize the predicted relationship between morphological and functional properties in the push-pull network *the only difference between X and Y cells in these simulations is that the RF center of the X cell is 0.3 times as wide as the Y cell RF center*. This assumption is based on the experimental observation that the primary morphological difference between X (beta) and Y (alpha) is not simply overall receptive field size, but rather the relationship between receptive field center and surround widths: it has generally been noted (e.g., Sterling, 1990) that in the beta cell the receptive field center is significantly narrower than the surround, whereas in alpha cells the width of the receptive field center is much closer to that of the surround.

Fig. 6b shows the behavior of a 1-D population of X cells (middle graph) and Y cells (bottom graph) in response to temporal square-wave modulation of a sinusoidal grating (top graph). Time evolves along the t axis. The thick solid and dashed lines trace the input and activation of X and Y cells at 0° and 270°, respectively. These lines correspond to the individual traces in Fig. 6a, and help to visualize the null responses in X cells and on-off responses in Y cells.

Fig. 6 thus shows that on-off responses sometimes ensue in the push-pull shunting model when the input is preprocessed by a nonlinear mechanism for temporal adaptation. Specifically, the network's sensitivity to sustained and transient inputs is differentially affected by the choice of morphological parameters: the beta-like (X cell) receptive field profile makes the network more sensitive to sustained and less sensitive to transient input components. As a result the simulated X cells are not affected by the input nonlinearity and exhibit null responses at ±90° regardless of contrast or average luminance. The Y cells, on the other hand, are more sensitive to the transient component of photoreceptor output, and exhibit on-off responses at ±90°.

The appearance of on-off responses in the Y cells can be explained if, as suggested by Hochstein and Shapley (1976a), the initial transduction is asymmetrical in its response to positive or negative deflections. The simultaneous existence of contrast-independent null responses for X cells, as Enroth-Cugell and Robson (1966) suggested, further requires that the nonlinearity be in some way symmetrical about the mean luminance level. *The push-pull network satisfies both of these constraints by ensuring that amplitude modulation in the response is independent of average input level, even when the input is generated by a nonlinear mechanism whose response to increments is asymmetrical with respect to its response to decrements.*

13 On-off response in X cells

All simulations presented here use a parameter choice that leads to a larger gain on the modulated component for Y cells than for X cells. However, as evident in the network's response to sinusoidal gratings found in eq. (20), the gain of the modulated response component is not fixed for a given parameter choice, but depends on the spatial and temporal waveform. Hence nonlinear responses due to input asymmetry may be noticeable in X cells for certain stimulus configurations.

Indeed, the existence of nonlinear responses in X cells has been documented in the experimental literature. For example, Enroth-Cugell and Robson (1966) showed the experimentally measured response of a single X cell to introduction and withdrawal of a bipartite field: inspection of their Fig. 12 shows the existence of small on-off responses at the location of the contrast edge. Furthermore, the original figure caption (Enroth-Cugell and Robson, 1966, Fig. 12) indicates that a contrast of only 0.2 was used for this example, whereas other figures showing null phase experiments used a contrast of 0.32. This point has been noted by Hochstein and Shapley (1976a, p. 247).

In terms of the present model, an explanation of X cell on-off responses to a modulated contrast edge is that such a stimulus configuration enhances the asymmetry between the inputs reaching the two sides of a cell's RF, and can thus overcome the relatively small gain of the X cell modulated response component. This observation was confirmed by mathematical analysis of the push-pull network response to modulation of a bipartite field (Gaudiano, 1991a,c).

The effectiveness of edge sharpness in enhancing on-off responses can be seen in Fig. 7. The

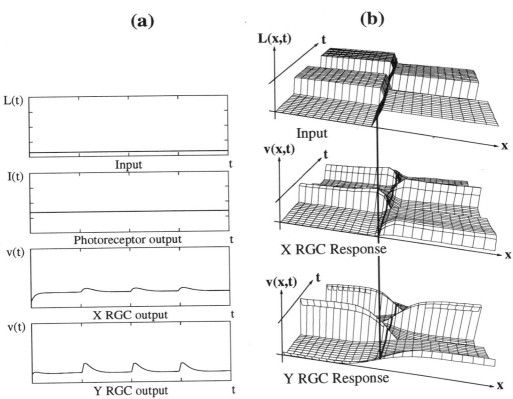

Figure 7: The response of simulated X and Y cells to a square-wave modulated bipartite field. All parameters are the same as in earlier figures. (a): Response of a photoreceptor, X, and Y cell located at contrast edge. X cell exhibits small on-off responses. (b): Response of 1-D populations of X and Y cells to the bipartite field. Thick solid line correspond to individual traces in (a).

activation of individual photoreceptors, X, and Y cells near the contrast edge are shown in Fig. 7a, confirming the existence of small on-off responses in the X cells, and more vigorous ones in Y cells. Note that the input and photoreceptor responses at the position shown in Fig. 7a are constant because the input in the middle of the contrast edge remains constant at the average luminance level. The on-off responses at that location arise from the nonlinearity in the activation of photoreceptors on the two sides of the RGC's receptive field (Gaudiano, 1991a,c).

This result is also illustrated in Fig. 7b, which shows the response of 1-D populations of simulated X and Y cells to the same square-wave modulated bipartite field. It is apparent that Y cells at the edge exhibit a large positive-going response at both contrast reversals. It is also apparent that, in this example, X cells at the contrast edge also show positive responses at both contrast reversals (on-off). All parameters are the same as those used for earlier simulations.

14 Spatial frequency and phase dependence of the on-off response

Hochstein and Shapley (1976a,b) extended the null phase test to include analysis of the temporal frequency components of X and Y cell response. They found that under most stimulus conditions X cell response is dominated by a fundamental temporal frequency component equal to the modulation frequency. In contrast, Y cell response exhibits second harmonic distortion, i.e., the response shows a mixture of fundamental and second harmonic (frequency doubling) components. The Y cell fundamental component exhibits linear spatial phase dependence similar to that of X cells, whereas the on-off, or second harmonic component is largely spatial phase-independent.

In general, the fundamental temporal frequency response component of Y cells falls off more rapidly at high spatial frequencies than that of X cells, owing to the Y cell's broader RF center. However, the Y cell second harmonic response persists at much higher spatial frequencies. Thus, there exists a range of moderately high spatial frequencies at which X cells exhibit a normal spatial phase-dependent response at the fundamental temporal frequency of modulation, whereas Y cells exhibit a pure phase-independent on-off (second harmonic) response.

This behavior is predicted by the push-pull shunting network, because the on-off response depends on photoreceptor asymmetry, which is propagated through the bipolar receptive fields. Mathematical analyses have shown (Gaudiano, 1991c) that the existence of on-off responses depends only partially on the spatial modulation function. For instance, a sinusoidal grating whose spatial frequency is too high for detection by a Y cell's center and surround RF components, can still cause asymmetrical responses in the bipolars, so that although unable to detect the spatial modulation, the Y cell can still exhibit on-off responses at all values of relative spatial phase.

The response of X and Y cells to a moderately high spatial frequency sinusoid is shown in Fig. 8a. All parameters are the same as those of Fig. 6b, but the spatial frequency is doubled. The figure shows that X cell response remains phase dependent, while Y cells exhibit on-off, or frequency doubling responses at all spatial phase values.

The model in its present form makes the simplifying assumption that the *cone→bipolar* connections are one-to-one, so that the second harmonic component of the response is expected to persist for the highest discriminable spatial frequencies. A more accurate model including larger bipolar receptive fields would show loss of the second harmonic response for high spatial frequencies that cannot be discriminated by the bipolars.

The properties of X and Y cell response to the classes of stimuli analyzed thus far should only change at extreme spatial frequency values if a more realistic bipolar cell model were used. However, a significant qualitative difference in RGC response may ensue for different classes of stimuli, particularly rapidly moving stimuli. Analysis of push-pull network behavior under these conditions will be the subject of future research.

15 Fading of stabilized images

The transduction mechanism presented in Section 11 is a form of nonlinear temporal adaptation. Temporal adaptation in general is a nonlinear process, in the sense that the output of such a mechanism is time-dependent. This point is very important because it bears on the issues discussed in Section 4. It is widely believed that photoreceptor and bipolar cell nonlinearities are not normally seen in X cells because these represent mild nonlinearities, such as a sigmoidal compression, and can thus be ignored at small signal values. However this argument only applies at steady state, that

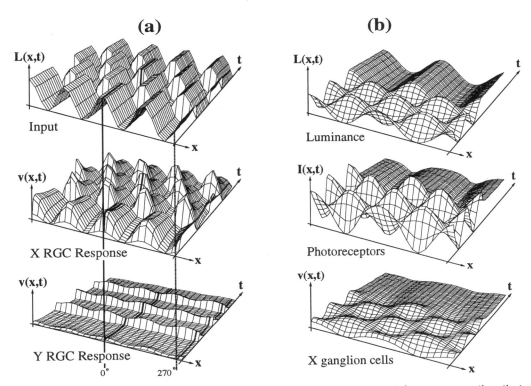

Figure 8: (a): Response of X and Y cell populations to a high spatial frequency grating that is square-wave modulated in time. All parameters are the same as in Fig. 6b, but the spatial frequency is doubled. (b): Responses of populations of photoreceptors and X cells to a sinusoidally modulated grating. Responses fade quickly when the image is stabilized.

is, once the photoreceptors have had time to adapt. In order to suppress nonlinearities in response to rapid temporal modulation, such as the square-wave modulation used by Enroth-Cugell and Robson (1966) and other authors, the photoreceptors would have to adapt *instantaneously*. Such instantaneous adaptation, aside from being unlikely in a biological setting, would lead to disastrous effects: if each photoreceptor were able to instantly adjust its output range in response to input fluctuations, then each photoreceptor would never be able to signal any temporal changes in luminance. Subsequent processing stages would then receive perfectly constant signals from all photoreceptors, and would thus be functionally blind!

To clarify this point I will show that the simple photoreceptor adaptation law embodied in equations (21)-(22) predicts that the ratio of photoreceptor activations in response to two distinct illumination levels approaches unity as the photoreceptors approach equilibrium. In other words, *stabilized images fade at the photoreceptors*. To see this, consider two spatially distinct points in a visual scene, one of intensity L_1, the other of intensity L_2. Let the ratio of these signals be $\theta = L_1/L_2$. Denote the response of two photoreceptors exposed to these signals by I_1 and I_2. If both photoreceptors have equilibrated to an initial signal L_0 when the new pattern is presented, eq. (26) shows that the respective responses at $t = 0^+$ are

$$I_1(t = 0^+) = \frac{L_1 FG}{F + HL_0} \tag{29}$$

$$I_2(t = 0^+) = \frac{L_2 FG}{F + HL_0}, \tag{30}$$

so that for the first few instants following instatement of the new input signals, the ratio of output signals is identical to the ratio of inputs:

$$\frac{I_1(t = 0^+)}{I_2(t = 0^+)} = \frac{L_1}{L_2} = \theta. \tag{31}$$

Consider now the equilibrium response approached by each photoreceptor:

$$I_1 \rightarrow \frac{L_1 FG}{F + HL_1} \quad \text{as} \quad t \rightarrow \infty, \tag{32}$$

$$I_2 \rightarrow \frac{L_2 FG}{F + HL_2} \quad \text{as} \quad t \rightarrow \infty. \tag{33}$$

Hence as the photoreceptors approach equilibrium, the ratio of output signals becomes

$$\frac{I_1}{I_2} = \frac{L_1}{L_2} \cdot \frac{F + HL_2}{F + HL_1}. \tag{34}$$

As long as HL_1 and HL_2 are larger than F, the resulting output ratio approaches unity, i.e., simultaneous contrast approaches zero and stabilized images fade, in general agreement with data on biological visual systems.

Figure 8b shows the behavior of 1-D populations of photoreceptors and RGCs in response to an input function consisting of a spatial sinusoidal grating that is sinusoidally modulated in time, and then suddenly stopped at the maximum contrast level. The network response shows that spatial contrast is almost completely lost when the image is stabilized. As is apparent from the figure, much of the contrast is lost at the photoreceptor layer.

As suggested in Section 4, instantaneous adaptation would always keep photoreceptors in approximate equilibrium, leading to inadequate transmission of *spatial* information.

The result shown here is based on a specific, simplified photoreceptor model. Similar results obtain with a more accurate model such as that presented by Carpenter and Grossberg (1981). In general this result should hold for *any* photoreceptor model that includes a mechanism for temporal adaptation, since fading should follow from the photoreceptor's ability to adapt to ambient light intensity. An analysis of this hypothesis is the subject of ongoing research.

16 Discussion

The work presented in this chapter is based on mathematical analyses of the spatiotemporal response of feed-forward shunting networks. The resulting model leads to qualitative simulations of some important retinal data, and proposes functional reasons for the underlying retinal circuits. One of the most compelling observations to support the validity of this model is perhaps captured in the paradoxical conclusion of Enroth-Cugell and Robson (1966) on the linearity of photoreceptors, as was discussed in Section 4. The ability to adapt to ambient illumination without the undesirable side effects of nonlinear image processing has obvious ecological value. The push-pull shunting network overcomes this problem by combining two copies of the incoming signal in such a way that the net input to the RGCs is always symmetrical about the average output level of a possibly nonlinear transducer, a property that can yield symmetrical null responses in X cells, and on-off responses in Y cells. The results presented here also explain why X cells are less sensitive than Y cells to nonlinear input transients, but may sometimes exhibit on-off responses.

In addition to simulating and clarifying certain experimental findings on retinal ganglion cells, the model shows that adaptation laws of the form used here lead to loss of perceived contrast in response to sustained images. In other words, stabilized images fade. The relationship of this finding to the biologically observed fading of stabilized images is the subject of ongoing work. Such loss of contrast appears to be an inescapable side effect of temporal adaptation. Hence, just as the original shunting network sacrifices temporal resolution in order to achieve spatial adaptation, so the photoreceptors sacrifice spatial resolution in order to achieve temporal adaptation.

Taken together with the results presented here, the above observations suggest that the retina provides an efficient means of maintaining high resolution in both space and time by sharing the load between three systems. The photoreceptors are able to optimize sensitivity in the temporal domain by adjusting their dynamic range in response to changes in local luminance levels, but in doing so they lose much of their ability to process sustained spatial information. Conversely, the center-surround receptive field mechanism and shunting dynamics in the RGCs can extract useful spatial information, but at least in some cases lead to degraded temporal resolution. Finally, the push-pull bipolar cells allow the RGCs to respond rapidly to increments and decrements of a nonnegative input signal, and in so doing they ensure that spatial RGC processing is not affected by photoreceptor nonlinearities.

The push-pull shunting network in its present form is a lumped model: no explicit equation is provided for the behavior of the bipolar cells, and no mention is made of horizontal and amacrine cells. However, the lateral interactions that are lumped in the model's equations must originate in horizontal and amacrine cells. More accurate, quantitative data fits will require a more detailed photoreceptor model, and an unlumped model of bipolar, horizontal and amacrine cells. This is particularly important for simulations that examine spatial and temporal frequency response, as the photoreceptor and bipolar cell response characteristics may be rate-limiting.

It is unclear at this time whether the transmitter gating law should be associated with photoreceptor behavior as was done here. From a functional perspective, the temporal adaptation should occur at the first stages of phototransduction. As indicated earlier, an enhanced version of this law has been used to model photoreceptor dynamics (Carpenter and Grossberg, 1981), and a heuristically derived law of form (22) has been implicated in the transduction of light into neural signals for the vertebrate rod (Pugh and Lamb, 1990). However, the transient overshoots and undershoots observed at and beyond the bipolar cell layer may be faster and more pronounced than those found in the photoreceptors. In order to improve response time, the same type of gated transmission law with overall faster dynamics may thus also take place at synaptic junctions in the inner and outer plexiform layers. For example, in a related model Öğmen and Gagné (1990) have suggested that this type of transmitter dynamics may be found in the synaptic connection from photoreceptors to large monopolar cells in the fly.

The push-pull model can be used to make strong predictions about properties of amacrine and horizontal cells, as well as bipolar cells. For example, if any portion of the RGC surround is carried by amacrine cells, then it is expected that the amacrine cells must also receive push-pull bipolar inputs. Also, the ability to selectively discount sustained input components should only arise *following* the bipolar cell layer. This is one of the strongest predictions of the model, because the entire development of this work is based on the assumption that it is the push-pull action at the bipolar level that allows the system to behave as a linear time-invariant system, and thus to *instantaneously* disregard sustained input components even though light is a nonnegative physical quantity. The model thus predicts that there should exist no classes of bipolar (or horizontal) cells that can give purely transient responses, a finding that receives some support in the experimental literature (e.g., chapter 4 of Dowling, 1987).

The existence of push-pull bipolar inputs to RGCs as proposed here receives some direct experimental support, though even this limited support has recently been questioned (Sterling, personal conversation). However, regardless of its validity as a model of specific retinal circuitry, the push-pull shunting network forces the conclusion that functional behavior of any cell cannot be qualitatively inferred by the type and distribution of the inputs it receives. This is clearly demonstrated by the X cell simulations in this chapter: the response of a simulated X cell appears linear in spite of nonlinear preprocessing and nonlinear membrane properties of the X cell itself.

By the same token, the nature and distribution of input cells of different types cannot be inferred by observing RGC behavior. For example this model shows that the lack of on-off responses in X cells does not preclude—but actually requires—the existence of inputs from both depolarizing and hyperpolarizing bipolar cells.

In light of these observations, I suggest that it is possible that push-pull bipolar inputs are ubiquitous in the retina, but have not been reported because their existence is functionally elusive.

Note that a constraint similar to the one found for RGCs should also apply to other neural processing stages. For instance, RGC output signals rely on axonal transmission (through action potentials) and are thus relegated to nonnegative values. Hence faithful spatiotemporal processing by cortical cells should require convergence of on-center and off-center RGCs. A similar mechanism has been suggested to explain the linearity of cortical simple cell responses in spite of floor nonlinearities (e.g., thresholds) in RGC output (Emerson et al., 1987). The model presented in this chapter extends that prediction by allowing for arbitrary preprocessing nonlinearities, and by suggesting that the same mechanism may apply to other cortical cell classes.

There are still many insights to be gained from the mathematical form of the push-pull network. It is apparent that different choices of dynamical and morphological parameters can significantly affect model behavior. I should point out that although simulated X and Y cells were made to only differ by a single parameter, this is obviously an oversimplification. For more accurate results the center and surround amplitudes (C and S), and the shunting saturation terms (B and D) should be manipulated in conjunction with the RF center and surround widths (σ_C and σ_S) to achieve a desired functional behavior. The single-parameter change is meant to emphasize the ability to affect RGC function by means of simple parametric changes within a single formal model.

Another unique aspect of this model is that it was not originally derived to simulate RGC data, but was instead formulated as a means of simultaneously achieving accurate spatial and temporal processing in feed forward shunting networks. This is a complementary approach to models that are grounded in experimental data, and often designed to reproduce retinal behavior by joining many detailed models of each element or layer of the retina (e.g., Rodieck, 1965; Siminoff, 1991; Sterling et al., 1987; Werblin, 1991). While the present model purports to explain the reason for adopting a certain architecture, the experimentally derived models contain very important data on many anatomical, physiological, and pharmacological properties of individual cells in the retina. Future research combining the push-pull architecture and details from such experimental models will undoubtedly yield more accurate results and a clearer understanding of retinal function.

References

Barlow, H.B. (1953). Summation and inhibition in the frog retina. *Journal of Physiology*, **119**: 69–88.

Boycott, B.B. and Wässle, H. (1974). The morphological types of ganglion cells of the domestic cat's retina. *Journal of Physiology*, **240**: 397–419.

Bullock, D. and Grossberg, S. (1988). Neural dynamics of planned arm movements: Emergent invariants and speed-accuracy properties during trajectory formation. *Psychological Review*, **95**, 49-90.

Carpenter, G.A. and Grossberg, S. (1981). Adaptation and trasnsmitter gating in vertebrate photoreceptors. *Journal of Theoretical Neurobiology*, **1**: 1-42.

Cleland, B.G. and Levick, W.R. (1974). Brisk and sluggish concentrically organized ganglion cells in the cat's retina. *Journal of Physiology*, **240**: 421–456.

Ditchburn, R.W., and Ginsborg, B.L. (1952). Vision with a stabilized retinal image. *Nature*, **170**, 36–37.

Dowling, J.E. (1987). **The retina: an approachable part of the brain**. Belknap: Cambridge, USA.

Emerson, R.C. Korenberg, M.J., and Citron, M.C. (1989). Identification of intensive nonlinearities in cascade models of visual cortex and its relation to cell classification. In Marmarelis, V.Z. (Ed.) **Advanced methods of physiological system modeling**. New York: Plenum Press.

Enroth-Cugell, C. and Robson, J.G. (1966). The contrast sensitivity of retinal ganglion cells of the cat. *Journal of Physiology*, **187**: 517–552.

Freed, M.A. and Sterling, P. (1988). The ON-alpha ganglion cell of the cat retina and its presynaptic cell types. *Journal of Neuroscience*, **8** (7): 2303–2320.

Fukuda, Y., Hsiao, C.-F., Watanabe, M., and Ito, H. (1984). Morphological correlates of physiologically identified Y, X and W cells in the cat retina. *Journal of Neurophysiology*, **52**: 999–1013.

Furman, G.G. (1965). Comparison of models for subtractive and shunting lateral-inhibition in receptor-neuron fields. *Kybernetyk*, **2**: 257–274.

Gaudiano, P. (1991a). Neural network models for spatio-temporal visual processing and adaptive sensory-motor control. Unpublished doctoral dissertation, Boston University.

Gaudiano, P. (1991b). A Unified Neural Network Model of Spatio-Temporal Processing in X and Y Retinal Ganglion Cells. I: Analytical Results. *Biological Cybernetics*, in press (1992).

Gaudiano, P. (1991c). A Unified Neural Network Model of Spatio-Temporal Processing in X and Y Retinal Ganglion Cells. II: Temporal Adaptation and Simulation of Experimental Data. *Biological Cybernetics*, in press (1992).

Grossberg, S. (1968). Some physiological and biochemical consequences of psychological postulates. *Proceedings of the National Academy of Sciences*, **60**: 758–765.

Grossberg, S. (1970). Neural Pattern Discrimination. *Journal of Theoretical Biology*, **27**: 291–337.

Grossberg, S. (1973). Contour enhancement, short-term memory, and constancies in reverberating neural networks. *Studies in Applied Mathematics*, **52**: 217-257.

Grossberg, S. (1980). Intracellular mechanisms of adaptation and self-regulation in self-organizing networks: the role of chemical transducers. *Bulletin of Mathematical Biology*, **42**: 365–396.

Grossberg, S. (1983). The quantized geometry of visual space: The coherent computation of depth, form, and lightness. *Behavioral and Brain Sciences*, **6**: 625–692.

Grossberg, S. (1988). Nonlinear neural networks: Principles, mechanisms, and architectures. *Neural Networks*, **1**: 17-61.

Hochstein, S. and Shapley, R.M. (1976a). Quantitative analysis of retinal ganglion cell classifications. *Journal of Physiology*, **262**: 237–264.

Hochstein, S. and Shapley, R.M. (1976b). Linear and nonlinear spatial subunits in Y cat retinal ganglion cells. *Journal of Physiology*, **262**: 265–284.

Hodgkin, A.L. (1964). **The conduction of the nervous impulse**. Liverpool: Liverpool University Press.

Kolb, H., Nelson, R. and Mariani, A. (1981). Amacrine cells, bipolar cells and ganglion cells of the cat retina: a Golgi study. *Vision Research*, **21**: 1081–1114.

Kuffler, S.W. (1953). Discharge patterns and functional organization of the mammalian retina. *Journal of Physiology*, **16**: 37–68.

McGuire, B.A., Stevens, J.K., and Sterling, P. (1986). Microrcircuitry of beta ganglion cells in cat retina. *Journal of Neuroscience*, **6**: 907–918.

Öğmen, H., and Gagné, S. (1990). Neural models for sustained and ON-OFF units of insect lamina. *Biological Cybernetics*, **63**: 51–60.

Pugh, E.N. Jr, and Lamb, T.D. (1990). Cyclic GMP and calcium: the internal messenger of excitation and adaptation in vertebrate photoreceptors. *Vision Research*, **30** (12): 1923–1948.

Rodieck, R.W. (1965). Quantitative analysis of cat retinal ganglion cell response to visual stimuli. *Vision Research*, **5**, 583–601.

Saito, H. (1983). Morphology of physiologically identified X-, Y- and W-type retinal ganglion cells of the cat. *Journal of Comparative Neurology*, **221**: 279–288.

Schiller, P. (1982). Central connections of the retinal On and Off pathways. *Nature*, **297**: 580–583.

Slaughter, M.M, and Miller, R.F. (1982). 2-Amino-4-phosphonobutyric acid: a new pharmacological tool for retina research. *Science*, **211**, 182–185.

Sperling, G. (1970). Model of visual adaptation and contrast detection. *Perc. & Psychophys.*, **8** (3): 143–157.

Sperling, G. and Sondhi, M.M. (1968). Model for visual luminance discrimination and flicker detection. *Journal of the Optical Society of America*, **58**: 1133–1145.

Sterling, P. (1990). Retina. In Shepherd, G.M. (Ed.) **The synaptic organization of the brain**. New York: Oxford University Press, third edition. Chapter 6, 170–213.

Werblin, F. (1971). Adaptation in a vertebrate retina: Intracellular recordings in Necturus. *Journal of Neurophysiology*, **34**: 228–241.

Werblin, F. (1991). Synaptic connections, receptive fields, and patterns of activity in the tiger slamander retina. *Investigative Ophthalmology and Visual Science*, **32** (3), 459–483.

Werblin, F. and Dowling, J.E. (1969). Organization of the retina of the mudpuppy, *Necturus maculosus*. II. Intracellular recording. *Journal of Neurophysiology*, **32**: 339–355.

NEURODYNAMICS OF REAL-TIME IMAGE VELOCITY EXTRACTION

David A. Fay and Allen M. Waxman[†]

Abstract

We describe our recent work on real-time implementations of early vision neural computations on the PIPE video-rate, parallel computer. In particular, neural networks based on shunting dynamics that perform light adaptation, spatial contrast enhancement, and image feature velocity extraction have been developed and implemented to process live imagery (256x256 pixels, 8-bit data) at 30 frames/sec. We demonstrate how moving edge features generate progressing waves of activity whose phase velocities are easy to obtain, and match those of the features themselves. The estimated velocities, which are oriented normal to the edge features, are then grouped to generate coherent local flow fields. Such flows are useful for segmenting time-varying imagery into separate objects, tracking the moving objects, and deducing their 3D structure and motion. These computations simulate spatiotemporal processing in the retina and motion pathway of the brain.

1. Introduction

One primary task performed by the retina is the detection of temporal change on the visual field. Psychophysical experiments by Yarbus (1967) have shown that objects kept stationary on the retina will fade within several seconds. This implies the retina adapts to stationary background illumination. Dowling (1987) has conducted electrophysiological measurements which show that, though photoreceptors respond rapidly to light spanning less than three orders of magnitude, they adapt over longer timescales to a range of illumination exceeding six orders of magnitude. The inherent computations have the effect of an adaptive offset and automatic gain control at each location on the visual field. Spatial change on the visual field is detected by the bipolar cells, which have center-surround receptive fields (cf. Fig. 4.17 in Dowling, 1987). On-center cells are excited by photoreceptors in their vicinity and inhibited by nearby horizontal cells which spatially average the surround on a slower timescale than the center. The off-center bipolar cells are excited and inhibited in opposite fashion. Together, these bipolar cells yield a rectified output that enhances areas of high spatial contrast. This combination of temporal and spatial change detection results in the enhancement of dynamic edge features.

A next step along the transient branch of the visual pathway involves the detection of motion and the extraction of velocity from moving edges and whole patterns. Barlow, Hill & Levick (1964) and Barlow & Levick (1965) have shown that some ganglion cells in rabbit retina respond selectively to both direction and speed of moving stimuli. Neurons tuned to

---

[†]This report is based on studies performed at Lincoln Laboratory, a center for research operated by the Massachusetts Institute of Technology. This work was sponsored, in part, by the Department of the Air Force. The authors are with the Machine Intelligence Technology Group at Lincoln Laboratory.

edge speed as well as direction have been found in area MT of macaque monkey brain by Maunsell & Van Essen (cf., their Figures 1, 5, & 6, 1983). Related studies in the cat have been described by Orban (1985). Detected edge motion is normal to the edge, and should not be confused with 2D pattern motion. The motion of an edge can be ambiguous, while the motion of a pattern is usually not. Movshon *et al.* (1985), using electrophysiological studies in area 17 of cat brain and area MT of macaque monkey brain, have found neurons sensitive to each of these classes of feature motion. Nakayama's (1985) review provides an excellent introduction to both the psychophysical and neurophysiological findings related to short-range visual motion sensitivity and its biological applications. Many of these issues have also been addressed (though rather unsatisfactorily) in the context of computer vision algorithms. Below, we develop and demonstrate dynamic neural network formulations for image velocity extraction, and real-time implementations, motivated by known physiology.

2. Neural Computation

The computational model for our network was inspired by Grossberg's dynamical formalism (cf., Grossberg, 1988, for an excellent overview), and utilizes a number of network interactions first developed by him for temporal adaptation, transient detection, and spatial contrast sharpening (cf., Chapters 1 & 5 of Grossberg, 1987; and Chapter 3 of Carpenter & Grossberg, 1991). Each node in the network has capacitor-like dynamics similar to the structure of cell membranes with variable conductances. In our network, light adaptation is achieved using an inhibitory interneuron whose temporal dynamics is slow compared to photodetector response time. Spatial contrast enhancement is achieved via a rectified on-center/off-surround feedforward network. The velocity of the resulting moving edges are found using a dynamic formulation of the Convected Activation method (Waxman, Wu & Bergholm, 1988). This method exploits the phase velocity of activity waves generated by moving edge features, in order to estimate the feature velocities. This speed estimate is invariant to the illuminant, as opposed to commonly used methods based on correlation techniques and its variants (Reichardt, 1961; van Santen & Sperling, 1985; Adelson & Bergen, 1985; Horn & Schunck, 1987; Anandan, 1989) which relate image velocity directly to moving intensity patterns. In fact, it is well known that moving intensity gradients alone (i.e., linear intensity profiles) do not generate a percept of motion (Nakayama, 1985). A method analogous to our 1988 formulation of Convected Activation, but based on the motion of phase contours resulting from spatiotemporal Gabor filtering of intensity (Fleet & Jepson, 1990) also suffers from problems related to the illuminant itself. The illuminant must be discounted, both in the amplitude and phase domains of any preprocessing, for otherwise it generates motion artifacts.

Only normal velocity of moving edges is provided by our network; a consequence of the familiar *aperture problem* (Marr & Ullman, 1981). To recover velocity fields of entire patterns, we use the Velocity Functional Method (Waxman & Wohn, 1985, 1988; Wohn & Waxman, 1990), which groups the estimated normal flows into coherent local flow fields, and also supports a flow segmentation process. Spatial derivatives of the flow are obtained directly by this approach, they do not result from any local differencing of velocity measurements (in contrast to Nakayama's (1985) assertions). This information is useful for segmenting time-varying imagery, detecting and tracking moving objects, determining time to collision with objects, and recovering local 3D structure and motion in the case of rigid bodies in motion.

These neural computations have been implemented both on a Sun Microsystems SPARC2 workstation, and on the PIPE[1] video-rate, parallel computer. PIPE is an 8-stage MIMD machine for real-time image processing, performing over one billion 8-bit operations per second (Kent, Shneier & Lumia, 1985). Temporal computations are implemented using the integral solutions of the differential equations that govern the neurodynamics via an exponentially fading memory. Spatial computations are performed by convolving the images with masks. Examples of these computations are described and illustrated below. We also compare the precision of these 8-bit computations with a floating-point implementation, and demonstrate the adequacy of the real-time results.

3. Network Dynamics

In this section we develop the network dynamics governing processing for 1D imagery. Extension to 2D imagery is straightforward. The complete network is shown in Figure 1 (illustrated in cross-section for simplicity). Temporal adaptation to scene illumination is performed by the bottom three layers of the network. The input image is received by the photodetection layer, the output of which excites an inhibitory interneuron with a slow time constant. The next layer is inhibited by the interneuron and excited by the photodetectors, yielding a light-adapted output. (These can also be thought of as chemical interactions within individual photoreceptors.) Denoting photodetector activity by $I_i(t)$, and interneuron activity by $\tilde{I}_i(t)$, the interneuron dynamics is modeled by

$$\frac{d\tilde{I}_i}{dt} = -\lambda \tilde{I}_i + I_i \quad , \tag{1a}$$

which implies a slow averaging of the input intensity according to

$$\tilde{I}_i(t) = \tilde{I}(0)e^{-\lambda t} + \int_0^t I_i(\tau)e^{-\lambda(t-\tau)}d\tau \quad . \tag{1b}$$

Denoting activity of the light-adapted nodes by $J_i(t)$, their dynamics is governed by a shunting short-term memory model,

$$\frac{dJ_i}{dt} = -\mu J_i + \kappa(1 - J_i)I_i - \kappa(1 + J_i)\tilde{I}_i \quad . \tag{2a}$$

The first term corresponds to a passive decay to the zero resting potential with a rate constant $\mu \gg \lambda$. The second and third terms represent the excitatory and inhibitory inputs, shunted (i.e., gated) by the adapted node activities so as to generate a fixed dynamic range $-1 < J_i < 1$. Thus, shunting acts as an automatic gain control. The rapidly adapted

---

[1] PIPE is manufactured by ASPEX, Inc. of New York. Illustrations of the PIPE architecture in this paper are presented with permission from ASPEX.

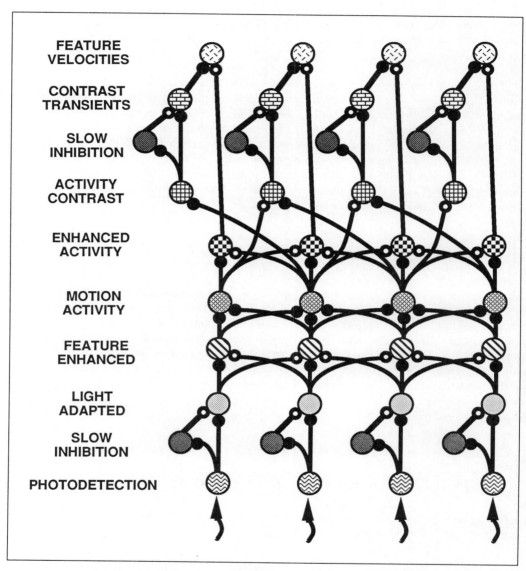

Figure 1: Cross-section of the dynamic network architecture illustrating the multiple layers of processing involved in light adaptation, edge enhancement, and edge velocity extraction.

equilibrium derived from (2a) is

$$J_i = \frac{I_i - \tilde{I}_i}{\frac{\mu}{\kappa} + I_i + \tilde{I}_i} \quad , \tag{2b}$$

that is, a normalized change relative to the slowly varying background. A detection threshold on J_i implies a Weber-Fechner law for just noticeable differences (JNDs) in intensity I_i.

The light-adapted activity is fed to the next layer via center-surround connections to spatially enhance dynamic edges. The dynamics of an on-center/off-surround edge enhancing node E_i is also modeled by shunting dynamics with decay rate α and normalized Gaussian weighted connections G_{ji}^+ and G_{ki}^-,

$$\frac{dE_i}{dt} = -\alpha E_i + \beta(1 - E_i)\sum_j G_{ji}^+ J_j - \beta(1 + E_i)\sum_k G_{ki}^- J_k \quad . \tag{3a}$$

The scale of the inhibitory Gaussian is chosen larger than that of the excitatory Gaussian. Rapid equilibration of (3a) yields a normalized edge response

$$E_i = \frac{DoG * J_i}{\frac{\alpha}{\beta} + SoG * J_i} \quad , \tag{3b}$$

where DoG is the difference of excitatory and inhibitory Gaussian weights, SoG is their sum, and $*$ implies a convolution. The rate constants are constrained such that $\frac{\alpha}{\beta} > 2$. This is required to prevent the denominator from becoming zero even when the term $SoG * J_i$ reaches its minimum value of -2. (Recall that J_i can approach -1.) An expression complimentary to (3a) governs the off-center/on-surround edge nodes. In order to provide insensitivity to direction of contrast, the output from these dual node layers can either be sent to different motion systems or combined into a single fully rectified response. Typically, we utilize a rectified output of the off-center node layer in order to prevent a double response near the edge location. Example results from each of the layers involved in creating the dynamic edges are shown in Figure 2. The upper-left quadrant contains an original image from a sequence showing a moving box with a star pattern. The time-averaged interneuron image is shown in the upper-right quadrant. The lower-left quadrant displays the resulting light-adapted imagery produced from the original and time-averaged images. The results of the edge enhancement is shown in the lower-right quadrant. These results are obtained from our PIPE implementation described below. These dynamic edge features feed the following velocity extraction layers.

The edge nodes convey their signals to the next layer, only if their activity exceeds a threshold. This thresholded binary output θ_i, excites the spatiotemporal activity layer A_i via a Gaussian fan-out, as shown in Figure 1. This layer also has an exponentially fading memory, and is governed by the slow dynamics characterized by rate constant λ,

$$\frac{dA_i}{dt} = -\lambda A_i + \sum_j G_{ji}^{(\sigma)} \theta_j \quad , \tag{4}$$

Figure 2: An example output from the real-time PIPE computations showing light adaptation and edge enhancement results. (A) Original live image from the sequence; (B) time-averaged interneuron layer; (C) light-adapted layer; (D) edge-enhanced layer.

where σ is the chosen scale of the excitatory fan-out. A similar dynamics was exploited by Grossberg & Rudd (1989) in their model of long-range apparent motion.

Consider the idealized problem of an isolated 1D edge in motion, which can be represented as a unit delta function as follows:

$$E(x_i, t) = \delta[x_i - x_e(t)] \quad , \tag{5a}$$

$$x_e(t) = \nu t \quad , \tag{5b}$$

where x_e represents the location of the edge, moving at unknown velocity ν, at time t. Equation (4) can be written in this simple case as

$$\frac{dA_i}{dt} = -\lambda A_i + C e^{-(x_i - x_e)^2 / 2\sigma^2} \quad , \tag{6}$$

where $C = \frac{1}{\sqrt{2\pi}\sigma}$ in order to normalize the Gaussian fan-out. The integral form of (6) emphasizes the fading memory,

$$A_i(t) = A_i(0)e^{-\lambda t} + Ce^{-\lambda t}\int_0^t e^{\lambda \tau}e^{-(\nu\tau - x_i)^2/2\sigma^2}d\tau \quad . \qquad (7)$$

Transforming the integral yields an expression for the activity in terms of the complimentary error function,

$$A_i(t) = C\sqrt{\frac{\pi}{2}}\frac{\sigma}{\nu}e^{\frac{\lambda}{\nu}[(x_i - \nu t) + \lambda\sigma^2/2\nu]}erfc\left[\frac{1}{\sqrt{2}\sigma}(x_i - \nu t + \lambda\sigma^2/\nu)\right] \quad . \qquad (8)$$

Consider this expression in the reference frame of the moving edge, and let $s \equiv (x_i - \nu t)/\sigma$ represent a scaled distance variable and $\rho \equiv \sigma\lambda/\nu$ represent a speed ratio (i.e., the ratio of characteristic speed $\sigma\lambda$ to edge speed ν). Then equation (8) can be rewritten as

$$A_i(t) = A_i(x_i - \nu t) \equiv A_i(s) = \frac{1}{2\nu}e^{\rho(s + \rho/2)}erfc\left[\frac{1}{\sqrt{2}}(s + \rho)\right] \quad . \qquad (9)$$

This implies that moving edges generate *moving waves of activity* that preserve their shape (over the timescales of nearly constant velocity). The phase velocity of each wave gives the speed of the corresponding moving edge; the direction of motion is normal to the edge. These concepts motivated the name *Convected Activation Profiles* in our original formulation (Waxman, Wu, & Bergholm, 1988).

Figure 3 shows two plots of activity waves and their gradients as functions of s and ρ. Increasing the value of ρ causes a decrease in the length of the "tail" due to the exponentially fading memory, as well as a broadening of the Gaussian curve. In Figure 3, only the decrease in the size of the tail is evident because the abscissa is scaled by σ, the size of the Gaussian fan-out. To accurately extract the velocity of an edge, the value of ρ should be kept constant. In order to best accomplish this, λ should be set at a fixed value while σ is varied proportionally to ν, the unknown velocity. This is needed to ensure that the characteristic speed is appropriately matched to the edge speed, and so can provide a valid estimation. The plots of activity gradients in Figure 3 help illustrate why a good match is required. Both plots contain a linear regime trailing the location of the moving edge (at $s = 0$). This linear region must be large enough to contain the displacement of the moving gradient. If σ is too small, the edge displacement in one frame-time will fall outside the linear region, corrupting the velocity estimation. On the other hand, if σ is too large the chance of edge interactions increases, that is, edges are never really isolated on all scales. Since the velocity estimates to be obtained are only accurate for isolated edges, the Gaussian fan-out should be kept as small as possible. Also, because we don't know how to set the characteristic speed *a priori*, there must in general be multiple motion channels, each tuned to a different speed. A competition between the motion channels activated at each location will yield the best match between the characteristic speed and the edge speed, and thus the best velocity estimate. This need for multiple channels is reminiscent of the speed tuned cells found in monkey and cat brains. Figure 4 shows moving activity waves as generated in real-time on the PIPE. The top image shows the spatiotemporal activity A_i of a thick

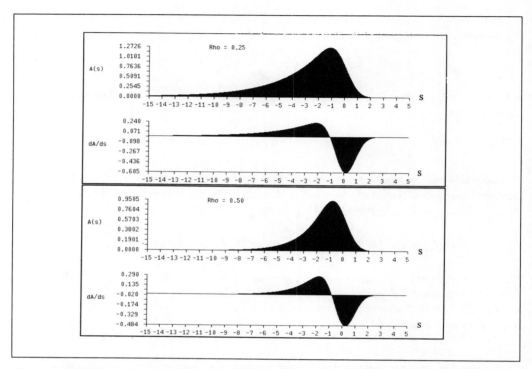

Figure 3: Plots showing examples of the activity waves from equation (9) and their gradients for two values of the dimensionless speed ratio ρ.

vertical black bar moving leftward against a white background. The bottom image shows a graph of a horizontal cut through the center of the top image. This graph illustrates the wave shape and the effects of both Gaussian fan-out and fading memory.

The next step is to extract the phase velocity of the activity waves, and hence, the velocities of the dynamic edges. Since the wave is convected, it will look the same at time t and at time $t + \Delta t$ (where Δt represents a time interval), though it will be displaced by $v\Delta t$ (where v is the velocity estimate). Describing the activity wave as a spatiotemporal function yields

$$A(x_i, t) = A(x_i + v\Delta t, t + \Delta t) \quad . \tag{10a}$$

Expanding the right hand side in a Taylor series to $O(\Delta t^2)$, we find

$$A(x_i, t) \cong A(x_i, t) + [A_t + vA_x]\Delta t + \left[A_{tt} + 2vA_{xt} + v^2 A_{xx}\right]\frac{\Delta t^2}{2} + \ldots \quad , \tag{10b}$$

where the subscripts denote partial derivatives evaluated at (x_i, t). Next we can cancel the term on the left with the 0th-order term on the right. We then divide by Δt and take the limit $\Delta t \to 0$; we obtain

$$A_t + vA_x = 0 \quad , \tag{11}$$

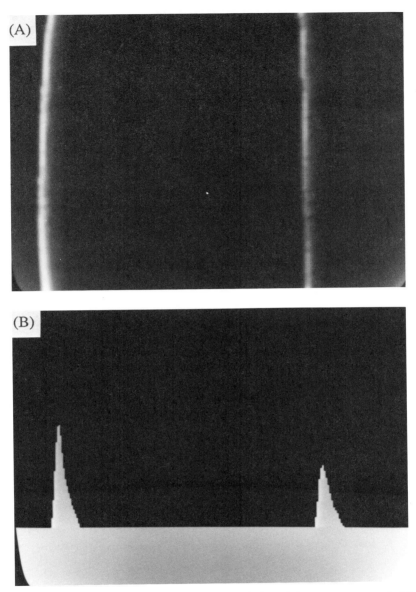

Figure 4: Example output from the PIPE showing (A) activity waves generated by a thick black bar moving to the left against a white background, and (B) a horizontal cut through the activity image illustrating the wave shapes.

which yields
$$v(x_i, t) = -\frac{A_t}{A_x} \quad . \tag{12}$$
However, this expression for the convected invariance of the activity profile will not result in an adequate velocity estimate. The reason being that we want to use this expression near edge locations, where the partials A_t and A_x approach zero. Although the edge location is not exactly at the maximum of the activity wave, due to the exponentially fading memory,

it is near enough that the velocity estimates would be unreliable. Thus, the terms of $O(\Delta t)$ would not provide a useful constraint near an edge. Since each order of equation (10b) must vanish identically, we can instead consider the terms of $O(\Delta t^2)$ in (10b) and again take the limit $\Delta t \to 0$; we find

$$A_{tt} + 2vA_{xt} + v^2 A_{xx} = 0 \quad , \tag{13a}$$

which can be rewritten as

$$\frac{\partial}{\partial t}[A_t + vA_x] + v[A_{xt} + vA_{xx}] = 0. \tag{13b}$$

For the reasons of convected shape embodied in equation (11) above, the first term in brackets vanishes for all times both near the edge and away from it. This leaves the expression

$$A_{xt} + vA_{xx} = 0 \quad , \tag{14a}$$

which can also be rewritten as

$$\left\{\frac{\partial}{\partial t} + v\frac{\partial}{\partial x}\right\} A_x = 0 \quad . \tag{14b}$$

This form of the equation shows that the activity gradient is also convected along with the moving edge. As was mentioned earlier, the velocity estimate is valid as long as the finite displacement in one frame-time falls within the linear regime of the gradient. The actual velocity estimate is derived from equation (14a), and as can be seen in Figure 3 is found trailing immediately behind the location of the dynamic edge (within the linear regime of the gradient),

$$v(x_i, t) = -\left[\frac{A_{xt}}{A_{xx}}\right] \quad . \tag{15}$$

In the simple case of an isolated edge, one can substitute expression (8) in the right-hand side of equation (15), and upon performing the differentiation one finds the velocity estimate $v = \nu$, the actual edge speed.

The derivative ratio in equation (15) suggests the use of shunting dynamics to govern the activity at the feature velocity layer in Figure 1. Let the activity gradient transient T_i be defined as

$$T_i(t) \equiv \frac{\partial^2}{\partial t \partial x} A_i(x_i - \nu t) \quad , \tag{16}$$

which can be found using slow inhibitory interneurons (cf. Fig. 1). Also define the contrast enhanced activity (a sustained signal) S_i as

$$S_i(t) \equiv -\frac{\partial^2}{\partial x^2} A_i(x_i - \nu t) \quad . \tag{17}$$

This can be found using on-center/off-surround connections similar to the ones used to enhance the dynamic edges (cf. Fig. 1). The feature velocity nodes V_i thus receive excitation from the contrast transient layer and shunted inhibition from the enhanced activity layer as shown in Figure 1, and governed by

$$\frac{dV_i}{dt} = -\gamma V_i + \frac{1}{\sigma\lambda}[T_i]^+ - V_i[S_i]^+ \quad . \tag{18a}$$

The first term provides a passive decay at a rate determined by γ, while the second and third terms provide rectified excitation and inhibition, respectively. The excitatory term is normalized by the characteristic speed $\sigma\lambda$, to provide the means for a later competition across different scales. The inhibitory term is shunted in order to provide a lower bound for the activity ($0 \leq V_i$). After rapid equilibration, the normalized feature velocity estimate is given by

$$V_i = (\frac{1}{\sigma\lambda})\left\{\frac{[T_i]^+}{\gamma + [S_i]^+}\right\} \quad . \tag{18b}$$

Since the feature velocity nodes are driven by rectified inputs, they will only provide a response in the "forward direction" (i.e., the direction to which the nodes are tuned to respond). This is similar to the directional selectivity *(preferred vs. null direction)* displayed by cells in rabbit retina, cat area 18, and monkey MT. The need for competition across orientation in 2D imagery, as well as scale, is created by this selectivity.

To extend these results to 2D imagery, we first have to establish certain restrictions:

- The edge must be locally straight on a scale $O(\sigma)$ - the size of the Gaussian fan-out to the spatiotemporal activity layer.

- The edge should not rotate much within the spatiotemporal neighborhood σ/λ at any point along the edge.

- The edge velocity should be constant over the timescale $(1/\lambda)$ of the spatiotemporal activity layer.

If these restrictions are not adhered to, the velocity estimates will represent average properties across a small neighborhood. Next, introduce at each edge contour a local coordinate system (ξ, η), where ξ is oriented perpendicular to that edge, then the estimated velocity (15) will represent the local normal component of flow v_n. This can be written as

$$v_n(\xi, \eta, t) = -\left[\frac{A_{\xi t}}{A_{\xi\xi}}\right] \quad , \tag{19}$$

which is better written in coordinate-free vector notation since there are many different local neighborhoods. Replacing A_ξ with ∇A and $A_{\xi\xi}$ with $\nabla^2 A$ (noting that $A_\eta = 0$ along an edge contour), and the normal speed v_n with the vector of normal velocity $\mathbf{v_n}$ (with x and y components) yields,

$$\mathbf{v_n}(x, y, t) = -\left[\frac{(\nabla A)_t}{\nabla^2 A}\right] \quad . \tag{20}$$

Figure 5 shows the extraction of the normal components of edge velocity from a model aircraft moving upward and to the left. The velocity is coded in gray, with white representing positive values and black representing negative values. The upper-left quadrant shows the original image. The extracted edges are shown in the upper-right. The lower-left shows the horizontal component of edge velocity, where motion to the left yields a negative velocity and rightward motion is positive. The vertical component is displayed in the lower-right

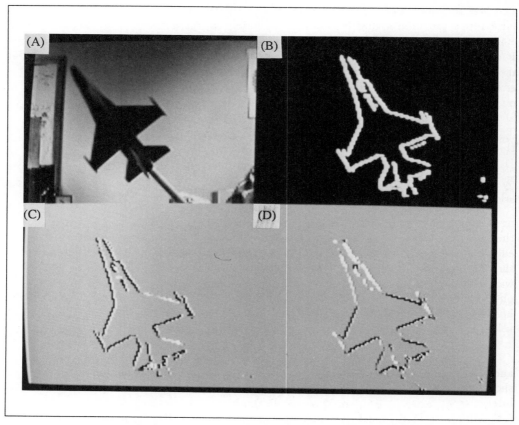

Figure 5: Example PIPE output showing edge velocity extraction from a model aircraft moving upward and to the left. (A) Original image from sequence; (B) light-adapted and edge-enhanced (thresholded); (C) horizontal component of normal velocity; (D) vertical component of normal velocity.

quadrant, with upward motion yielding negative velocity and downward yielding positive. This speed and direction coding is clearly noticeable along the wings of the aircraft.

Below we discuss implementations of our network on a Sun Microsystems SPARC2 workstation, and on the PIPE computer which performs the light adaptation and spatial contrast enhancement (creating dynamic edges) as well as the extraction of normal velocity from edges in real-time.

4. Implementations and Results

Many approximations had to be made in order to implement our network on the PIPE. For example, Gaussian fan-out, spatial and temporal derivatives, and exponential fading memory, were all modeled via small mask spatial convolutions and temporal finite differences and sums, restricted to 8-bit precision. So, in addition to implementing these computations on the PIPE, we simulated the PIPE's model of the network on a SPARC2 workstation. This allowed us to easily check the network's performance when limited to only 8-bit precision. The 8-bit results were compared to a floating-point implementation so that we could assess the loss in accuracy due to this fixed precision. For simplicity, the network was modeled in 1D, with each layer consisting of 400 nodes. The PIPE implementation supports 2D image sequence processing in real-time.

An example of a simple input to the network, and various layer outputs, is shown in Figure 6 from our SPARC2 implementation. The five plots on the left display the outputs from the light adaptation and spatial contrast enhancement processes that extract the dynamic edge features from a moving input pattern. On the right side there are five plots which show the activity wave and its derivatives, which are used to find the phase velocity of the wave. The output of some of the final velocity nodes are shown in detail in Figure 7. This plot shows the responses in the vicinity of the edge location. *The valid velocity estimate will always be found in the small linear regime just behind the edge.* In this example the actual velocity $\nu = 1.0$, fan-out $\sigma = 1.94$, and decay rate constant $\lambda = 0.25$.

The parameters which most influence the network's performance are the constants σ and λ that determine the characteristic speed in the spatiotemporal activity layer. Table 1 shows some example floating-point speed estimates for different actual velocities of a simple intensity discontinuity, given $\lambda = 0.25$ and $1.50 \leq \sigma \leq 2.29$. For speeds $\nu = 1.0$ and $\nu = 2.0$, a choice of $\sigma \approx 1.90$ will produce an accurate velocity estimate, but for $\nu = 3.0$ a larger $\sigma \approx 2.08$ is needed. This illustrates the need for larger Gaussian fan-outs in order to accurately estimate higher velocities. However, if σ is too large, the network will underestimate the unknown velocity. This is due to a shifting of the linear regime (as seen in Figure 3) which causes the location of the edge to move away from the center of the linear regime. As it does so, the slope of the curve decreases, resulting in the underestimation of speed. A second problem with utilizing too large a fan-out, is the possible interactions between edges moving at different speeds. This will cause the net activity profile to change its shape over time, in conflict with the underlying principle of *convected activation* here.

Table 2 shows some example results from the network when it uses *artificially thickened edges*. The edges are thickened by spreading activity to the two neighboring nodes of an edge, such that an edge that was one node wide will become three nodes wide, providing more input to the spatiotemporal activity layer. This increase in input yields more activity, thus widening the wave and indirectly widening the linear regime of its gradient. By comparing the velocity estimates of Table 2 with those from Table 1, it is clear that thickening the edges has the same effect as increasing the size of σ. This is useful for extracting high velocities when hardware limits the size of the Gaussian fan-out that can be used. In particular, a

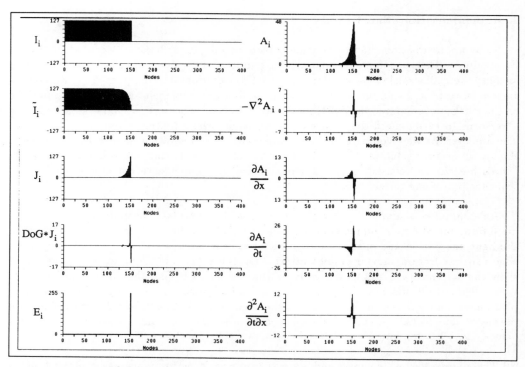

Figure 6: Example outputs from the network layers obtained in the SPARC2 implementation: An intensity discontinuity I_i moving to the right at 1 pixel per frame. $\sigma = 1.94$ (corresponding to a discrete 15 pixel Gaussian fan-out), rate constant $\lambda = 0.25$.

discrete 9-pixel Gaussian is satisfactory over the range $1.0 \leq \nu \leq 3.0$.

Next we performed simulations on the SPARC2 workstation using 8-bit precision throughout, in order to test how this limitation affects the resulting velocity estimates. Table 3 gives results for $1.50 \leq \sigma \leq 1.94$. As can be seen, the small σ values were good enough for estimating a speed $\nu = 1.0$, but performed poorly at high speeds. This indicates the need for a larger linear regime in the activity gradient, created either by increasing the size of σ, or by thickening the edges. Thickening the edges is quite simple, and with limited precision is essential. The results are shown in Table 4. Now the network is able to not only accurately estimate speeds of $\nu = 1.0$, but also higher velocities with $\nu = 2.0$ and $\nu = 3.0$ using this range of σ values. In fact, a simple 9-pixel fan-out yields excellent results, as long as the edges are at least three pixels wide.

The network was then implemented on a PIPE computer which can processes live imagery at 60 fields/sec. PIPE is an 8-stage MIMD parallel machine that is used for real-time image processing, capable of performing 1.2 billion arithmetic operations per second on 8-bit data. In addition to the 8 Modular Processing Stages, there is a video interface stage

N	σ	ν		
		1.0	2.0	3.0
9	1.50	1.088	2.547	8.700
11	1.66	1.044	2.253	4.937
13	1.80	1.017	2.084	3.845
15	1.94	0.998	1.976	3.327
17	2.06	0.985	1.902	3.026
19	2.18	0.975	1.850	2.831
21	2.29	0.968	1.811	2.695

Table 1: Speed estimates from the SPARC2 floating-point implementation of the network, for an intensity discontinuity moving at speed ν. Velocity estimates are given for various degrees of fan-out characterized by σ (or size N of the discrete Gaussian).

N	σ	ν		
		1.0	2.0	3.0
9	1.50	0.986	1.925	3.000
11	1.66	0.980	1.886	2.929
13	1.80	0.974	1.848	2.814
15	1.94	0.968	1.814	2.706

Table 2: Speed estimates from the SPARC2 floating-point implementation of the network with "artificially thickened edges." A discrete 9-pixel Gaussian fan-out yields satisfactory results for speeds $1.0 \leq \nu \leq 3.0$.

N	σ	ν		
		1.0	2.0	3.0
9	1.50	1	3	7
11	1.66	1	4	6
13	1.80	1	3	5
15	1.94	1	2	9

Table 3: Speed estimates from the SPARC2 8-bit implementation of the network with "normal edges" (one or two pixels wide).

N	σ	ν		
		1.0	2.0	3.0
9	1.50	1	2	3
11	1.66	1	2	3
13	1.80	1	2	3
15	1.94	1	2	2

Table 4: Speed estimates from the SPARC2 8-bit implementation of the network with "artificially thickened" edges (three or four pixels wide).

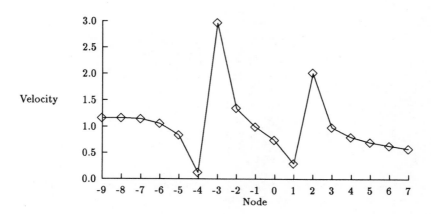

Figure 7: Example plot showing the output from the velocity extraction layer in the vicinity of the edge (located at node 0) obtained in the SPARC2 implementation. The valid velocity estimate is at node -1; the actual edge speed is $\nu = 1.0$, moving to the right.

for receiving data (256x256 8-bit pixel fields) from a camera or video tape machine, an input stage which receives data from the video interface stage, and an output stage for storing results which may be further processed or displayed. There is also a special purpose processing stage called ISMAP which supports real-time histogramming and other sorting operations. Figure 8 shows a block diagram of the PIPE, displaying the connections between the different processing stages.

The Modular Processing Stages are interconnected by a variety of pathways. A forward path provides pipelined processing by connecting each stage's output to the input of the next stage. A recursive path connects a stage's output to its own input allowing feedback. A connection from one stage's output to the input of a previous stage is also available. Finally, a set of six video buses provides connections from one stage to any of the 8 stages. Each of the processing stages contains neighborhood operators, arithmetic/logic units, single and two-input look-up tables, 3x3 (9x1) convolvers, and image buffers. A VME bus interface supports data transfer between all buffers and a host computer. A block diagram of a Modular Processing Stage is shown in Figure 9.

In order to implement the network on the PIPE, utilizing its capabilities to the fullest, we had to make many approximations to the network's dynamics. Grayscale imagery from a CCD camera provides input to the network, representing the activity at the photodetector layer. The temporal averaging of the interneuron is performed by the input stage ALU adding 1/4 of the incoming input intensity to 3/4 of the time-averaged image intensity stored during the previous frame-time of 1/30 of a second. This result is later subtracted from the next input yielding the light-adapted image. This image is then thresholded with

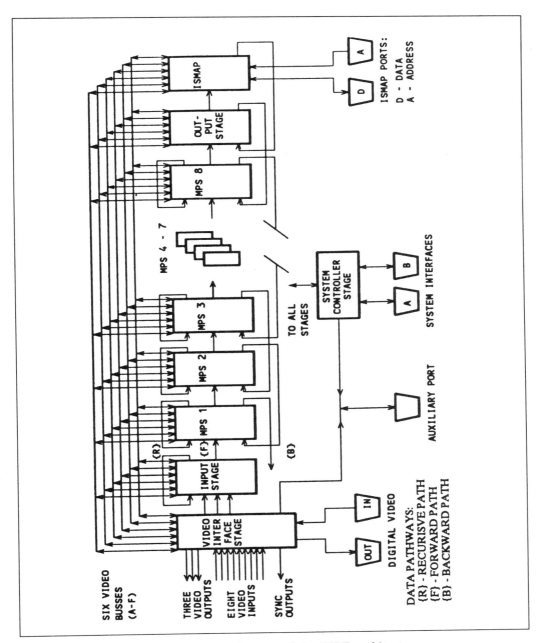

Figure 8: Block diagram of the PIPE architecture.

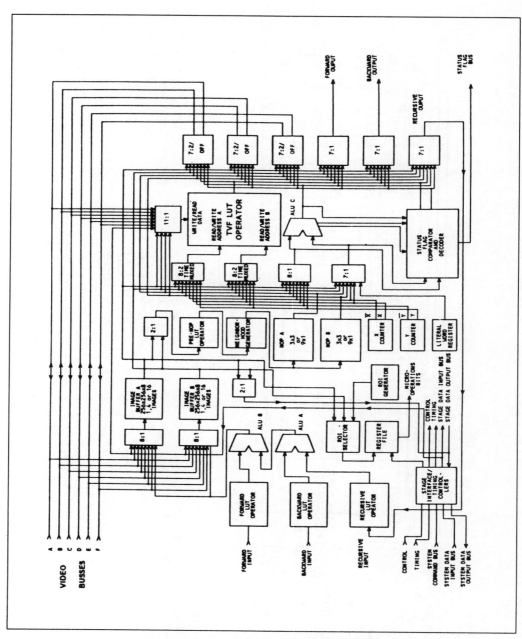

Figure 9: Block diagram of a Modular Processing Stage on PIPE.

a threshold-linear function (by using a look-up table) to remove residual noise. The spatial contrast enhancement is performed on the light-adapted image by utilizing the neighborhood operators to convolve the image with 3x3 masks. In this case we use a cascade of 3x3 Gaussian masks to produce a 7x7 Gaussian. From this we subtract a 3x3 Gaussian to create an off-center difference-of-Gaussians (DoG), and also used them to normalize the contrast approximating the equilibrium DoG/SoG. Binary edges are then created by thresholding this normalized contrast. This processing takes up four stages of the PIPE, performing about 400 million arithmetic operations per second. Example results of these computations are shown in Figure 10. The upper-left quadrant shows the original image, in which Fay (on the left)

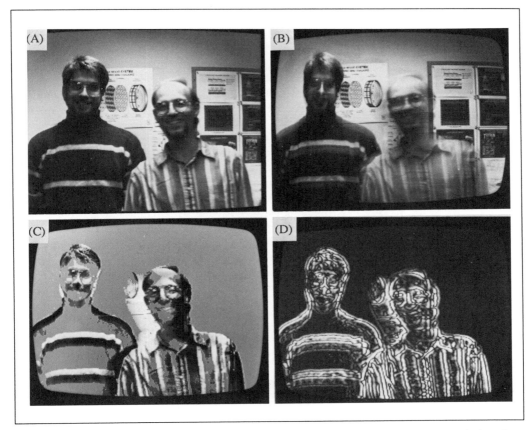

Figure 10: Example output from the PIPE real-time implementation, showing light adaptation and edge enhancement with Fay moving upwards and Waxman moving to the right. (A) Original live imagery; (B) time-averaged interneuron imagery; (C) light-adapted imagery; (D) edge-enhanced imagery.

is moving upward and Waxman (on the right) is moving to the right. In the upper-right the time-averaged interneuron image is shown; notice the blurring caused by the body motions.

The light-adapted imagery is shown in the lower-left quadrant. The stationary background has faded away, as have the large homogeneous regions, such as the dark areas in Fay's sweater. The enhanced edges of the light-adapted image are displayed in the lower-right. Strong responses are given along the edges of the stripes and around the silhouettes of the moving bodies.

Activity waves are generated by spatially smoothing the binary edge image with a 9x9 Gaussian (again created by cascading 3x3 Gaussian masks), then adding 1/4 of that image (pixel by pixel) to 3/4 of the activity image from the previous frame (approximating an exponentially fading memory in time). An example activity image is shown in Figure 4, as well as a horizontal cut through the middle of the image showing the profile of the moving activity wave.

In order to extract the phase velocity of these moving waves via equation (20), their spatial and temporal derivatives are calculated. The term $\nabla^2 A$ is estimated by simply convolving the activity image with a 3x3 Laplacian mask. The term $\partial \nabla A / \partial t$ is generated by first estimating the temporal derivative (by subtracting the activity image of the previous frame from that of the current), then convolving the resulting image with a pair of Sobel gradient masks. Finally, the velocity estimate is computed by 'dividing' each component image of $-\partial \nabla A / \partial t$ by $\nabla^2 A$ and scaling the result (via a two-input look-up table) so that it is in a visible range on the display. The total network takes up all eight stages of the PIPE, which in this case performs about 800 million arithmetic operations per second. Example results of the velocity extraction are shown in Figure 11. Again the upper-left shows the original imagery, in this case Fay is moving to the left while Waxman is moving to the right. The upper-right shows the enhanced edges, notice again that the stationary background is not visible. The horizontal velocity component of the edge motions is shown in the lower left, clearly displaying the directional selectivity (leftward motion is coded dark and rightward motion is coded light). The lower-right quadrant shows the vertical component of edge motion (downward is light, upward is dark); there is still a response due to the sloping edges though the object motion is in the horizontal direction. Another example of speed variations along sloping edges is illustrated in Figure 5, where object motion is upward and to the left (edge speed being coded in gray). Recovering whole pattern motion from edge motions is addressed in the next section.

Finally, we should note that our PIPE implementation supports velocity estimation at multiple scales (hence, supporting high edge speed estimation). This is achieved by first creating a multi-resolution pyramid of each image in the sequence, and processing the pyramid through the same network shown in Figure 1. The lower resolution imagery results in lower edge density (hence, less edge interactions via fan-out on the effectively larger scale), and supports high speed sensitivity, much like peripheral vision. By combining these speed estimates at different resolutions across the visual field, we are able to simulate the velocity sensitivity of a non-uniform retina.

5. Full Flow Recovery via the Velocity Functional Method

The theory and implementations described above comprise a real-time method for the quantitative extraction of speed along moving edge contours. It is rather obvious that image motion *along contours* is not perceptible; only 'normal flow' *perpendicular to contours* is

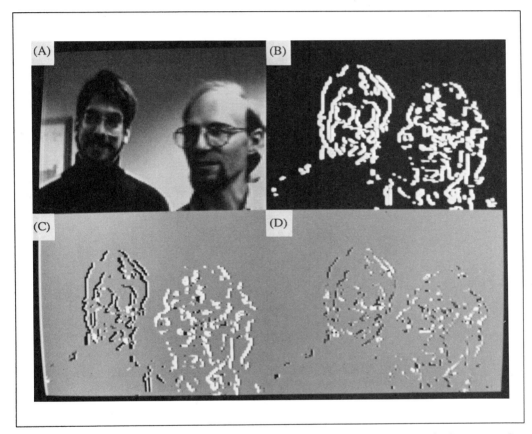

Figure 11: Example output from the PIPE implementation showing velocity extraction from Fay (moving left) and Waxman (moving right): (A) Original image from live sequence; (B) light-adapted and edge-enhanced imagery; (C) horizontal component of edge motion coded in gray (rightward is light, leftward is dark); (D) vertical component of edge motion coded in gray (downward is light, upward is dark).

perceptible and measurable. This has been termed the *aperture problem* (Marr & Ullman, 1981). To recover whole pattern motion from component motion perpendicular to sparsely located edge contours, it is necessary to combine information over neighborhoods containing many edges. Thus, it is not surprising that receptive field sizes of MT cells that respond to pattern motion are indeed large. A variety of methods have been proposed for the recovery of full flow or pattern motion from normal flow measurements, but most are based on heuristics such as locally constant velocity or nearly constant velocity (interpreted by some as *smoothness* of the flow field). The *Velocity Functional Method* developed by Waxman & Wohn (1985) provides a means to recover local flow fields based solely on the notion of flow analyticity, and hence, yields a means of detecting flow discontinuities where the analyticity breaks down (Waxman & Wohn, 1988; Wohn & Waxman, 1990). For completeness, we summarize this method here, show how spatial derivatives of flow are obtained directly, and illustrate the classes of neighborhood deformation that are captured by the approach. Conceptually, the visual field of moving edges is decomposed into overlapping neighborhoods of *receptive fields* corresponding to pattern motion cells. Within each neighborhood the flow is assumed analytic, and is modeled by polynomials (deviations from analyticity are readily detected, and lead to flow segmentation). Waxman and Wohn have argued that each component of full flow (e.g., horizontal and vertical in a vector formulation) is adequately modeled by second-order polynomials with respect to angular distance from the center of the neighborhood. This representation is useful because: (1) it models exactly the flow induced by a planar surface in rigid body motion; (2) it captures the effects of surface curvature on the flow for non-planar surfaces in rigid body motion; (3) it can model the local flow induced by non-rigid surfaces in motion; (4) it captures local flow variations such as divergence, spin, deformation and higher-order perspective effects; and (5) provides the information necessary to recover 3D surface structure and 3D rigid body motions in closed form. This polynomial model is easily extended to incorporate temporal variations in the image flow, providing a kind of temporal filtering of pattern motion (Wohn & Waxman, 1990; also see Nakayama, 1985). Defining the local flow model in terms of second-order polynomials in image coordinates (measured in radians) can be expressed as:

$$v_x(x,y) = \sum_{i=0}^{2} \sum_{j=0}^{2} v_x^{(i,j)} \frac{x^i}{i!} \frac{y^j}{j!} \quad , \tag{21}$$

$$v_y(x,y) = \sum_{i=0}^{2} \sum_{j=0}^{2} v_y^{(i,j)} \frac{x^i}{i!} \frac{y^j}{j!} \quad , \tag{22}$$

where the summation indices are limited by $i + j \leq 2$. The twelve unknown coefficients in these polynomials $v_x^{(i,j)}$ and $v_y^{(i,j)}$, are equivalent to the partial derivatives of the local flow through second-order, and are to be determined from the sparse measurements of normal flow, i.e., edge speed as obtained from equation (20). For each edge contour point at which we are able to measure the local edge orientation (and hence, local normal direction $\mathbf{n}(x,y)$) and speed, we can impose a constraint that the model flow, when projected onto the edge normal, yield the measured normal flow. This is equivalent to the constraint

$$v_n(x,y) = \sum_{i=0}^{2} \sum_{j=0}^{2} [v_x^{(i,j)} n_x(x,y) + v_y^{(i,j)} n_y(x,y)] \frac{x^i}{i!} \frac{y^j}{j!} \quad . \tag{23}$$

Given a minimum of twelve independent edge fragments in a neighborhood, these linear equations can be solved for the local flow coefficients. In general, one utilizes all edge speed data in a neighborhood, and solves equation (23) in the least-squares sense. Such quadratic error minimization problems can be mapped quite directly to the determination of adaptive weights on ADALINES (adaptive linear neurons) using the Widrow-Hoff LMS adaptation algorithm (Widrow & Sterns, 1985). In this case the weights are interpreted as the pattern velocity and its first and second spatial derivatives (unpublished work by Baxter & Waxman, 1989). The zeroth-order coefficients are all that is required for target tracking, the first-order terms are useful for determining target spin and time to collision, and the second-order terms are necessary to determine 3D surface curvature and target rigid body motion in 3D. We note that the size of the residual error in the least-squares is useful for segmenting individual neighborhoods, whereas comparison of flow models between adjacent neighborhoods can drive segmentation between neighborhoods.

Finally, we note that once the flow coefficients $\mathbf{v}^{(i,j)}$ have been determined from local edge speed data, equations (21-22) determine a dense flow field throughout the neighborhood. However, we have suggested that it is these flow coefficients themselves that characterize the evolving metrical structure of the time-varying imagery. In particular, they represent a decomposition of local image deformations onto a basis of primitive deformations that are revealed through the deforming edge contours. In effect, each term of the polynomial series corresponds to such a primitive neighborhood deformation, and can be visualized by deforming a regular grid by a unitary amount of each term in isolation. This is illustrated in Figure 12 for initially square and polar grids (from Wohn & Waxman, 1990).

As we have noted above, these flow coefficients can be used to segment flow fields and recover 3D information about local surfaces moving as rigid bodies (Waxman & Wohn, 1988; Wohn & Waxman, 1990).

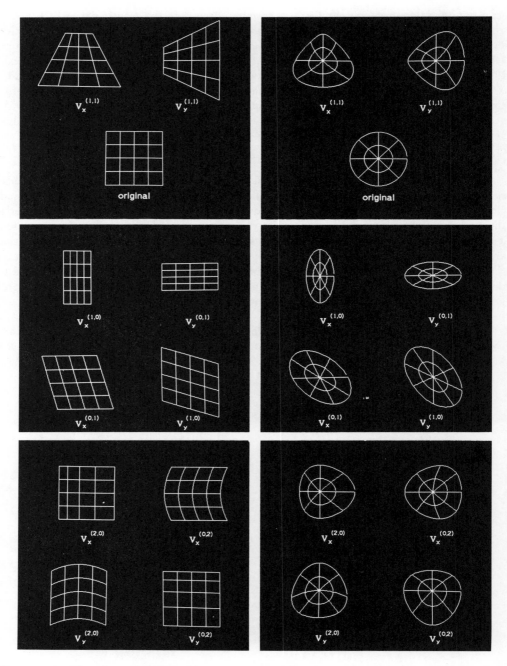

Figure 12: The class of deformations represented by the terms of the second-order flow model, illustrated for initially square (left) and polar (right) grids. The zeroth-order terms (not shown) displace the neighborhood without changing its shape.

REFERENCES

Adelson, E.H., and Bergen, J.R. (1985). Spatio-temporal energy models for the perception of motion. *Journal of the Optical Society of America A 2*, 284-299.

Anandan, P. (1989). A computational framework and an algorithm for the measurement of visual motion. *International Journal of Computer Vision 2*, 283-310.

Barlow, H.B., Hill, R.M., and Levick, W.R. (1964). Retinal ganglion cells responding selectively to direction and speed of image motion in the rabbit. *Journal of Physiology 173*, 377-407.

Barlow, H.B., and Levick, W.R. (1965). The mechanism of directionally selective units in rabbit's retina. *Journal of Physiology 178*, 477-504.

Carpenter, G.A., and Grossberg, S. (1991). **Pattern Recognition by Self-Organizing Neural Networks**. See in particular Chapter 3, *Neural pattern discrimination* (originally published in 1970 by S. Grossberg). Cambridge, MA: MIT Press.

Dowling, J.E. (1987). **The Retina: An Approachable Part of the Brain**. Cambridge, MA: Harvard Press.

Fleet, D.J., and Jepson, A.D. (1990). Computation of component image velocity from local phase information. *International Journal of Computer Vision 5*, 77-104.

Grossberg, S. (1987). **The Adaptive Brain, II: Vision, Speech, Language, and Motor Control**. See in particular Chapter 1, *The quantized geometry of visual space: The coherent computation of depth, form, and lightness* (originally published in 1983 by S. Grossberg); and Chapter 5, *Adaptation and transmitter gating in vertebrate photoreceptors* (originally published in 1981 by G.A. Carpenter & S. Grossberg). Amsterdam: Elsevier/North-Holland.

Grossberg, S. (1988). Nonlinear neural networks: Principles, mechanisms, and architectures. *Neural Networks 1*, 17-61.

Grossberg, S., and Rudd (1989). A neural architecture for visual motion perception: Group and element apparent motion. *Neural Networks 2*, 421-450.

Horn, B.K.P., and Schunck, B.G. (1981). Determining optical flow. *Artificial Intelligence 17*, 185-203.

Kent, E.W., Shneier, M.O., and Lumia, R. (1985). PIPE: Pipelined image processing engine. *Journal of Parallel and Distributed Computing 2*, 50-78.

Marr, D., and Ullman, S. (1981). Directional selectivity and its use in early visual processing. *Proceedings of the Royal Society of London B211*, 151-180.

Maunsell, J.H.R., and Van Essen, D.C. (1983). Functional properties of neurons in middle temporal visual area of the macaque monkey. I. Selectivity for stimulus direction, speed, and orientation. *Journal of Neurophysiology 49*, 1127-1147.

Movshon, J.A., Adelson, E.H., Gizzi, M.S., and Newsome, W.T. (1985). The analysis of moving visual patterns. In **Pattern Recognition Mechanisms** (C. Chagras, R. Gattass, & C. Gross, editors), 116-151, New York: Springer-Verlag.

Nakayama, K. (1985). Biological image motion processing: A review. *Vision Research 25*, 625-660.

Orban, G.A. (1985). Velocity tuned cortical cells and human velocity discrimination. In **Brain Mechanisms of Vision** (D. Ingle, M. Jeannerod, & D. Lee, editors), 371-388, Berlin: Martinus Nijhoff/Springer-Verlag.

Reichardt, W. (1961). Autocorrellation: a principle in motion perception. In **Sensory Communication** (W. Rosenblith, editor), 303-317, New York: Wiley Press.

Van Santen, J.P.H., and Sperling, G. (1985). Elaborated Reichardt detectors. *Journal of the Optical Society of America A2*, 300-321.

Waxman, A.M., and Wohn, K. (1985). Contour evolution, neighborhood deformation, and global image flow: Planar surfaces in motion. *International Journal of Robotics Research 4*, 95-108.

Waxman, A.M. and Wohn, K. (1988). Image flow theory: A framework for 3-D inference from time-varying imagery. Chapter 3 of **Advances in Computer Vision** (C. Brown, editor), Hillsdale, NJ: Lawrence Erlbaum Associates.

Waxman, A.M., Wu, J., and Bergholm, F. (1988). Convected activation profiles and the measurement of visual motion. *Proceedings of the IEEE Conference on Computer Vision and Pattern Recognition*, 717-723, Ann Arbor, MI.

Widrow, B., and Stearns, S.D. (1985). **Adaptive Signal Processing**. Englewood Cliffs, NJ: Prentice-Hall.

Wohn, K., and Waxman, A.M. (1990). The analytic structure of image flows: Deformation and segmentation. *Computer Vision, Graphics, and Image Processing 49*, 127-151.

Yarbus, A.L. (1967). **Eye Movements and Vision**. New York: Plenum Press.

WHY DO PARALLEL CORTICAL SYSTEMS EXIST FOR THE PERCEPTION OF STATIC FORM AND MOVING FORM? †

Stephen Grossberg‡
Center for Adaptive Systems
Boston University
111 Cummington Street
Boston, MA 02215

ABSTRACT

This article analyses computational properties that clarify why the parallel cortical systems $V1 \to V2$, $V1 \to MT$, and $V1 \to V2 \to MT$ exist for the perceptual processing of static visual forms and moving visual forms. The article describes a symmetry principle, called FM Symmetry, that is predicted to govern the development of these parallel cortical systems by computing all possible ways of symmetrically gating sustained cells with transient cells and organizing these sustained-transient cells into opponent pairs of on-cells and off-cells whose output signals are insensitive to direction-of-contrast. This symmetric organization explains how the static form system (Static BCS) generates emergent boundary segmentations whose outputs are insensitive to direction-of-contrast and insensitive to direction-of-motion, whereas the motion form system (Motion BCS) generates emergent boundary segmentations whose outputs are insensitive to direction-of-contrast but sensitive to direction-of-motion. FM Symmetry clarifies why the geometries of static and motion form perception differ; for example, why the opposite orientation of vertical is horizontal (90°), but the opposite direction of up is down (180°). Opposite orientations and directions are embedded in gated dipole opponent processes that are capable of antagonistic rebound. Negative afterimages, such as the MacKay and waterfall illusions, are hereby explained, as are aftereffects of long-range apparent motion. These antagonistic rebounds help to control a dynamic balance between complementary perceptual states of resonance and reset. Resonance cooperatively links features into emergent boundary segmentations via positive feedback in a CC Loop, and reset terminates a resonance when the image changes, thereby preventing massive smearing of percepts. These complementary preattentive states of resonance and reset are related to analogous states that govern attentive feature integration, learning, and memory search in

---

† Reprinted with permission from *Perception & Psychophysics*, **49**, 117–141. ©1991 The Psychonomic Society.

‡ Supported in part by the Air Force Office of Scientific Research (AFOSR 90-0175), the Army Research Office (ARO DAAL-03-88-K-0088), DARPA (AFOSR 90-0083), and Hughes Research Labs (S1-903136).

Acknowledgements: Thanks to Cynthia Bradford and Diana Meyers for their valuable assistance in the preparation of the manuscript and illustrations.

Adaptive Resonance Theory. The mechanism used in the $V1 \to MT$ system to generate a wave of apparent motion between discrete flashes may also be used in other cortical systems to generate spatial shifts of attention. The theory suggests how the $V1 \to V2 \to MT$ cortical stream helps to compute moving-form-in-depth and how long-range apparent motion of illusory contours occurs. These results collectively argue against vision theories that espouse independent processing modules. Instead, specialized subsystems interact to overcome computational uncertainties and complementary deficiencies, to cooperatively bind features into context-sensitive resonances, and to realize symmetry principles that are predicted to govern the development of visual cortex.

1. The Motion Boundary Contour System

This article contributes further evidence for a new theory of biological motion perception that was outlined in Grossberg (1987b) and quantitatively specified and analyzed in Grossberg and Rudd (1989a, 1989b, 1990a, 1990b) and in Grossberg and Mingolla (1990a, 1990b, 1990c). The new theory consists of a neural architecture called a *Motion Boundary Contour System*, or *Motion BCS*. The Motion BCS consists of several parallel copies, such that each copy is activated by a different range of receptive field sizes. Each copy is further subdivided into hierarchically organized subsystems: a *Motion Oriented Contrast Filter*, or *MOC Filter*, for preprocessing moving images; and a Cooperative-Competitive Feedback Loop, or CC Loop, for generating coherent emergent boundary segmentations of the filtered signals.

These results have provided a computational explanation for the cortical stream $V1 \to V2$ that joins the areas V1 and MT of visual cortex. A previous model of static form perception, summarized below, modeled aspects of the parallel $V1 \to V2$ cortical stream. Evidence for the MOC Filter includes its ability to explain a variety of classical and recent data about short-range and long-range apparent motion, motion capture, induced motion, and cortical cell properties that have not yet been explained by alternative models. Grossberg and Rudd (1989c) have, moreover, shown how the main properties of other motion perception models can be assimilated into different parts of the Motion BCS design.

2. Why is not a Static Form Perception System Sufficient?

The Motion BCS model provides a computationally precise answer to the following perplexing question. It is well known that some regions of visual cortex are specialized for motion processing, notably region MT (Albright, Desimone, and Gross, 1984; Maunsell and van Essen, 1983; Newsome, Gizzi, and Movshon, 1983; Zeki, 1974a, 1974b). On the other hand, even the earliest stages of visual cortex precessing, such as the simple cells in area V1, are sensitive to changes in stimulus intensity and to direction-of-motion (De Valois, Albrecht, and Thorell, 1982; Heggelund, 1981; Hubel and Wiesel, 1962, 1968, 1977; Tanaka, Lee and Creutzfeldt, 1983). Why has evolution gone to the trouble to generate specialized regions such as MT, when even the simple cells of V1 are already change-sensitive and direction-sensitive? What computational properties are achieved by MT that are not already available in V1 and its prestriate projections V2 and V4? In response to this question, many scientists reply that a motion system needs larger receptive fields. This may be true, but cannot be the heart of the answer, because V1, V2, and V4 also possess multiple receptive field sizes.

Our answer, along with a new theory of motion form perception, came into view after

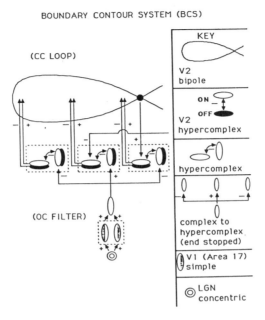

Figure 1. The static Boundary Contour System circuit described by Grossberg and Mingolla (1985b). The circuit is divided into an oriented contrast-sensitive filter (SOC Filter) followed by a cooperative-competitive feedback network (CC Loop). Multiple copies of this circuit are used, each corresponding to a different range of receptive field sizes of the SOC Filter. The depicted circuit has been used to analyse data about monocular vision. A binocular generalization of the circuit has also been described (Grossberg, 1987b; Grossberg and Marshall, 1989).

some unexpected implications of our previous theory of static form perception were noticed. This latter theory has been called FACADE Theory, because its visual representations are predicted to multiplex together properties of Form-And-Color-And-DEpth in prestriate cortical area V4. FACADE Theory describes the neural architecture of two subsystems, the Boundary Contour System (BCS) and the Feature Contour System (FCS), whose properties are computationally complementary (Grossberg, Mingolla, and Todorović, 1989). The BCS generates an emergent 3-D boundary segmentation of edges, texture, shading, and stereo information at multiple spatial scales (Grossberg, 1987a, 1987b, 1990; Grossberg and Marshall, 1989; Grossberg and Mingolla, 1985a, 1985b, 1987). The FCS compensates for variable illumination conditions and fills-in surface properties of brightness, color, and depth among multiple spatial scales (Cohen and Grossberg, 1984; Grossberg, 1987a, 1987b; Grossberg and Mingolla, 1985a; Grossberg and Todorović, 1988).

The BCS provided a new computational rationale as well as a model of the neural circuits governing classical cortical cell types such as simple cells, complex cells, and hypercomplex cells. The theory also predicted a new cell type, the bipole cell (Cohen and Grossberg, 1984; Grossberg and Mingolla, 1985a) whose properties have been supported by neurophysiological experiments (von der Heydt, Peterhans, and Baumgartner, 1984; Peterhans and von der Heydt, 1989).

This BCS model, now called the *Static* BCS model, consists of several parallel copies, such that each copy is activated by a different range of receptive field sizes, as in the Motion BCS. Also as in the Motion BCS, each Static BCS copy is further subdivided into two hierarchically organized systems (Figure 1): a *Static Oriented Contrast Filter*, or *SOC Filter*, for preprocessing quasi-static images (the eye never ceases to jiggle in its orbit); and a Cooperative-Competitive Feedback Loop, or CC Loop, for generating coherent emergent boundary segmentations of the filtered signals. Thus the Motion BCS and Static BCS models share many common design features. This important fact, which is not evident in other form and motion theories, enables us to view both models as variations on a common architectural design for visual cortex. A great conceptual simplification is afforded by the fact that variations on a common design can now be used to explain large data bases about form and motion perception that have heretofore been treated separately.

3. Joining Sensitivity to Direction-of-Motion with Insensitivity to Direction-of-Contrast

Analysis of the SOC Filter design revealed that one of its basic properties made it unsuitable for motion processing; namely, the output of the SOC Filter cannot effectively process the direction-of-motion of a moving figure. This deficiency arises from the way in which the SOC Filter becomes insensitive to direction-of-contrast at its complex cell level. Insensitivity to direction-of-contrast of the SOC Filter's complex cells enables the CC Loop of the Static BCS, which involves feedback interactions between hypercomplex cells and bipole cells (Figure 1), to generate boundary segmentations along scenic contrast reversals (Figures 2 and 3).

The simple cells at the first BCS level are, however, sensitive to direction-of-contrast (Figure 4). The activities of like-oriented simple cells that are sensitive to opposite directions-of-contrast are rectified before they generate outputs to their target complex cells. Because the complex cells pool outputs from both directions-of-contrast, they are themselves insensitive to direction-of-contrast.

Figure 4 shows a single pair of simple cells generating inputs to each complex cell. Such an arrangement is not sufficient in general. For example, Grossberg (1987b) and Grossberg and Marshall (1989) have shown that multiple simple cells may input to each complex cell. The number of converging simple cells is predicted to covary in a self-similar manner with the size of the simple cell receptive fields, and then to trigger nonlinear contrast-enhancing competition at the complex cell level, in order to explain basic data about binocular vision such as the size-disparity correlation and binocular fusion and rivalry.

Inspection of the (simple cell)-to-(complex cell) interaction in Figure 4 shows that a vertically-oriented complex cell could respond, say, to a dark-light vertical edge moving to the right *and* to a light-dark vertical edge moving to the left. Thus the process whereby complex cells become insensitive to direction-of-contrast has rendered them insensitive to direction-of-motion in the SOC Filter.

The main design problem leading to a MOC Filter is to make the minimal changes in the SOC Filter that are needed to model an oriented, contrast-sensitive filter whose outputs are insensitive to direction-of-contrast—a property that is just as important for moving images

Figure 2. Long vertical and horizontal boundaries are detected despite regular contrast reversals in defining the grid of alternating black and white squares. (From "Perception of surface curvature and direction of illumination from patterns of shading" by J. Todd and E. Mingolla, 1983, *Journal of Experimental Psychology: Human Perception and Performance*, 9, 4, 583-595, by American Psychological Association Inc. Printed with permission.)

Figure 3. A reverse-contrast Kanizsa square: The BCS is capable of completing illusory boundaries between the vertical dark-light and light-dark contrasts of the pac man figures. This boundary completion, or emergent segmentation, process enables the BCS to detect boundaries along contrast reversals, as in Figure 2.

as for static images—yet is also *sensitive* to direction-of-motion—a property that is certainly essential in a motion perception system. This modification (Figure 5) involves introduction of an extra degree of computational freedom which achieves several important properties in one stroke. These covarying properties are summarized in Table 1. The simple cells at the input end of the MOC Filter are sensitive to direction-of-contrast and to stimulus orientation. They are also monocular (except for ocular dominance column overlap) and interact via spatially short-range interactions. The cells at the output end of the MOC Filter play a role analogous to SOC Filter complex cells. In the MOC Filter, however, the extra degree of freedom renders these cells insensitive to direction-of-contrast and sensitive to direction-of-motion. In addition, although their preferred orientation is perpendicular to their preferred direction-of-motion, these "motion complex cells" also respond to other stimulus orientations that move in the same preferred direction (Grossberg and Mingolla, 1990b, 1990c). Such a difference between sensitivity to static orientation versus sensitivity to motion direction is also found between cortical cells in V1 and in MT, respectively (Albright, 1984; Albright, Desimone, and Gross, 1984; Maunsell and van Essen, 1983; Newsome, Gizzi, and Movshon, 1983). In addition, the MOC Filter output cells are binocular and have large receptive fields that permit them to engage in the long-range spatial interactions that subserve many apparent motion percepts (Grossberg and Rudd, 1989c, 1990b), as illustrated in Section 15.

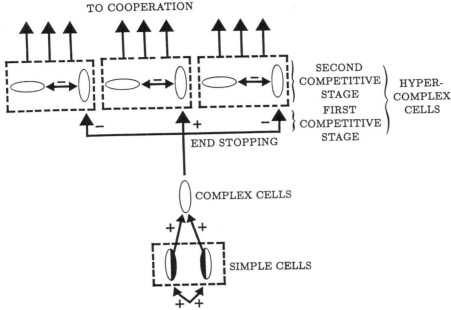

Figure 4. Early stages of SOC Filter processing: At each position exist cells with elongated receptive fields (simple cells) of various sizes which are sensitive to orientation, amount-of-contrast, and direction-of-contrast. Pairs of such cells sensitive to like orientation but opposite directions-of-contrast (lower dashed box) input to cells (complex cells) that are sensitive to orientation and amount-of-contrast but not to direction-of-contrast (white ellipses). These cells, in turn, excite like-oriented cells (hypercomplex cells) corresponding to the same position and inhibit like-oriented cells corresponding to nearby positions at the first competitive stage. At the second competitive stage, cells corresponding to the same position but different orientations (higher-order hypercomplex cells) inhibit each other via a push-pull competitive interaction.

4. Why is not a Motion Form Perception System Sufficient?

A further analysis of the Static BCS and Motion BCS poses a new puzzle. This puzzle arises because it seems that the Motion BCS has stronger computational properties than the Static BCS. Why, then has Nature not opted to use only a Motion BCS? Why has not the Static BCS fallen by the wayside of evolution due to disuse? In particular, if Nature could design a MOC filter that is sensitive to direction-of-motion and insensitive to direction-of-contrast, then why did the SOC Filter evolve, in which insensitivity to direction-of-contrast comes only at the cost of a loss of sensitivity to direction-of-motion? This question is perplexing given the facts that animals are usually in relative motion with respect to their visual environment, and that simple cells in V1 are already sensitive to direction-of-motion.

An answer to this question can be derived by first noting that the SOC Filter design that was described by Grossberg and Mingolla (1985b), Grossberg (1987b), and Grossberg and Marshall (1989) is incomplete. This design omits the processes that would be needed to

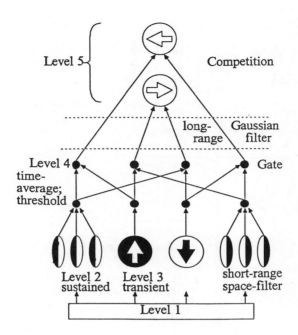

Figure 5. The MOC Filter: Level 1 registers the input pattern. Level 2 consists of sustained response cells with oriented receptive fields that are sensitive to direction-of-contrast. Level 3 consists of transient response cells with unoriented receptive fields that are sensitive to direction-of-change in the total cell input. Level 4 cells combine sustained cell and transient cell signals to become sensitive to direction-of-motion and sensitive to direction-of-contrast. Level 5 cells combine Level 4 cells via a long-range filter to become sensitive to direction-of-motion and insensitive to direction-of-contrast.

make the SOC Filter sensitive to transient changes in the input pattern. An analysis of how to correct this omission leads herein to an enhanced FACADE Theory in which the Static BCS and the Motion BCS may be viewed as parallel subsystems of a single total system. I predict herein that this total system unfolds during the development of the visual cortex as an expression of an underlying symmetry principle, called FM Symmetry (F = form, M = motion). Many manifestations of symmetry are familiar from our daily experiences with the physical world, and symmetry principles provide an important predictive and explanatory tool in the modern physical sciences. Here I suggest that the static form perception and motion form perception systems are not independent modules that obey different rules. Rather, they express two sides of a unifying organizational principle that is predicted to control the development of visual cortex.

TABLE 1

LEVELS OF MOTION OC FILTER

LEVEL 1:   INPUT PATTERN

LEVEL 2:   SUSTAINED RESPONSE CELLS
Time-averaged and shunted signals from rectified outputs of spatially filtered oriented receptive fields

LEVEL 3:   TRANSIENT RESPONSE CELLS
Rectified outputs of time-averaged and shunted signals from unoriented change-sensitive cells

LEVEL 4:   LOCAL MOTION DETECTORS
Pairwise gating of sustained and transient response combinations
Sensitive to direction-of-contrast
Sensitive to direction-of-motion

LEVEL 4 → 5   LONG-RANGE GAUSSIAN FILTER

LEVEL 5:   MOTION-DIRECTION DETECTORS
Contrast-enhancing competition
Insensitive to direction-of-contrast
Sensitive to direction-of-motion

5. A Symmetry Principle for Cortical Development: Sustained-Transient Gating, Opponent Processing, and Insensitivity to Direction-of-Contrast

This symmetry principle is predicted to control the simultaneous satisfaction of three constraints; namely,

(1) multiplicative interaction, or gating, of all combinations of sustained cell and transient cell output signals to form four sustained-transient cell types;

(2) symmetric organization of these sustained-transient cell types into two opponent on-cell and off-cell pairs, such that

(3) output signals from all the opponent cell types are independent of direction-of-contrast.

Multiplicative gating of sustained cells and transient cells is shown below to generate change-sensitive receptive field properties of oriented on-cells and off-cells within the Static BCS, and direction-sensitive cells within the Motion BCS. The constraint that output signals be independent of direction-of-contrast enables both the Static BCS and the Motion BCS to generate emergent boundary segmentations along image contrast reversals.

The previous discussion suggests how the static form and motion form systems may both arise. This discussion does not, however, disclose how these systems control different perceptual properties whose behavioral usefulness has preserved their integrity throughout the evolutionary process. The following behavioral implications of the symmmetry principle will be explained herein:

6. Different Geometries for Perception of Static Form and Motion Form

We are all so familiar with the different geometries for processing static *orientations* and motion *directions* that we often take them for granted. For example, we all take for granted that the opposite *orientation* of "vertical" is "horizontal," a difference of 90°; yet the opposite *direction* of "up" is "down," a difference of 180°. Why are the perceptual symmetries of static form and motion form different?

7. Negative Afterimages via Antagonistic Rebound

A clue is provided by considering how the 90° and 180° symmetries are reflected in percepts of negative afterimages. These symmetries suggest an opponent organization whereby orientations that differ by 90° are grouped together, whereas directions that differ by 180° are grouped together. Negative aftereffects illustrate a key property of this opponent organization. For example, after sustained viewing of a radial input pattern, looking at a uniform field triggers a percept of a circular afterimage (MacKay, 1957). The orientations within the input and the circular afterimage differ from each other by 90°. After sustained viewing of a downwardly moving image, looking at a uniform field triggers a percept of an upwardly moving afterimage, as in the waterfall illusion (Sekuler, 1975). The directions within the downward input and the upward afterimage differ from each other by 180°.

In summary, the geometries of both static form perception and motion form perception include an opponent organization in which offset of the input pattern after sustained viewing triggers onset of a transient *antagonistic rebound*, or activation of the opponent channel.

8. Resonance versus Reset: Cooperative Feature Linking without Destructive Smearing

Antagonistic rebound within opponent channels is needed to control the complementary perceptual processes of *resonance* and *reset*. Within the CC Loop (Figure 1), positive feedback signals between the hypercomplex cells and bipole cells can cooperatively link similarly oriented features at approximately colinear locations into emergent boundary segmentations (Grossberg and Mingolla, 1985a, 1985b, 1987). Several neurophysiological laboratories (Eckhorn et al., 1988; Gray et al., 1989) have supported the prediction that such cooperative linking occurs between cortical representations of similarly oriented features. In the original Grossberg and Mingolla computer simulations of this phenomenon, a lumped version of the CC Loop was used in which only "fast" variables were included, for simplicity. In this approximation, the cooperative linking process approaches an equilibrium configuration through time (Figure 6). My student David Somers and I have shown, however, that unlumping the CC Loop model by introducing "slow" variables enables emergent segmentations to generate resonant standing waves in which cooperatively linked features oscillate in phase with one another (Grossberg and Somers, 1990). The equilibrium configurations are a limiting, or singular, approximation of these standing waves.

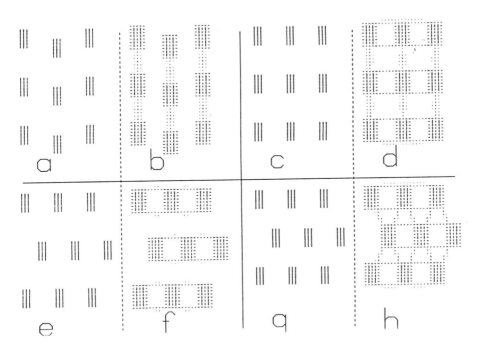

Figure 6. Computer simulations of processes underlying textural grouping: The length of each line segment is proportional to the activation of a network node responsive to one of 12 possible orientations. Parts a, c, e, and g display the activities of oriented cells that input to the CC Loop. Parts b, d, f, and h display equilibrium activities of oriented cells at the second competitive stage of the CC Loop. A pairwise comparison of (a) with (b), (c) and (d), and so on, indicates the major groupings sensed by the network. (From "Neural dynamics of surface perception: Boundary webs, illuminants, and shape-from-shading" by S. Grossberg and E. Mingolla, 1987, *Computer Vision, Graphics, and Image Processing*, **37**, 133. Copyright 1987 by Academic Press. Reprinted with Permission.)

In Grossberg (1976, 1978) it was predicted that perceptual codes in the visual cortex would express themselves either as standing waves—if enough slow variables were operative—or as (approximate) equilibria if they were not. Both the Eckhorn and Singer laboratories whose work is referenced above were aware of this prediction before carrying out their experiments.

Within the BCS, such a resonant segmentation, whether in standing wave or equilibrium form, derives from the positive feedback interactions between hypercomplex on-cells and bipole cells of the CC Loop. These positive feedback interactions selectively amplify and sharpen the globally "best" cooperative grouping and provide the activation for inhibiting less favored groupings. The positive feedback interactions also subserve the coherence, hysteresis, and structural properties of the emergent segmentations.

The positive feedback can, however, maintain itself for a long time after visual inputs terminate. Thus the very existence of cooperative linking could seriously degrade perception by maintaining long-lasting positive afterimages, or smearing, of every percept.

Although some smearing can occur, it is known to be actively limited by inhibitory processes that are triggered by changing images (Hogben and DiLollo, 1985). Herein I suggest how antagonistic rebounds between opponently organized on-cells and off-cells can actively inhibit CC Loop resonances when the input pattern changes. This inhibitory process *resets* the resonance and enables the CC Loop to flexibly establish new resonances in response to rapidly changing scenes.

In summary, the symmetry principle that is predicted to control the parallel development of the static form and motion form systems enables these systems to rapidly reset their resonant segmentations in response to rapidly changing inputs.

9. Resonance and Reset Control Stable Autonomous Learning

The control of complementary states of resonance and reset during emergent segmentation within the BCS appears to be a special case of a more general principle of brain design. Adaptive Resonance Theory, or ART, clarifies how the brain can continue to learn new perceptual and cognitive recognition codes throughout life without experiencing unselective forgetting of previously learned, but still effective, recognition codes (Carpenter and Grossberg, 1987a, 1987b, 1988, 1990; Grossberg, 1976, 1980, 1982, 1987c, 1988). The ability to stably preserve previous memories while engaging in rapid new learning is called a solution to the *stability-plasticity dilemma*.

ART solves the stability-plasticity dilemma by suggesting how the brain autonomously controls complementary states of resonance and reset. Within ART, the resonant state focuses attention upon predictive groupings of perceptual features. This attentive resonant state also triggers learning of new recognition codes or selective refinement of previously learned recognition codes. The reset event drives a rapid search, or hypothesis testing, cycle for more appropriate recognition codes when top-down learned expectations do not adequately match bottom-up perceptual data. This hypothesis testing process prevents unselective forgetting of recognition codes from occurring by rapidly initiating reset and search before the bottom-up data can be associated with the incorrect recognition category. Within ART, an opponent organization capable of antagonistic rebound helps to trigger the reset events that drive the search process, and the same model of opponent processing, called a *gated dipole*, is used in both the BCS and ART reset schemes (Section 22).

BCS resonance and reset play a functional role similar to ART resonance and reset. In fact, the prediction that standing waves of cooperatively linked features should exist in visual cortex was made in the context of ART resonance (Grossberg, 1976, 1978). It remains to directly test whether the standing waves reported by Eckhorn *et al.* (1988) and Gray *et al.* (1989) utilize BCS bipole cells, as suggested here, or more general ART adaptive filters. This can be accomplished by combining the tests of Eckhorn *et al.* (1988) and Gray *et al.* (1989) with those of von der Heydt *et al.* (1984). In both the BCS and ART, the resonance subserves a perceptual event that may be attentively modulated. In particular, top-down learned ART expectations from, say, inferotemporal cortex are predicted to attentively modulate BCS segmentations that are predicted to arise in cortical area V2 (Grossberg, 1987a, Figure 2; Grossberg and Mingolla, 1985b, Figure 1). In ART, resonance triggers learning. I hypothesize that BCS resonance also controls a learning process. In the BCS, such learning would enable appropriate synaptic linkages to be selected and stably learned between bipole

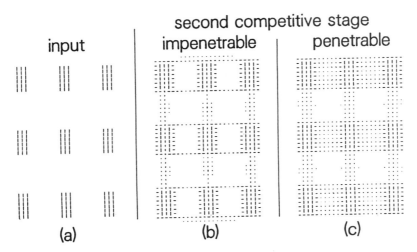

Figure 7. Computer simulations of processing underlying textural grouping. The length of each line segment is proportional to the activation of a network node responsive to one of 12 possible orientations. Part (a) displays the activity of oriented cells that input to the CC Loop. Part (b) displays the groupings sensed by our actual model network. Part (c) displays the resulting flooding of boundary activity that occurs when the model's mechanism for spatial impenetrability is removed. Reprinted with permission from Grossberg and Mingolla (1986).

cells and hypercomplex cells of the CC Loop. This type of learning would be driven by statistical regularities, such as edges, curves, and angles, in the visual environment. Marshall (1990) has, in fact, already used ART learning principles to train the synapses of a motion segmentation network. His model does not, however, use a CC Loop and cannot yet explain various data which CC Loop dynamics can explain. It remains an open problem to demonstrate adaptive tuning of CC Loop synapses during a resonant boundary segmentation.

10. Combining Rapid Reset and Spatial Impenetrability Predictions

The network design that controls rapid reset of a CC Loop resonance also constrains which combinations of features can resonate together, and thereby helps to structure the geometry of perceptual space. In particular, rapid reset of a resonating segmentation uses on-cells and off-cells of a given orientation that generate excitatory inputs and inhibitory inputs, respectively, to bipole cells of the same orientation. When on-cells lose their input, an antagonistic rebound activates off-cells that inhibit bipole cells and terminate the resonance.

Grossberg and Mingolla (1985b) have shown that the same mechanism can also generate the property of *spatial impenetrability* whereby emergent segmentations, during the resonance phase, are prevented from penetrating figures whose boundaries are built up from non-colinear orientations (Figure 7). In particular, in a cartoon drawing of a person standing in a grassy field, the horizontal contours where the ground touches the sky do not generate horizontal emergent boundaries that cut the person's vertical body in half. This property follows from the fact that vertical on-cells inhibit vertical off-cells, which disinhibit horizontal off-cells. The horizontal off-cells, in turn, inhibit horizontal bipole cells, and thereby

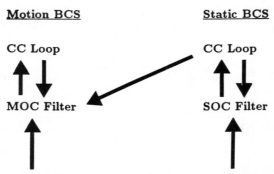

Figure 8. Model analog of $V1 \to V2 \to MT$ pathway: Stereo-sensitive emergent segmentations from the Static CC Loop help to select the depthfully correct combinations of motion signals in the MOC Filter.

undermine horizontal segmentations that might otherwise have penetrated the vertical figure.

The hypothesis that these reset and impenetrability mechanisms are one and the same may be tested by variants of the prediction that sudden offset of a previously sustained figure that contains many vertically oriented lines may facilitate, rather than block, the propagation of horizontal emergent boundary segmentations between the horizontally oriented lines that surround the location of the figure on both sides. Such a facilitation would be due to antagonistic rebounds that activate horizontal orientations when the vertical orientations terminate. Then these horizontal orientations could cooperate with the horizontal orientations of the background to facilitate formation of a horizontal segmentation that spans the region where the vertical figure had been.

11. Perception of Moving-Form-in-Depth: The $V1 \to V2 \to MT$ Pathway

An additional consequence of the symmetry principle clarifies why an indirect cortical pathway $V1 \to V2 \to MT$ from V1 to MT exists in addition to the direct $V1 \to MT$ pathway (DeYoe and van Essen, 1988). Outputs from the MOC Filter sacrifice a measure of orientational specificity in order to effectively process direction-of-motion. However, precisely oriented binocular matches are important in the selection of cortical cells that are tuned to the correct binocular disparities (von der Heydt, Hänny, and Dürsteler, 1981). The Static BCS can carry out such precise oriented matches; the Motion BCS cannot. This fact suggests that a pathway from the Static BCS to the Motion BCS exists in order to help the Motion BCS to generate its motion segmentations at correctly calibrated depths.

Such a pathway needs to arise after the level of BCS processing at which cells capable of binocular fusion are chosen and binocularly rivalrous cells are suppressed. This occurs within the hypercomplex cells and bipole cells of the Static BCS (Grossberg, 1987b; Grossberg and Marshall, 1989), hence within the model analog of prestriate cortical area V2 (Figure 8). Thus the existence of a pathway from V2 and/or V4 to MT is consistent with the different functional roles of the Static BCS and the Motion BCS.

According to this reasoning, the $V2 \to MT$ pathway should occur at a processing stage prior to the one at which several orientations are pooled into a single direction-of-motion within each spatial scale. Thus, the pathway ends in the MOC Filter at a stage no later than Level 4 in Figure 5. Such a pathway would join like orientations within like spatial

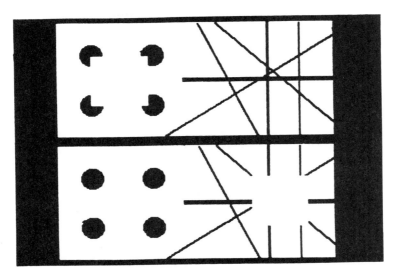

Figure 9. Images used to demonstrate that apparent motion of illusory figures arises through interactions of the static illusory figures, not from the inducing elements. Frame 1 (row 1) is temporally followed by Frame 2 (row 2). (From "Apparent motion of subjective surfaces" by V.S. Ramachandran, 1985, *Perception*, **14**, 129. Copyright 1985 by Pion Ltd. Reprinted by permission.)

scales between the Static BCS and the Motion BCS. It could thereby enhance the activation within the Motion BCS of those spatial scales and orientations that are binocularly fused within the Static BCS.

12. Apparent Motion of Illusory Figures

This interpretation of the $V2 \to MT$ pathway helps to explain the percept of apparent motion of illusory figures—a type of "doubly illusory" percept. Ramachandran, Rao, and Vidyasagar (1973) and Ramachandran (1985) have, for example, studied this phenomenon using the display summarized in Figure 9. Frame 1 of this display generates the percept of an illusory square using a Kanizsa figure. Frame 2 generates the percept of an illusory square using a different combination of image elements. When Frame 2 is flashed on after Frame 1 shuts off, the illusory square is seen to move continuously from its location in Frame 1 to its location in Frame 2. Because matching of image elements between the two frames is impossible, the experiment demonstrates that the illusory square, not the image elements that generate it, is undergoing apparent motion.

This phenomenon can be explained using the pathway from the CC Loop of the Static BCS to Level 4 of the MOC Filter. The CC Loop is capable of generating an illusory square in response to Frame 1 and Frame 2 (Grossberg and Mingolla, 1985b; Van Allen and Kolodzy, 1987). Successive inputs to Level 4 of the MOC Filter can induce continuous apparent motion if they are properly timed and spatially arranged (Grossberg and Rudd, 1989c, 1990b), as I will indicate below in Section 15. When this happens, the two static illusory squares can induce a continuous wave of apparent motion at Level 5 of the MOC

Filter.

This explanation of apparent motion of illusory figures can be used to test whether the $V2 \to MT$ pathway plays the role suggested above. One possible approach is to train a monkey to respond differently when the two illusory figures appear to move and when they do not. Then a (reversible) lesion of V2 or of the $V2 \to MT$ pathway should abolish the former behavior but not the latter.

13. Augmenting the Static BCS

The design of the Motion BCS, and the symmetry principle which combines the Static BCS and Motion BCS into a unified theory, both came into view by noting and correcting incomplete features of the Static BCS model that was introduced by Grossberg and Mingolla (1985b, 1987). These features can be understood by inspecting Figure 1.

A. Insensitivity to Direction-of-Motion

As shown in Figure 1, although the simple cells of the Static BCS are sensitive to direction-of-contrast, or contrast polarity, the complex cells are rendered insensitive to direction-of-contrast by receiving inputs from pairs of simple cells with opposite direction-of-contrast. Such a property is also true of the simple cells and complex cells in area V1 (DeValois, Albrecht, and Thorell, 1982; Poggio, Motter, Squatrito, and Trotter, 1985; Thorell, DeValois, and Albrecht, 1984).

This property is useful for extracting boundary structure that is independent of contrast polarity, as in Figures 2 and 3. As remarked in Section 3, however, the output of the SOC Filter is unable to differentiate direction-of-motion. The complex cell depicted in Figure 1 can, for example, respond to a vertical light-dark contrast moving to the right, and to a vertical dark-light contrast moving to the left. Because the complex cell can respond to image contrasts that move in opposite directions, it is insensitive to direction-of-motion. A key property of the Motion BCS model is that it possesses an MOC Filter that joins the property of insensitivity to direction-of-contrast, which is also needed to process moving segmentations, with sensitivity to direction-of-motion.

This analysis suggested that a fundamental computational property achieved by a motion segmentation system, such as MT, is to generate output signals that maintain insensitivity to direction-of-contrast without sacrificing sensitivity to direction-of-motion. The fact that a modest change of the Static BCS leads to a Motion BCS that can explain a large body of data concerning motion perception provides additional support for both the Static BCS model and the Motion BCS model by showing that both models may be considered variations on a single neural architectural theme.

Although this modification of the Static BCS is computationally modest, it is based upon a conceptually radical departure of FACADE Theory from previous vision models. Indeed, the property of insensitivity to direction-of-contrast in the Static BCS reflects one of the fundamental new insights of FACADE Theory. Insensitivity to direction-of-contrast is possible within the BCS because all boundary segmentations within the BCS are perceptually invisible. Visibility is a property of the FCS, whose computations are sensitive to direction-of-contrast. A vision theory built up from independent processing modules could not articulate the heuristics or the mechanisms of the Motion BCS because it could not articulate the fact that the BCS and FCS are computationally *complementary* subsystems of a single larger system, rather than being independent modules for the processing of form and color.

In order to overcome the inability of the Static BCS to process direction-of-motion, it is sufficient to design a MOC Filter, which realizes the minimal variation of SOC Filter design that is capable of generating outputs that are *insensitive* to direction-of-contrast and *sensitive* to direction-of-motion. The design of such a MOC Filter is described in Section 14.

B. Insensitivity to Input Transients

In Figure 1, the simple cells of the SOC Filter are modelled as oriented sustained-response cells. Sustained-response cells can respond with a constant output to a constant input. In contrast, simple cells *in vivo* are sensitive to transient changes in input patterns, including changes due to moving images.

When the SOC Filter is modified to be sensitive to image transients, it may be compared with the MOC Filter, which is obviously also sensitive to image transients. Such a comparison led to the discovery of the symmetry principle, and to the realization that both filters may be viewed as parallel halves of a larger system design.

The answer to how the SOC Filter computes image transients was suggested by another incomplete feature of the original SOC Filter model:

C. No Simple and Complex Off-Cells

The simple cells and complex cells in Figure 1 are all on-cells; they are activated when external inputs turn on. No simple or complex off-cells are represented. In contrast, the hypercomplex cells in Figure 1 include both on-cells and off-cells. This asymmetry in the network raises the question of how to design simple off-cells and complex off-cells to interact with the hypercomplex off-cells.

It turns out that problems (B) and (C), which seem to be two distinct problems, have the same solution. Multiplicative coupling of transient on-cells and transient off-cells with oriented sustained on-cells define simple on-cells and off-cells, as well as complex on-cells and off-cells, that are sensitive to image transients.

In summary, including off-cells and the property of sensitivity to image transients in the SOC Filter leads to a more symmetric SOC Filter model which, when compared with the MOC Filter model, reveals a deeper principle of symmetric design. The remainder of this article shows how problems (A)–(C) may be solved, and describes how these solutions lead to the data implications summarized in Sections 1–12. On a first reading, the reader may skip to Section 15 to follow the qualitative analysis of how the SOC Filter may be augmented and the FM Symmetry described.

14. Design of a MOC Filter

This section suggests a solution to the problem raises in Section 13A. The equations for a one-dimensional MOC Filter were described in Grossberg and Rudd (1989c). The MOC Filter's five processing levels are described qualitatively below for the more general two-dimensional case. The equations used for the 1-D theory are also described to provide a basis for rigorously defining the augmented SOC Filter and FM Symmetry.

Level 1: Preprocess Input Pattern

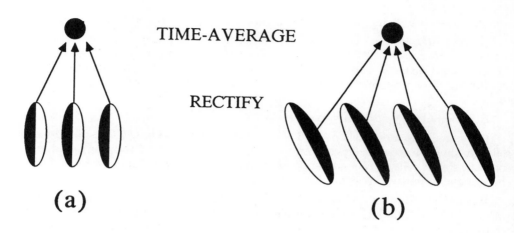

Figure 10. The sustained cell short-range filter combines several spatially contiguous receptive fields of like orientation via a spatial filter with a fixed directional preference. The orientation perpendicular to the direction is preferred, but non-orthogonal orientations can also be grouped in a prescribed direction.

The image is preprocessed before activating the MOC Filter. For example, it is passed through a shunting on-center off-surround net to compensate for variable illumination, or to "discount the illuminant" (Grossberg and Todorović, 1988).

In the 1-D theory, I_i denotes the input at position i.

Level 2: Sustained Cell Short-Range Filter

Four operations occur here, as illustrated in Figure 10.

(1) **Space-Average:** Inputs are processed by individual sustained cells with oriented receptive fields.

(2) **Rectify:** The output signal from a sustained cell grows with its activity above a signal threshold.

(3) **Short-Range Spatial Filter:** A spatially alligned array of sustained cells with like orientation and direction-of-contrast pool their output signals to activate the next cell level. This spatial pooling plays the role of the short-range motion limit $D_{\max}$ (Braddick, 1974). The breadth of spatial pooling scales with the size of the simple cell receptive fields (Figures 10a and 10b). Thus "$D_{\max}$" is not independent of the spatial frequency content of the image (Anderson and Burr, 1987; Burr, Ross, and Morrone, 1986; Nakayama and Silverman, 1984, 1985), and is not a universal constant.

The direction of spatial pooling may not be perpendicular to the oriented axis of the sustained cell receptive field (Figure 10b). The target cells are thus sensitive to a movement *direction* that may not be perpendicular to the sustained cell's preferred orientation.

(4) **Time-Average:** The target cell time-averages the directionally-sensitive inputs that it receives from the short-range spatial filter. This operation has properties akin to the

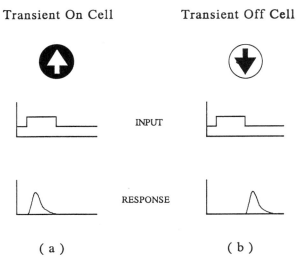

Figure 11. The transient cell filter consists of on-cells which react to input increments and off-cells which react to input decrements.

"visual inertia" during apparent motion that was reported by Anstis and Ramachandran (1987); see Figure 19.

In the 1-D theory, only horizontal motions are considered. It therefore suffices to consider two types of such cells that filter the input pattern I_i, one of which responds to a light-dark luminance contrast (designated by L, for left) and the other of which responds to a dark-light luminance contrast (designated by R, for right). Output pathways from like cells converge (Figure 10) to generate inputs J_{iL} and J_{iR} at each position i. The activity x_{ik} of the ith target cell at Level 2 obeys a membrane equation

$$\frac{d}{dt} x_{ik} = -A x_{ik} + (1 - B x_{ik}) J_{ik}, \qquad (1)$$

where $k = L, R$, which performs a time-average of the input J_{ik}.

Level 3: Transient Cell Filter:

In parallel with the sustained cell filter, a transient cell filter reacts to input increments (on-cells) and decrements (off-cells) with positive outputs (Figure 11). This filter uses four operations too:

(1) **Space-Average:** This is accomplished by a receptive field that sums inputs over its entire range.

(2) **Time-Average:** This sum is time-averaged to generate a gradual growth and decay of total activation.

(3) **Transient Detector:** The on-cells are activated when the time-average increases (Figure 11a). The off-cells are activated when the time-average decreases (Figure 11b).

(4) **Rectify:** The output signal from a transient cell grows with its activity above a signal threshold.

In the 1-D theory, the activities of the transient cells were computed as the rectified time derivatives of an unoriented space-time average x_i of the input pattern I_i. The time derivative is given by the membrane equation

$$\frac{d}{dt}x_i = -Cx_i + (D - Ex_i)\sum_j I_j F_{ji}, \qquad (2)$$

where F_{ji} is the unoriented spatial kernel that represents a transient cell receptive field.

Positive and negative half-wave rectifications of the time derivative were performed independently by defining

$$y_i^+ = \max\left(\frac{d}{dt}x_i - \Gamma, 0\right), \qquad (3)$$

and

$$y_i^- = \max\left(\Omega - \frac{d}{dt}x_i, 0\right), \qquad (4)$$

where Γ and Ω are constant thresholds. The activity y_i^+ models the response of a transient on-cell; and the activity y_i^- models the response of a transient off-cell.

Level 4: Sustained-Transient Gating Yields Direction-of-Motion Sensitivity and Direction-of-Contrast Sensitivity

Maximal activation of a Level 2 sustained cell filter is caused by image contrasts moving in either of two directions that differ by 180°. Multiplicative gating of each Level 2 sustained cell output with a Level 3 transient cell on-cell or off-cell removes this ambiguity (Figure 12). For example, consider a sustained cell output from vertically oriented light-dark simple cell receptive fields that are joined together in the horizontal direction by the short-range spatial filter (Figure 10a). Such a sustained cell output is maximized by a light-dark image contrast moving to the right or to the left. Multiplying this Level 2 output with a Level 3 transient on-cell output generates a Level 4 cell that responds maximally to motion to the right.

In the 1-D theory, there are two types of sustained cells (corresponding to the two antisymmetric directions-of-contrast), and also two type of transient cells (the on-cells and the off-cells). Consequently, there are four types of gated responses that can be computed. Two of these produce cells that are sensitive to local rightward motion: the $(L, +)$ cells that respond to $x_{iL}y_i^+$, and the $(R, -)$ cells that respond to $x_{iR}y_i^-$. The other two produce cells which are sensitive to local leftward motion: the $(L, -)$ cells that respond to $x_{iL}y_i^-$, and the $(R, +)$ cells that respond to $x_{iR}y_i^+$. All of these cells inherit a sensitivity to the direction-of-contrast of their inputs from the Level 2 sustained cells from which they are constructed.

The cell outputs from Level 4 are sensitive to direction-of-contrast. Level 5 consists of cells that pool outputs from Level 4 cells which are sensitive to the same direction-of-motion but to opposite directions-of-contrast.

Level 5: Long-Range Spatial Filter and Competition

Outputs from Level 4 cells sensitive to the same direction-of-motion but opposite directions-of-contrast activate individual Level 5 cells via a long-range spatial filter that has a Gaussianly profile across space (Figure 13). This long-range filter groups together Level 4 cell

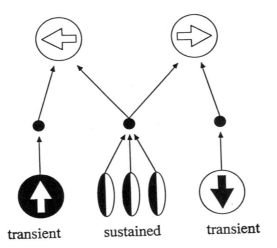

Figure 12. Sustained-transient gating generates cells that are sensitive to direction-of-motion as well as to direction-of-contrast.

outputs that are derived from Level 3 short-range filters with the same directional preference but different simple cell orientations. Thus the long-range filter provides the extra degree of freedom that enables Level 5 cells to function as "direction" cells, rather than "orientation" cells.

The long-range spatial filter broadcasts each Level 4 signal over a wide spatial range in Level 5. Competitive, or lateral inhibitory, interactions within Level 5 contrast-enhance this input pattern to generate spatially sharp Level 5 responses. A winner-take-all competitive network (Grossberg, 1973, 1982) can transform even a very broad input pattern into a focal activation at the position that receives the maximal input. The winner-take-all assumption is a limiting case of how competition can restore positional localization. More generally, we suggest that this competitive process partially contrast-enhances its input pattern to generate a motion signal whose breadth across space increases with the breadth of its inducing pattern. A contrast-enhancing competitive interaction has also been modeled at the complex cell level of the SOC Filter (Grossberg, 1987b; Grossberg and Marshall, 1989). The Level 5 cells of the MOC Filter are, in other respects too, computationally homologous to the SOC Filter complex cells.

In the 1-D theory, we define the transformation from Level 4 to Level 5 by letting

$$r_i = x_{iL} y_i^+ + x_{iR} y_i^-, \tag{5}$$

and

$$l_i = x_{iL} y_i^- + x_{iR} y_i^+, \tag{6}$$

be the total response of the local right motion and left motion detectors, respectively, at position i of Level 4. Signal r_i increases if either a light-dark or a dark-light contrast pattern moves to the right. Signal l_i increases if either a light-dark or a dark-light contrast pattern moves to the left.

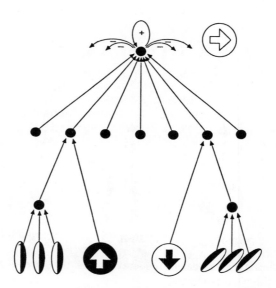

Figure 13. The long-range spatial filter combines sustained-transient cells with the same preference for direction-of-motion, including cells whose sustained cell inputs are sensitive to opposite directions-of-contrast and to different orientations.

These local motion signals are assumed to be filtered independently by a long-range operator with a Gaussian kernel

$$G_{ji} = H \exp[-(j-i)^2/2K^2], \tag{7}$$

which defines the input fields of the Level 5 cells. Thus, there exist two types of direction sensitive cells at each position i of Level 5. The activity at i of the right-motion sensitive cell is given by

$$R_i = \sum_j r_j G_{ji}, \tag{8}$$

and the corresponding activity of the left-motion sensitive cell is given by

$$L_i = \sum_j l_j G_{ji}. \tag{9}$$

We assume that contrast-enhancing competitive, or lateral inhibitory, interactions within Level 5 generate the activities which encode motion information. In the simplest case, the competition is tuned to select that population whose input is maximal, as in

$$x_i^{(R)} = \begin{cases} 1 & \text{if } R_i > R_j, j \neq i \\ 0 & \text{otherwise,} \end{cases} \tag{10}$$

and
$$x_i^{(L)} = \begin{cases} 1 & \text{if } L_i > L_j, j \neq i \\ 0 & \text{otherwise.} \end{cases} \quad (11)$$

In the simulations summarized below, the above assumption was made for simplicity. The functions $x_i^{(R)}$ and $x_i^{(L)}$ change through time in a manner that idealizes parametric properties of many apparent motion phenomena. See Grossberg and Rudd (1989c, 1990b) for details. More generally, we suggest that the competitive process idealized by (10) and (11) performs a partial contrast enhancement of its input pattern and thereby generates a motion signal whose breadth across space increases with the breadth of its inducing pattern.

The total MOC Filter design is summarized in Figure 5.

15. Continuous Motion Paths from Spatially Stationary Flashes

The model equations listed in Section 14 provide an answer to long-standing questions in the vision literature concerning why individual flashes do not produce a percept of long-range motion, yet long-range interaction between spatially discrete pairs of flashes can produce a spatially sharp percept of continuous motion. Such apparent motion phenomena are a particularly useful probe of motion mechanisms because they describe controllable experimental situations in which nothing moves, yet a compelling percept of motion is generated. For example, two brief flashes of light, separated in both time and space, create an illusion of movement from the location of the first flash to that of the second when the spatiotemporal parameters of the display are within the correct range (Figure 14a).

Outstanding theoretical issues concerning apparent motion include the resolution of a trade-off that exists between the long-range spatial interaction that is needed to generate the motion percept, and the localization of the perceived motion signal that smoothly interpolates the inducing flashes. If a long-range interaction between the flashes must exist in order to generate the motion percept, then why is it not perceived when only a single light is flashed? Why are not outward waves of motion-carrying signals induced by a single flash? What kind of long-range influence is generated by each flash, yet only triggers a perceived motion signal when at least two flashes are activated? What kind of long-range influence from individual flashes can generate a smooth motion signal between flashes placed at variable distances from one another? How does the motion signal speed up to smoothly interpolate flashes that occur at larger distances but the same time lag (Kolers, 1972)? How does the motion signal speed up to smoothly interpolate flashes when they occur at the same distance but shorter time lags (Kolers, 1972)?

Variants of apparent motion include *phi motion*, or the *phi phenomenon*, whereby a "figureless" or "objectless" motion signal propagates from one flash to another; *beta motion*, whereby a well-defined form seems to move smoothly and continuously from one flash to the other; *gamma motion*, the apparent expansion at onset and contraction at offset of a single flash of light; *delta motion*, or backward motion from a more intense second flash to a less intense first flash; and *split motion*, or simultaneous motion paths from a single first flash to a simultaneous pair of second flashes (Bartley, 1941; Kolers, 1972).

Another well-known apparent motion display, originally due to Ternus (1926/1950), illustrates the fact that not only the existence of a motion percept, but also its figural identity, may depend on subtle aspects of the display, such as the interstimulus interval, or ISI, between the offset of the first flash and the onset of the second flash (Figure 14b). In the Ternus

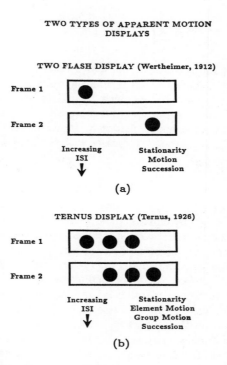

Figure 14. Two types of apparent motion displays in which the two frames outline the same region in space into which the dots are flashed at successive times: In (a), a single dot is flashed, followed by an interstimulus interval (ISI), followed by a second dot. At small ISI's, the two dots appear to flicker in place. At longer ISI's, motion from the position of the first dot to that of the second is perceived. (b) In the Ternus display, three dots are presented in each frame such that two of the dots in each frame occupy the same positions. At short ISI's, all the dots appear to be stationary. At longer ISI's the dots at the shared positions appear to be stationary, while apparent motion occurs from the left dot in Frame 1 to the right dot in Frame 2. At still longer ISI's, the three dots appear to move from Frame 1 to Frame 2 as a group.

display, a cyclic alternation of two stimulus frames gives rise to competing visual movement percepts. In Frame 1, three elements are arranged in a horizontal row on a uniform background. In Frame 2, the elements are shifted to the right in such a way that the positions of the two leftwardmost elements in Frame 2 are identical to those of the two rightwardmost elements in Frame 1. Depending on the stimulus conditions, the observer will see either of two bistable motion percepts. Either the elements will appear to move to the right as a group between Frames 1 and 2 and then back again during the second half of a cycle of the display or, alternatively, the leftwardmost element in Frame 1 will appear to move to the location of the rightwardmost element in Frame 2, jumping across two intermediate elements which

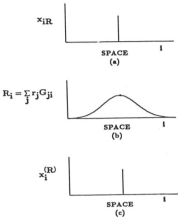

Figure 15. Spatial response of the MOC Filter to a point input. (a) Sustained activity of a Level 2 cell. (b) Total input pattern to Level 5. (c) Contrast-enhanced response at Level 5. (From "A neural architecture for visual motion perception: Group and element apparent motion" by S. Grossberg and M.E. Rudd, 1989, *Neural Networks*, 2, p. 429. Copyright 1989 by Pergamon Press. Reprinted by permission.)

appear to remain stationary. The first percept is called "group" motion, and the second percept "element" motion. At short ISI's there is a tendency to observe element motion. At longer ISI's, there is a tendency to observe group motion.

Remarkably, formal analogs of all these and many other motion phenomena occur at Level 5 of the motion MOC Filter in response to sequences of flashes presented to Level 1. Intuitively, a signal for motion will arise when a continuous wave of activation connects the locations corresponding to the flashes; that is, when a connected array of the functions $x_i^{(R)}$, $x_{i+1}^{(R)}$, $x_{i+2}^{(R)}$, ... are activated sequentially through time, or alternatively the functions $x_i^{(L)}$, $x_{i-1}^{(L)}$, $x_{i-2}^{(L)}$, ... are activated sequentially through time. Each activation $x_i^{(R)}$, or $x_i^{(L)}$, represents the peak, or maximal activity, of a broad spatial pattern of activation across the network.

The broad activation pattern (Figure 15b) is generated by the long-range Gaussian filter G_{ji} in (7) in response to a spatially localized flash to Level 1 (Figure 15a). The sharply localized response function $x_i^{(R)}$ is due to the contrast-enhancing action of the competitive network within Level 5 (Figure 15c). A stationary localized $x_i^{(R)}$ response is hereby generated in response to a single flashing input.

In contrast, suppose that two input flashes occur with the following spatial and temporal separations. Let the positions of the flashes be $i = 1$ and $i = N$. Let the activity $r_1(t)$ in (5) caused by the first flash start to decay as the activity $r_N(t)$ in (5) caused by the second flash starts to grow. Suppose, moreover, that the flashes are close enough that their spatial

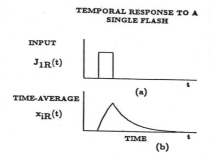

Figure 16. Temporal response of sustained response cells to a point input: (a) The input is presented for a brief duration at location 1. (b) The activity of the sustained response cell gradually builds up after input onset, then decays after input offset. (c) Growth of the input pattern to Level 5 through time with transient cell activity held constant. The activity pattern retains a Gaussian shape centered at the location of the input. (From "A neural architecture for visual motion perception: Group and element apparent motion" by S. Grossberg and M.E. Rudd, 1989, *Neural Networks*, **2**, p. 429. Copyright 1989 by Pergamon Press. Reprinted by permission.)

patterns $r_1 G_{1i}$ and $r_N G_{Ni}$ overlap. Then the total input

$$R_i = r_1 G_{1i} + r_N G_{Ni} \tag{12}$$

to the ith cell in Level 5 can change in such a way that the maximum value of the spatial pattern $R_i(t)$ through time, namely $x_i^{(R)}(t)$ in (10), first occurs at $i = 1$, then $i = 2$, then $i = 3$, and so on until $i = N$. A percept of continuous motion from the position of the first flash to that of the second will result.

This basic property of the MOC Filter is illustrated by the computer simulations from Grossberg and Rudd (1989c) summarized in Figures 16–18. Figure 16 depicts the temporal response to a single flash at position 1 of Level 1. The sustained cell response at position 1 of Level 2 undergoes a gradual growth and decay of activation (Figure 16b), although the position of maximal activation in the input to Level 5 does not change through time (Figure 16c). The temporal decay of activation in Figure 16b may be compared with the "visual inertia" by Anstis and Ramachandran (1987, Figure 6) during their experiments on apparent motion (Figure 19).

Figure 17 illustrates an important implication of the fact that the Level 2 cell activations persist due to temporal averaging after their Level 1 inputs shut off. If a flash at position 1 is followed, after an appropriate delay, by a flash at position N, then the sustained response

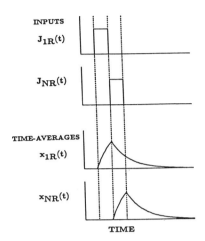

Figure 17. Temporal response of the sustained response cells at Level 2 to two successive point inputs. One input is presented briefly at location 1, followed by a second input at location N. For an appropriately timed display, the decaying response at position 1 overlaps the rising response at position N. (From "A neural architecture for visual motion perception: Group and element apparent motion" by S. Grossberg and M.E. Rudd, 1989, *Neural Networks*, 2, p. 429. Copyright 1989 by Pergamon Press. Reprinted by permission.)

to the first flash [e.g., $x_{1R}(t)$] can decay while the response to the second flash [e.g., $x_{NR}(t)$] grows.

Assume for simplicity that the transient signals defined by equations (3) and (4) are held constant and consider how the waxing and waning of sustained cell responses control the motion percept. Then the total input pattern R_i to Level 5 can change through time in the manner depicted in Figure 18. Each row of Figure 18a illustrates the total input to Level 5 caused, at a prescribed time t, by $x_{1R}(t)$ alone, by $x_{NR}(t)$ alone, and by both flashes together. Successive rows plot these functions at equally spaced later times. As $x_{1R}(t)$ decays and $x_{NR}(t)$ grows, the maximum value of $R_i(t)$ moves continuously to the right. Figure 18b depicts the position $x_i^{(R)}(t)$ of the maximum value at the corresponding times.

16. Feature Integration and Spatial Attention Shifts

The conditions under which such a travelling wave of activation can occur are proved in Grossberg and Rudd (1989c) to be quite general. The phenomenon can arise whenever a decaying trace of one activation adds to an increasing trace of a second activation via spatially long-range Gaussian receptive fields before the sum is contrast-enhanced. Such a travelling wave may, for example, subserve certain shifts in spatial attention (Eriksen and Murphy, 1987; LaBerge and Brown, 1989; Remington and Pierce, 1984). It remains for future analyses to determine whether discrete jumps of spatial attention and continuous shifts of attention may receive a unified analysis in terms of the same formal constraints that explain

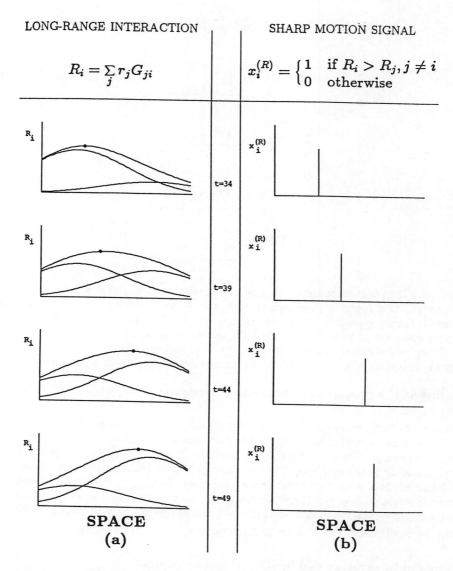

Figure 18. MOC Filter simulation in response to a two flash display. Successive rows correspond to increasing times: (a) The two lower curves in each row depict the total input to Level 5 caused by each of the two flashes. The input due to the left flash decreases while the input due to the right flash increases. The total input due to both flashes is a travelling wave whose maximum value moves from the location of the first flash to that of the second flash. (b) Position of the contrast-enhanced response at Level 5. (From "A neural architecture for visual motion perception: Group and element apparent motion" by S. Grossberg and M.E. Rudd, 1989, *Neural Networks*, **2**, p. 425. Copyright 1989 by Pergamon Press. Reprinted by permission.)

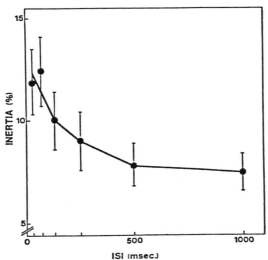

Figure 19. Strength of visual inertia as a function of the timing of dots that prime the direction of a subsequent apparent motion. (From "Visual inertia in apparent motion" by S. Anstis and V.S. Ramachandran, 1987, *Vision Research*, **27**, p. 759. Copyright 1987 by Pergamin Press. Reprinted by permision.)

how discrete flashes or continuous apparent motion are perceived.

Within the general theoretical framework of FACADE Theory and ART, mechanisms of "preattentive feature integration" by cooperative BCS segmentation and FCS filling-in, "attentive feature integration" through resonance with top-down learned ART expectations after reset and search terminate, and shifts in spatial attention due to mechanisms similar in formal properties to preattentive mechanisms of motion perception—come together in a unified computational theory. Such a theory provides an alternative framework to Treisman's seminal account of feature integration (Treisman and Gelade, 1980; Treisman and Souther, 1985) whose conceptual difficulties and the demands of new data have gradually led to qualitative theoretical approaches more in harmony with the quantitative mechanisms of FACADE Theory and ART (Duncan and Humphreys, 1989; Nakayama and Silverman, 1986; Pashler, 1987; Triesman and Gormican, 1988).

17. Design of Simple On-Cells and Off-Cells

This section suggests a solution to the problems raised in Sections 13B and 13C. As noted there, hypercomplex cells in Figure 1 are organized into opponent on-cells and off-cells, yet the SOC Filter explicitly depicts only pathways to the hypercomplex on-cells from the simple on-cells via complex on-cells. Moreover, all of these cells are of sustained cell type. Interactions with simple off-cells, complex off-cells, and transient cells are not described. It is now shown how multiplicative gating of sustained cells with transient on-cells and transient off-cells solves both of these problems, and reveals the modified SOC Filter and the MOC Filter as parallel subsystems of a symmetric total system.

In Figures 1 and 4, pairs of like-oriented simple cells that are sensitive to opposite directions-of-contrast input to a single complex cell, which is insensitive to direction-of-

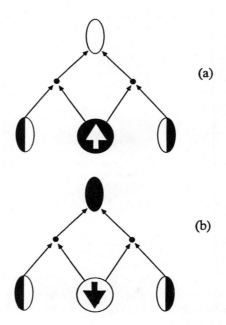

Figure 20. (a) A complex/orientation/on cell: Pairs of rectified sustained cells with opposite direction-of-contrast are gated by rectified transient on-cells to generate simple sustained-transient on-cells before the gated responses are added. (b) A complex/orientation/off cell: Pairs of rectified sustained cells with opposite direction-of-contrast are gated by rectified transient off-cells to generate simple sustained-transient off-cells before the gated responses are added.

contrast. Our task is to preserve this fundamental property while rendering the simple cells sensitive to image transients and defining both on-cells and off-cells of simple and complex type. To define simple on-cells that are sensitive to image transients, let a transient on-cell, as defined in equation (3), multiply each sustained cell in the pair of like-oriented cells depicted in Figures 1 and 4. This gives rise to a pair of like-oriented simple on-cells (Figure 20a) that are sensitive to opposite directions-of-contrast and are activated when properly oriented and positional inputs are turned on. In the 1-D model notation of equations (1) and (2), these cell responses are defined by $x_{iL}y_i^+$ and $x_{iR}y_i^+$, rather than x_{iL} and x_{iR} alone. As in Figures 1 and 4, a pair of simple on-cells with like-orientation but opposite direction-of-contrast inputs to a complex on-cell, as in Figure 20a. The complex on-cell is defined by summing the rectified outputs of the simple on-cells, as in the equation

$$c_i^+ = x_{iL}y_i^+ + x_{iR}y_i^+. \tag{13}$$

Likewise, a pair of simple off-cells can be defined by gating the pair of like-oriented sustained cells in Figure 1 with a transient off-cell, as defined in equation (4). The pair of simple off-cell responses is thus defined by $x_{iL}y_i^-$ and $x_{iR}y_i^-$, rather than x_{iL} and x_{iR} alone. The off-cells are activated when properly oriented and positioned inputs shut off. Such a pair of

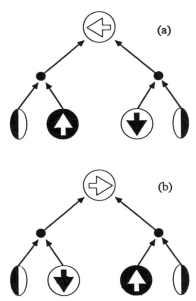

Figure 21. (a) A complex/direction/left cell: Pairs of rectified sustained cells with opposite direction-of-contrast are gated by pairs of rectified transient on-cells and off-cells, before the gated responses are added. (b) A complex/direction/right cell: Same as in (a), except sustained cells are gated by the opposite transient cell.

simple off-cells is depicted in Figure 20b, where it gives rise to a complex off-cell through the interaction

$$c_i^- = x_{iL} y_i^- + x_{iR} y_i^-. \qquad (14)$$

By construction, both complex on-cells and off-cells are insensitive to direction of contrast.

Let the complex on-cell in Figure 20a input to hypercomplex on-cells as in Figure 1. In a similar fashion, let the complex off-cell in Figure 20b input to the hypercomplex off-cells in Figure 1. The process of gating sustained cells by transient cells to generate on-cells and off-cells in the Static BCS thus makes the overall design of this architecture more symmetric by showing how simple and complex on-cells and off-cells fit into the scheme.

18. FM Symmetry

The above refinement of the SOC Filter merely adds sensitivity to image transients in a manner consistent with Figure 1. Having done so, a comparison of the modified SOC Filter with the MOC Filter reveals the FM Symmetry principle that was introduced in Section 5. FM Symmetry is embodied in the following set of four equations, the first two from the MOC Filter, and the last two from the enhanced SOC Filter.

Left-Direction Motion Complex On-Cell

$$r_i = x_{iL} y_i^+ + x_{iR} y_i^- \qquad (5)$$

Right-Direction Motion Complex On-Cell

$$l_i = x_{iL} y_i^- + x_{iR} y_i^+ \qquad (6)$$

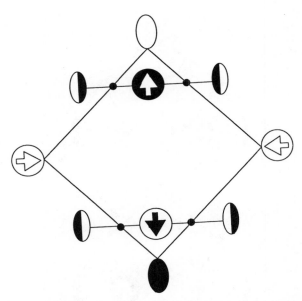

Figure 22. FM Symmetry: Symmetric unfolding of pairs of opponent orientation cells and opponent direction cells whose outputs are insensitive to direction-of-contrast. The gating combinations from Figures 20 and 21 are combined to emphasize their underlying symmetry.

Vertical-Orientation Static Complex On-Cell

$$c_i^+ = x_{iL}y_i^+ + x_{iR}y_i^+ \tag{13}$$

Vertical-Orientation Static Complex Off-Cell

$$c_i^- = x_{iL}y_i^- + x_{iR}y_i^- \tag{14}$$

These equations describe all possible ways of symmetrically gating an opponent pair (x_{iL}, x_{iR}) of sustained cells with transient cells to generate two opponent pairs, (c_i^+, c_i^-) and (r_i, l_i), of output signals that are insensitive to direction-of-contrast. One opponent pair of outcomes (c_i^+, c_i^-) contains cell pairs that are insensitive to direction-of-motion, but sensitive to either the onset or the offset of an oriented contrast difference. These cells may be called complex/orientation/on cells and complex/orientation/off cells, respectively, as in equations (13) and (14). They belong to the SOC Filter.

The other opponent pair of outcomes (r_i, l_i) contains the MOC Filter cell pairs, schematized in Figure 21, that are sensitive to opposite directions-of-motion. These cells may be called complex/direction/left cells and complex/direction/right cells, as in equations (5) and (6). When both sets of pairs are combined into a single symmetric diagram, the result is shown in Figure 22. Figure 22 suggests that parallel, but interdependent, streams of static form and motion form processing arise in visual cortex because the cortex develops by computing all possible sustained-transient output signals that are independent of direction-of-contrast and organized into opponent on-cells and off-cells. Experimental tests of this prediction will require a coordinated analysis of cell types and processing levels.

19. 90° Orientations: From V1 to V2

An important consequence of the abstract symmetry described in Figure 22 is the familiar fact from daily life that opposite static orientations are 90° apart, whereas opposite motion directions are 180° apart, as summarized in Section 6.

The 90° symmetry of opposite orientations may be explained by the way in which perpendicular *end cuts* are generated at the hypercomplex cells of the Static BCS, as analysed in Grossberg and Mingolla (1985b). This perpendicularity property arises from the fact that the opponent feature of a complex/orientation/on cell is a complex/orientation/off cell. This mechanism is reviewed below for completeness. For those familiar with the end cut concept, the 90° symmetry may tersely be summarized as follows: Suppose that a vertical line end excites a complex/vertical/on cell in Figure 1. Suppose that the end stopped competition inhibits hypercomplex/vertical/on cells at positions beyond the line end. Hypercomplex/horizontal/on cells at these positions are thereby activated, and generate a horizontally oriented end cut. In addition, hypercomplex/horizontal/off cells at these positions are inhibited by the opponent interaction. As a result, a net excitatory input is generated from the horizontally oriented hypercomplex cells to the horizontally oriented bipole cells of the CC Loop at that position. These excitatory end cut inputs cooperate across positions to generate a horizontal emergent segmentation, that is perpendicular to the vertical line, along the entire line end.

The more complete summary below of how end cuts are generated also provides an occasion for summarizing various data and predictions based upon the Grossberg and Mingolla (1985b) prediction that end cuts exist. Readers familiar with this analysis can skip to Section 21.

20. End Cuts: Cortical Simple Cells, Complex Cells, and Hypercomplex Cells as a Module for Hierarchical Resolution of Uncertainty

In order to effectively build up boundaries, the SOC Filter must be able to determine the orientation of a boundary at every position. To accomplish this, the cells at the first stage of the SOC Filter possess orientationally tuned simple cell receptive fields. Such simple cells, or cell populations, are selectively responsive to orientations that activate a prescribed small region of the retina, and whose orientations lie within a prescribed band of orientations with respect to the retina. A collection of such orientationally tuned cells is assumed to exist at every network position, such that each cell type is sensitive to a different band of oriented contrasts within its prescribed small region of the scene, as in the hypercolumn model, which was developed to explain the responses of simple cells in area V1 of the striate cortex (Hubel and Wiesel, 1977).

These oriented receptive fields are oriented *local contrast* detectors, rather than edge detectors. This property enables them to fire in response to a wide variety of spatially nonuniform image contrasts, including edges, spatially nonuniform densities of unoriented textural elements, and spatially nonuniform densities of surface gradients. Thus by sacrificing a certain amount of spatial resolution in order to detect oriented local contrasts, these masks achieve a general detection characteristic which can respond to edges, textures, and surfaces.

The fact that the receptive fields of the SOC Filter are *oriented* greatly reduces the number of possible groupings into which their target cells can enter. On the other hand, in order to detect oriented local contrasts, the receptive fields must be elongated along their

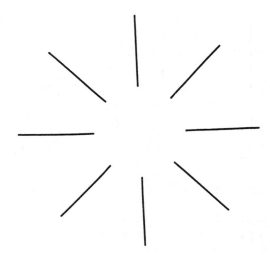

Figure 23. An Ehrenstein figure: A bright circular disk is perceived even though all white areas of the image are equally luminant. It is suggested that end cuts formed perpendicular to the line ends by the SOC Filter cooperate within the CC Loop to form a circular illusory boundary. This boundary separates two regions within the FCS whose filled-in activity levels differ. This difference is perceived as a difference in brightness.

preferred axis of symmetry. Then the cells can preferentially detect differences of average contrast across this axis of symmetry, yet can remain silent in response to differences of average contrast that are perpendicular to the axis of symmetry. Such receptive field elongation creates greater positional uncertainty about the exact locations within the receptive field of the image contrasts which fire the cell. This positional uncertainty becomes acute during the processing of image line ends and corners.

Oriented receptive fields cannot easily detect the ends of thin scenic lines or corners (Grossberg and Mingolla, 1985b). This property illustrates a basic uncertainty principle which says: Orientational "certainty" implies positional "uncertainty" at the ends of scenic lines whose widths are neither too small nor too large with respect to the dimensions of the oriented receptive field. If no BC signals are elicited at the ends of lines, however, then in the absence of further processing within the BCS, Boundary Contours will not be activated to prevent color and brightness signals from flowing out of line ends within the FCS during filling-in. Many percepts would hereby become badly degraded by featural flow. Thus, orientational certainty implies a type of positional uncertainty, which is unacceptable from the perspective of featural filling-in requirements.

Later processing stages within the BCS are needed to recover both the positional and orientational information that are lost in this way so that the boundaries at line ends and corners can be completed before they are mapped into the FCS to control filling-in of surface brightness, color, and depth. Grossberg and Mingolla (1985b) have called the process which completes the boundary at a line end an *end cut*. End cuts actively reconstruct the line end at a processing stage higher than the oriented receptive field much as they do to form a

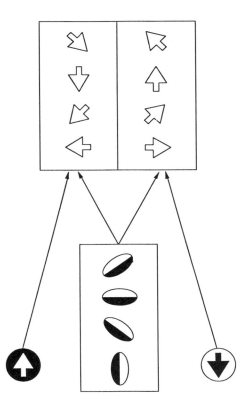

Figure 24. Orientation and direction hypercolumns: A single hypercolumn of orientation cells (say in $V1$) can give rise to a double hypercolumn of opponent direction cells (say in MT) through gating with opponent pairs of transient cells.

circular Ehrenstein figure (Figure 23). In order to emphasize the paradoxical nature of this process, we say that *all line ends are illusory*. Interactions between simple cells, complex cells, and hypercomplex cells were predicted to generate these perpendicular end cuts.

The processing stages that are hypothesized to generate end cuts are summarized in Figure 4. First, oriented simple cell receptive fields of like position and orientation, but opposite direction-of-contrast, generate rectified output signals that summate at the next processing stage to activate complex cells whose receptive fields are sensitive to the same position and orientation as themselves, but are insensitive to direction-of-contrast. These complex cells maintain their sensitivity to *amount* of oriented contrast, but not to the *direction* of this oriented contrast, in order to generate boundary detectors which can detect the broadest possible range of luminance or chromatic contrasts, in particular boundaries capable of bridging contrast-reversals.

The rectified output from the complex cells activates a second filter which is composed of two successive stages of spatially short-range competitive interaction whose net effect is to generate end cuts (Figure 4). First, a cell of prescribed orientation excites like-oriented cells corresponding to its location and inhibits like-oriented cells corresponding to nearby locations at the next processing stage. In other words, an on-center off-surround organization of like-oriented cell interactions exists around each perceptual location. This mechanism is analogous to the neurophysiological process of *end stopping*, whereby hypercomplex cell receptive fields are fashioned from interactions of complex cell output signals (Hubel and Wiesel, 1965; Orban, Kato, and Bishop, 1979). The outputs from this competitive mechanism interact with the second competitive mechanism. Here, hypercomplex cells compete that represent different orientations, notably perpendicular orientations, at the same perceptual location. This competition defines a push-pull opponent process. If a given orientation is excited, then its perpendicular orientation is inhibited. If a given orientation is inhibited, then its perpendicular orientation is excited via disinhibition.

The combined effect of these two competitive interactions generate end cuts as follows. The strong vertical activations along the edges of a scenic line inhibit the weak vertical activations near the line end. These inhibited vertical activations, in turn, disinhibit horizontal activations near the line end, thereby generating a horizontal end cut that is perpendicular to its inducing vertical line end. Thus the positional uncertainty generated by orientational certainty is eliminated at a subsequent processing level by the interaction of two spatially short-range competitive mechanisms which convert complex cells into two distinct populations of hypercomplex cells. This analysis of end cuts suggests that the simple cells, complex cells, and hypercomplex cells function as a unitary network for achieving hierarchical resolution of uncertainty.

These model mechanisms have successfully predicted and helped to explain a variety of neural and perceptual data. For example, the Grossberg and Mingolla (1985b) complex cell model is in accord with the complex cell model that was independently derived by Spitzer and Hochstein (1985) from their neurophysiological experiments on cats. The full BCS model also clarifies, in a manner that goes beyond the Spitzer-Hochstein model, why many complex cells receive inputs from several different classes of color opponent cells in the lateral geniculate nucleus (De Valois, Albrecht, and Thorell, 1982). This convergence of opponent cells generates a chromatically broad-band boundary detector. The BCS model hereby clarifies how inputs to complex cells from simple cells with chromatically opponent receptive fields may be attenuated in response to isoluminant stimuli, without denying that BCS boundaries are sensitive to color inputs. This model of complex cells also needs to be extended to explain basic facts of binocular fusion and rivalry (Grossberg, 1987b; Grossberg and Marshall, 1989).

An important prediction of the theory anticipated the report by von der Heydt, Baumgartner, and Peterhans (1984) that cells in prestriate visual cortex respond to perpendicular line ends, whereas cells in striate visual cortex do not. These cell properties also helped to explain why color is sometimes perceived to spread across a scene, as in the phenomenon of neon color spreading (Grossberg, 1987a; Grossberg and Mingolla, 1985a; Redies and Spillmann, 1981), by showing how some of the boundaries that would otherwise have been generated by image contrasts may be inhibited by the competitive mechanisms underlying end cuts in response to certain scenes, thereby allowing colors to spread beyond these image contrasts. The end cut process also exhibits properties of hyperacuity which have been used (Grossberg, 1987a) to explain subsequent psychophysical data about spatial localization and hyperacuity (Badcock and Westheimer, 1985a, 1985b; Watt and Campbell, 1985). A similar

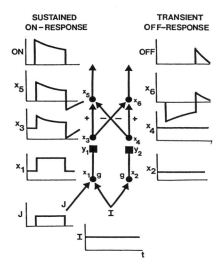

Figure 25. Example of a feedforward gated dipole: A sustained habituating on-response (top left) and a transient off-rebound (top right) are elicited in response to onset and offset, respectively, of a phasic input J (bottom left), when tonic arousal, I (bottom center), and opponent processing (diagonal pathways) supplement the slow gating actions (square synapses).

double filter model has been used to analyse data about texture segregation (Sutter, Beck, and Graham, 1989) in a way that supports the texture analyses of Grossberg and Mingolla (1985b). These latter texture segregation analyses also utilized the cooperative-competitive feedback interactions of the CC Loop (Figure 1) to generate emergent boundary segmentations, such as the Kanizsa square generated in response to the four pac-man figures in Figure 19.

21. 180° Opponent Directions from V1 to MT

The fact that opponent directions differ by 180°, rather than 90°, follows from the fact, summarized in Figure 21, that the opposite feature of a complex/direction cell is another complex/direction cell whose direction preference differs from it by 180°. When this latter property is organized into a network topography, one finds the type of direction hypercolumns that were described in MT by Albright, Desimone, and Gross (1984). A schematic explanation of how direction hypercolumns in MT may be generated from the orientation hypercolumns of V1 is shown in Figure 24. This explanation suggests that the pathways from V1 to MT combine signals from sustained cells and transient cells, as in Figure 21, in a different way than the pathways from V1 to V2, as in Figure 20.

22. Opponent Rebounds: Rapid Reset of Resonating Segmentations

A final refinement of the SOC Filter and MOC Filter design assumes that the opponent cell pairs shown in Figures 20 and 21 are capable of *antagonistic rebound*; that is, offset of

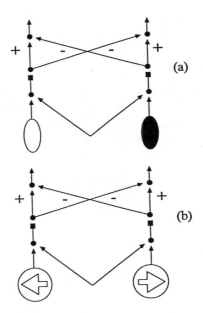

Figure 26. Opponent rebounds: When orientationally tuned complex cells in the SOC Filter are organized into gated dipole opponent circuits, as in (a), offset of a complex on-cell can transiently activate like-oriented complex off-cells, as well as perpendicular hypercomplex on-cells at the second competitive stage (see text). Offset of directionally tuned complex cells within the MOC Filter, as in (b), can transiently activate complex cells tuned to the opposite direction.

one cell in the pair after its sustained activation can trigger an antagonistic rebound that transiently activates the opponent cell in the pair. A minimal neural model of such an opponent rebound, illustrated in Figure 25, is called a *gated dipole* (Grossberg, 1972, 1982, 1988). Such an antagonistic rebound, when appropriately embedded in an SOC Filter or MOC Filter, can reset a resonating segmentation in response to rapid changes in the stimulus, as discussed in Section 8. For example, suppose that horizontally oriented hypercomplex cells in the SOC Filter are cooperating with horizontally oriented bipole cells to generate a horizontal boundary segmentation in the CC Loop (Figure 1) when the input pattern is suddenly shut off. In the absence of opponent processing, the positive feedback signals between the active hypercomplex on-cells and bipole cells could maintain the boundary segmentation for a long time after input offset, thereby causing massive smearing of the visual percept in response to rapidly changing scenes.

Suppose, however, that due to opponent processing by gated dipoles, offset of the horizontal complex on-cells can trigger an antagonistic rebound that activates the horizontal hypercomplex off-cells. The horizontal hypercomplex off-cells would then generate inhibitory signals to the horizontal bipole cells, as in Figure 1. These inhibitory signals would actively shut off the resonating segmentation, thereby preventing too much smearing from occurring. Such inhibitory signals from hypercomplex cells to bipole cells are predicted to be one of the inhibitory processes that control the amount of smearing caused by a moving image in the experiments of Hogben and DiLollo (1985).

This analysis of how antagonistic rebounds can reset a resonating segmentation leads to the prediction that gated dipoles occur at the complex cell level or the hypercomplex cell level (Figure 26) in the Static BCS and Motion BCS.

23. MacKay Afterimages, the Waterfall Effect, and Long Range MAE

The previous sections argued that some positive aftereffects may be partly due to a lingering resonance in a CC Loop, and some negative aftereffects may be partly due to an antagonistic rebound that resets such a resonance. Within the Static BCS, negative aftereffects tend to activate perpendicular segmentations via the same 90° symmetry of the SOC Filter that generates perpendicular end cuts (Section 20). Due to this symmetry, sustained inspection of a radial image can induce a circular aftereffect if a blank field is subsequently inspected (MacKay, 1957). In a similar fashion, it follows from the 180° symmetry of the MOC Filter, summarized in Figure 24, that sustained inspection of a waterfall can induce an upward-moving motion aftereffect (MAE) if a blank field is subsequently attended (Sekuler, 1975).

The assumption that a level of gated dipoles occurs at, or subsequent to, Level 5 of the MOC Filter also provides an explanation of how a long-range MAE can occur between the locations of two flashes that previously generated apparent motion between themselves (von Grünau, 1986). As discussed in Section 15, a wave of apparent motion is synthesized at Level 5 due to interactions of the flashes through the long-range Gaussian filter described in equations (5)–(11). The gated dipoles at, or subsequent to, Level 5 will habituate to the wave of apparent motion much as they would in response to a "real" motion signal expressed at Level 5.

24. Concluding Remarks: The Inadequacy of Independent Visual Modules

FACADE Theory clarifies that, whereas specialization of function surely exists during visual perception, it is not the type of specialization that may adequately be described by independent neural modules for the processing of edges, textures, shading, depth, motion, and color information. In particular, FACADE Theory provides an explanation of many data that do not support the modular approach advocated by Marr (1982).

A basic conceptual problem faced by a modular approach may be described as follows. Suppose that specialized modules capable of processing edges, or textures, or shading, etc. are available. Typically each of these modules is described using different mathematical rules that are not easily combined into a unified theory. Correspondingly, the modules do not respond well to visual data other than the type of data which they were designed to process. In order to function well, either the visual world which such a module is allowed to process must be restricted, whence the module could not be used to process realistic scenes; or a smart preprocessor is needed to sort scenes into parts according to the type of data that each module can process well, and to expose a module only to that part of a scene for which it was designed. Such a smart preprocessor would, however, embody a more powerful vision model than the modules themselves; hence would render the modules obsolete. In either case, modular algorithms do not provide a viable approach to the study of real-world vision.

The task of such a smart preprocessor is, in any case, more difficult that one of sorting scenes into parts which contain only one type of visual information. This is because each part of a visual scene often contains locally ambiguous information about edges, textures, shading, depth, motion, and color, all overlaid together. Humans are capable of using these multiple types of visual information cooperatively to generate an unambiguous 3-dimensional representation of Form-And-Color-And-DEpth; hence the term FACADE representation. The hyphens in "Form-And-Color-And-DEpth" emphasize the well-known fact that changes in perceived color can cause changes in perceived depth and form, changes in perceived depth can cause changes in perceived brightness and form, and so on. Every stage of visual processing *multiplexes* together several key properties of the scenic representation. It is a central task of biological vision theories to understand how the organization of visual information processing regulates which properties are multiplexed together at each processing stage, and how the stages interact to generate these properties.

FACADE Theory became possible through the discovery of several new uncertainty principles; that is, principles which show what combinations of visual properties cannot, in principle, be computed at a single processing stage (Grossberg, 1987a; Grossberg and Mingolla, 1985b). The theory has by now described how to design parallel and hierarchical interactions that can resolve these uncertainties using several processing stages. The hierarchical computations of end cuts to overcome orientational uncertainty is illustrative (Section 20). These interactions occur within and between two subsystems, the Boundary Contour System (BCS) and the Feature Contour System (FCS), whose computations are computationally complementary. In addition, principles of symmetry seem to govern the organization of these subsystems, as the designs of SOC Filter and MOC Filter illustrate. Resonance principles are also operative, as in the design of the CC Loop.

Issues concerning uncertainty principles, complementarity, symmetry, and resonance lie at the foundations of quantum mechanics and other physical theories. Mammalian vision systems are also quantum systems in the sense that they can generate visual percepts in response to just a few light quanta. How the types of uncertainty, complementarity, symmetry, and resonance that are resolved by biological vision systems for purposes of macroscopic perception may be related to concepts of uncertainty, complementarity, symmetry, and resonance in quantum mechanics or other physical theories is a theme of considerable importance for future research.

The themes of uncertainty, complementarity, symmetry, and resonance show the inadequacy of the modular and rule-based approaches from a deeper information theoretic perspective. Although the BCS, FCS, and their individual processing stages are computationally specialized, their *interactions* overcome computational uncertainties and complementary deficiencies to generate useful visual representations, rather than properties that may be computed by independent processing modules. Context-sensitive interactions also determine which combinations of positions, orientations, disparities, spatial scales, and the like will be cooperatively linked through resonance. Likewise, the symmetry principle that integrates static form and motion form properties cannot be stated as a property of independent modules for form or motion perception, because the Static BCS and the Motion BCS each process aspects of both form and motion, and the design of each of these networks can best be understood as parts of a single larger system, as in Figure 22.

Such an interactive theory precludes the sharp separation between formal algorithm and mechanistic realization that Marr (1982) proposed. How computational uncertainties can be overcome, how complementary processing properties can be interactively synthesized, and how particular combinations of multiplexed properties may resonate or be symmetrically organized are all properties of particular classes of mechanistic realizations. Many workers in the field of neural networks summarize this state of affairs by saying that "the architecture is the algorithm." Future tasks in theoretically understanding biological vision promise to require that we replace algorithmic rules and independent modules by architectural designs whose emergent properties constitute intelligence as we know it.

REFERENCES

Albright, T.D. (1984). Direction and orientation selectivity of neurons in visual area MT of the macaque. *Journal of Neurophysiology*, **52**, 1106–1130.

Albright, T.D., Desimone, R., and Gross, C.G. (1984). Columnar organization of directionally sensitive cells in visual area MT of the macaque. *Journal of Neurophysiology*, **51**, 16–31.

Anderson, S.J. and Burr, D.C. (1987). Receptive field size of human motion detection units. *Vision Research*, **27**, 621–635.

Anstis, S. and Ramachandran, V.S. (1987). Visual inertia in apparent motion. *Vision Research*, **27**, 755–764.

Badcock, D.R. and Westheimer, G. (1985a). Spatial location and hyperacuity: The centre/surround localization contribution function has two substrates. *Vision Research*, **25**, 1259–1267.

Badcock, D.R. and Westheimer, G. (1985b). Spatial location and hyperacuity: Flank position within the centre and surround zones. *Spatial Vision*, **1**, 3–11.

Bartley, S.H. (1941). **Vision, a study of its basis**. New York: D. Van Nostrand.

Braddick, O.J. (1974). A short range process in apparent motion. *Vision Research*, **14**, 519–527.

Burr, D.C., Ross, J., and Morrone, M.C. (1986). Smooth and sampled motion. *Vision Research*, **26**, 643–652.

Carpenter, G.A. and Grossberg, S. (1987a). A massively parallel architecture for a self-organizing neural pattern recognition machine. *Computer Vision, Graphics, and Image Processing*, **37**, 54–115.

Carpenter, G.A. and Grossberg, S. (1987b). ART 2: Stable self-organization of pattern recognition for analog input patterns. *Applied Optics*, **26**, 4919–4930.

Carpenter, G.A. and Grossberg, S. (1988). The ART of adaptive pattern recognition by a self-organizing neural network. *Computer*, **21**, 77–88.

Carpenter, G.A. and Grossberg, S. (1990). ART 3: Hierarchical search using chemical transmitters in self-organizing pattern recognition architectures. *Neural Networks*, **3**, 129–152.

Cohen, M.A. and Grossberg, S. (1984). Neural dynamics of brightness perception: Features, boundaries, diffusion, and resonance. *Perception and Psychophysics*, **36**, 428–456.

DeValois, R.L., Albrecht, D.G., and Thorell, L.G. (1982). Spatial frequency selectivity of cells in macaque visual cortex. *Vision Research*, **22**, 545–559.

DeYoe, E.A. and van Essen, D.C. (1988). Concurrent processing streams in monkey visual cortex. *Trends in Neuroscience*, **11**, 219–226.

Duncan, J. and Humphreys, G.W. (1989). Visual search and stimulus similarity. *Psychological Review*, **96**, 433–458.

Eckhorn, R., Bauer, R., Jordan, W., Brosch, M., Kruse, W., Munk, M., and Reitboeck, H.J. (1988). Coherent oscillations: A mechanism of feature linking in the visual cortex? *Biological Cybernetics*, **60**, 121–130.

Eriksen, C.W. and Murphy, T.D. (1987). Movement of attentional focus across the visual field: A critical look at the evidence. *Perception and Psychophysics*, **42**, 29–305.

Gray, C.M., Konig, P., Engel, A.K., and Singer, W. (1989). Oscillatory responses in cat visual cortex exhibit inter-columnar synchronization which reflects global stimulus properties. *Nature*, **338**, 334–337.

Grossberg, S. (1972). A neural theory of punishment and avoidance, II. Quantitative theory. *Mathematical Biosciences*, **15**, 253–285.

Grossberg, S. (1973). Contour enhancement, short-term memory, and constancies in reverberating neural networks. *Studies in Applied Mathematics*, **52**, 217–257.

Grossberg, S. (1976). Adaptive pattern classification and universal recoding, II: Feedback, expectation, olfaction, and illusions. *Biological Cybernetics*, **23**, 187–202.

Grossberg, S. (1978). A theory of visual coding, memory, and development. In E. Leeuwenberg and H. Buffart (Eds.), **Formal theories of visual perception**. New York: Wiley.

Grossberg, S. (1980). How does a brain build a cognitive code? *Psychological Review*, **87**, 1–51.

Grossberg, S. (1982). **Studies of mind and brain: Neural principles of learning, perception, development, cognition, and motor control**. Boston: Reidel Press.

Grossberg, S. (1987a). Cortical dynamics of three-dimensional form, color, and brightness perception, I: Monocular theory. *Perception and Psychophysics*, **41**, 87–116.

Grossberg, S. (1987b). Cortical dynamics of three-dimensional form, color, and brightness perception, II: Binocular theory. *Perception and Psychophysics*, **41**, 117–158.

Grossberg, S. (Ed.) (1987c). **The adaptive brain, II: Vision, speech, language, and motor control**. Amsterdam: Elsevier/North-Holland.

Grossberg, S. (Ed.) (1988). **Neural networks and natural intelligence**. Cambridge, MA: MIT Press.

Grossberg, S. (1990). 3-D vision and figure-ground separation by visual cortex. Submitted for publication.

Grossberg, S. and Marshall, J. (1989). Stereo boundary fusion by cortical complex cells: A system of maps, filters, and feedback networks for multiplexing distributed data. *Neural Networks*, **2**, 29–51.

Grossberg, S. and Mingolla, E. (1985a). Neural dynamics of form perception: Boundary completion, illusory figures, and neon color spreading, *Psychological Review*, **92**, 173–211.

Grossberg, S. and Mingolla, E. (1985b). Neural dynamics of perceptual grouping: Textures, boundaries, and emergent segmentations, *Perception and Psychophysics*, **38**, 141–171.

Grossberg, S. and Mingolla, E. (1986). Computer simulation of neural networks for perceptual psychology. *Behavior Research Methods, Instruments, and Computers*, **18**, 601–607.

Grossberg, S. and Mingolla, E. (1987). Neural dynamics of surface perception: Boundary webs, illuminants, and shape-from-shading. *Computer Vision, Graphics, and Image Processing*, **37**, 116–165.

Grossberg, S. and Mingolla, E. (1990a). Neural dynamics of motion segmentation: Direction fields, apertures, and resonant grouping. In M. Caudill (Ed.), **Proceedings of the international joint conference on neural networks**, January, I, 11–14. Hillsdale, NJ: Erlbaum Associates.

Grossberg, S. and Mingolla, E. (1990b). Neural dynamics of motion segmentation. In **Proceedings of Vision Interface '90**, Halifax, Nova Scotia, May 14-18.

Grossberg, S. and Mingolla, E. (1990c). Neural dynamics of motion segmentation: Direction fields, apertures, and resonant grouping. Submitted for publication.

Grossberg, S., Mingolla, E., and Todorović, D. (1989). A neural network architecture for preattentive vision. *IEEE Transactions on Biomedical Engineering*, **36**, 65-84.

Grossberg, S. and Rudd, M.E. (1989a). A neural architecture for visual motion perception: Group and element apparent motion. In M. Caudill (Ed.), **Proceedings of the international joint conference on neural networks**, June, **I**, 195-199. Piscataway, NJ: IEEE.

Grossberg, S. and Rudd, M.E. (1989b). Neural dynamics of visual motion perception: Group and element apparent motion. *Investigative Ophthalmology Supplement*, **30**, 73.

Grossberg, S. and Rudd, M.E. (1989c). A neural architecture for visual motion perception: Group and element apparent motion. *Neural Networks*, **2**, 421-450.

Grossberg, S. and Rudd, M.E. (1990a). Cortical dynamics of visual motion perception: Short- and long-range motion. *Investigative Ophthalmology Supplement*, **31**, 529.

Grossberg, S. and Rudd, M.E. (1990b). Cortical dynamics of visual motion perception: Short-range and long-range motion. Submitted for publication.

Grossberg, S. and Somers, D. (1990). Synchronized oscillations during cooperative feature linking in a cortical model of visual perception. Submitted for publication.

Grossberg, S. and Todorović, D. (1988). Neural dynamics of 1-D and 2-D brightness perception: A unified model of classical and recent phenomena. *Perception and Psychophysics*, **43**, 241-277.

Heggelund, P. (1981). Receptive field organization of simple cells in cat striate cortex. *Experimental Brain Research*, **42**, 89-98.

Hogben, J.H. and DiLollo, V. (1985). Suppression of visual persistence in apparent motion. *Perception and Psychophysics*, **38**, 450-460.

Hubel, D.H. and Wiesel, T.N. (1962). Receptive fields, binocular interaction and functional architecture in the cat's visual cortex. *Journal of Physiology*, **160**, 106-154.

Hubel, D.H. and Wiesel, T.N. (1965). Receptive fields and functional architecture in two nonstriate visual areas (18 and 19) of the cat, *Journal of Neurophysiology*, **28**, 229-289.

Hubel, D.H. and Wiesel, T.N. (1968). Receptive fields and functional architecture of monkey striate cortex. *Journal of Physiology*, **195**, 215-243.

Hubel, D.H. and Wiesel, T.N. (1977). Functional architecture of macaque monkey visual cortex. *Proceedings of the Royal Society of London (B)*, **198**, 1-59.

Kolers, P.A. (1972). **Aspects of motion perception**. Oxford: Pergamon Press.

LaBerge, D. and Brown, V. (1989). Theory of attentional operations in shape identification. *Psychological Review*, **96**, 101-124.

MacKay, D.M. (1957). Moving visual images produced by regular stationary patterns. *Nature*, **180**, 849-850.

Marr, D. (1982). **Vision**. San Francisco: Freeman.

Marshall, J. (1990). Self-organizing neural networks for perception of visual motion. *Neural Networks*, **3**, 45-74.

Maunsell, J.H.R. and van Essen, D.C. (1983). Response properties of single units in middle temporal visual area of the macaque. *Journal of Neurophysiology*, **49**, 1127-1147.

Nakayama, K. and Silverman, G.H. (1984). Temporal and spatial characteristics of the upper displacement limit for motion in random dots. *Vision Research*, **24**, 293-299.

Nakayama, K. and Silverman, G.H. (1985). Detection and discrimination of sinusoidal grating displacements. *Journal of the Optical Society of America*, **2**, 267-273.

Nakayama, K. and Silverman, G.H. (1986). Serial and parallel processing of visual feature conjunctions. *Nature*, **320**, 264-265.

Newsome, W.T., Gizzi, M.S., and Movshon, J.A. (1983). Spatial and temporal properties of neurons in macaque MT. *Investigative Ophthalmology and Visual Science*, **24**, 106.

Orban, G.A., Kato, H., and Bishop, P.O. (1979). Dimensions and properties of end-zone inhibitory areas in receptive fields of hypercomplex cells in cat striate cortex, *Journal of Neurophysiology*, **42**, 833-849.

Pashler, H. (1987). Detecting conjunctions of color and form: Reassessing the serial search hypothesis. *Perception and Psychophysics*, **41**, 191-201.

Peterhans, E. and von der Heydt, R. (1989). Mechanisms of contour perception in monkey visual cortex, II. Contours bridging gaps. *Journal of Neuroscience*, **9**, 1749-1763.

Poggio, G.F., Motter, B.C., Squatrito, S., and Trotter, Y. (1985). Responses of neurons in visual cortex (V1 and V2) of the alert macaque to dynamic random-dot stereograms. *Vision Research*, **25**, 397-406.

Ramachandran, V.S. (1985). Apparent motion of subjective surfaces. *Perception*, **14**, 127-134.

Ramachandran, V.S., Rao, V.M., and Vidyasagar, T.R. (1973). Apparent motion with subjective contours. *Vision Research*, **13**, 1399-1401.

Redies, C. and Spillmann, L. (1981). The neon color effect in the Ehrenstein illusion, *Perception*, **10**, 667-681.

Remington, R. and Pierce, L. (1984). Moving attention: Evidence for time-invariant shifts of visual selective attention. *Perception and Psychophysics*, **35**, 393-399.

Sekuler, R. (1975). Visual motion perception. In E.C. Carterette and M.P. Friedman (Eds.), **Handbook of perception, Volume V: Seeing.** New York: Academic Press.

Spitzer, H. and Hochstein, S. (1985). A complex-cell receptive field model, *Journal of Neurophysiology*, **53**, 1266-1286.

Sutter, A., Beck, J., and Graham, N. (1989). Contrast and spatial variables in texture segregation: Testing a simple spatial-frequency channels model. *Perception and Psychophysics*, **46**, 312-332.

Tanaka, M. Lee, B.B., and Creutzfeldt, O.D. (1983). Spectral tuning and contour representation in area 17 of the awake monkey. In J.D. Mollon and L.T. Sharpe (Eds.), **Colour vision.** New York: Academic Press, 1983.

Ternus, J. (1926/1950). Experimentelle Untersuchungen über phänomenale Identität. *Psychologische Forschung*, **7**, 81-136. Abstracted and translated in W.D. Ellis (Ed.), **A sourcebook of Gestalt psychology.** New York: Humanities Press, 1950.

Thorell, L.G., De Valois, R.L., and Albrecht, D.G. (1984). Spatial mapping of monkey V1 cells with pure color and luminance stimuli. *Vision Research*, **24**, 751–769.

Treisman, A. and Gelade, G. (1980). A feature integration theory of attention. *Cognitive Psychology*, **12**, 97–136.

Treisman, A. and Gormican, S. (1988). Feature analysis in early vision: Evidence from search asymmetries. *Psychological Review*, **95**, 15–48.

Treisman, A. and Souther, J. (1985). Search asymmetry: A diagnostic for preattentive processing of separable features. *Journal of Experimental Psychology: General*, **114**, 285–310.

van Allen, E.J. and Kolodzy, P.J. (1987). Application of a boundary contour neural network to illusions and infrared sensor imagery. In M. Caudill and C. Butler (Eds.), **Proceedings of the IEEE first international conference on neural networks, IV**, 193–197. Piscataway, NJ: IEEE Press.

von der Heydt, R., Peterhans, E., and Baumgartner, G. (1984). Illusory contours and cortical neuron responses. *Science*, **224**, 1260–1262.

von der Heydt, R., Hänny, P., and Dürsteler, M.R. (1981). The role of orientation disparity in stereoscopic perception and the development of binocular correspondence. In E. Grastyán and P. Molnár (Eds.), **Advances in physiological sciences, volume 16: Sensory functions**. Elmsford, NY: Pergamon Press.

von Grünau, M.W. (1986). A motion aftereffect for long-range stroboscopic apparent motion. *Perception and Psychophysics*, **40**, 31–38.

Watt, R.J. and Campbell, F.W. (1985). Vernier acuity: Interactions between length effects and gaps when orientation cues are eliminated. *Spatial Vision*, **1**, 31–38.

Zeki, S.M. (1974a). Functional organization of a visual area in the posterior bank of the superior temporal sulcus of the rhesus monkey. *Journal of Physiology (London)*, **236**, 549–573.

Zeki, S.M. (1974b). Cells responding to changing image size and disparity in the cortex of the rhesus monkey. *Journal of Physiology (London)*, **242**, 827–841.

NEURAL DYNAMICS OF VISUAL MOTION PERCEPTION: LOCAL DETECTION AND GLOBAL GROUPING

Stephen Grossberg[†] and Ennio Mingolla[‡]

Center for Adaptive Systems
and
Department of Cognitive and Neural Systems
Boston University
111 Cummington Street
Boston, Massachusetts 02215

Abstract

How do we perceive coherent patterns of motion in regions of our visual fields containing distributions of locally ambiguous motion signals? How does the visual system solve the twin problems of segmentation, the segregation of motion signals from nearby signals belonging to other objects with different trajectories, and grouping, the combining of relatively distant or separated and often locally unequal motion signals into coherent perceptual objects? A neural network model of global motion segmentation and grouping by visual cortex is here described. Called the Motion Boundary Contour System (BCS), the model clarifies how ambiguous local movements on a complex moving shape are actively reorganized into a coherent global motion signal. Unlike many previous researchers, we analyse how a coherent motion signal is imparted to all regions of a moving figure, not only to regions at which unambiguous motion signals exist. The model hereby suggests a solution to the global aperture problem. The Motion BCS describes how preprocessing of motion signals by a Motion Oriented Contrast Filter (MOC Filter) is joined to long-range cooperative grouping mechanisms in a Motion Cooperative-Competitive Loop (MOCC Loop) to control phenomena such as motion capture. The Motion BCS is computed in parallel with the Static BCS of Grossberg and Mingolla (1985a, 1985b, 1987). Homologous properties of the Motion BCS and the Static BCS, specialized to process movement directions and static orientations, respectively, support a unified explanation of many data about static form perception and motion form perception that have heretofore been unexplained or treated separately. It is shown how the Motion BCS can compute motion directions that may be synthesized from multiple orientations with opposite directions-of-contrast. Interactions of model simple cells, complex cells, hypercomplex cells, and bipole cells are described, with special emphasis given to new functional roles in direction disambiguation for end stopping at multiple processing stages and to the dynamic interplay of spatially short-range and long-range interactions.

[†] Supported in part by the Air Force Office of Scientific Research (AFOSR 90-0175), DARPA (AFOSR 90-0083), and the Office of Naval Research (ONR N00014-91-J-4100).

[‡] Supported in part by the Air Force Office of Scientific Research (AFOSR 90-0175).

Acknowledgements: The authors wish to thank Cynthia E. Bradford for her expert assistance in the preparation of the manuscript.

1. Introduction: The Aperture Problem and Motion Grouping

The unity and persistent identity of objects undergoing motion is so immediate that we may all too readily take for granted the subtle problems posed in discriminating such unity within the scintillating mosaic of visual stimulation. In the familiar example of trying to detect a leopard moving through a forest canopy, the change in the optic array over time is a jumble of contrast changes whose local components point in a variety of directions, due to the vagaries of the motions of rustling leaves and of the occlusions and disocclusions of markings on the leopard's body. For a leopard to be detected in such contexts, a key problem must be solved: How can the locally ambiguous local motion signals corresponding to the many parts of the leopard's body be *reorganized* into a coherent object whole with a unitary motion?

The *aperture problem* illustrates these issues in a vivid way. It inquires into how a straight edge or grating viewed through a circular aperture appears to be moving perpendicular to its orientation of contrast regardless of its true motion direction (Wallach, 1976). While this description of the aperture problem is at the molar perceptual level, any early visual cell with a localized receptive field, whether concentric or elongated, experiences its own "aperture problem," because its estimate of motion direction is a spatiotemporally weighted function of changes in local stimulation only. Adelson and Movshon (1982) introduced diagrams similar to Figure 1a to illustrate the ambiguity of local motion direction and speed from information confined to an aperture. In this figure, the length of arrows codes possible trajectories of the point A that would be consistent with the measured change of contrast over time of the cell in question. These authors use the term "velocity space" to emphasize that such a cell seems to be sensitive to the normal component of velocity. A complementary way of thinking about this situation is illustrated in Figure 1b, in which the length of arrows is roughly proportional to the cell's "prior probability distribution" for interpreting changing stimulation as occurring in one of several directions. The direction perpendicular to the cell's receptive field's orientation is locally preferred. However, a cell with an oriented receptive field, such as a simple cell, may be stimulated by a moving edge that is not perfectly aligned with its receptive field's dark-to-light contrast axis. Within a hypercolumn of cells that are tuned to similar position, spatial frequency, contrast, and temporal parameters but vary in preferred orientation (Hubel and Wiesel, 1977), the distribution of local motion signals across cells tuned to all orientations at a given position would favor the direction perpendicular to the orientation of the edge.

The barberpole illusion (Wallach, 1976) shows that these local preferences can be readily overridden by global factors related to the forms within which stationary and moving segments of a display are located (Figure 2). This illusion indicates that, to the extent that locally *unambiguous* motion signals exist at line ends or corners, those unambiguous signals can somehow enforce an interpretation of motion direction consistent with their own throughout the length of a contour bounded by those ends or corners (Figure 3). Clearly some long-range grouping process subsequent to the early generation of local motion signals is at work. This process acts to cooperatively "choose and sharpen" relatively large domains of ambiguous signals using relatively localized, but unambiguous, signals.

In a recent psychophysical study of motion perception, Mingolla, Todd, and Norman (1992) created displays containing spatial arrays of many small circular apertures. For each display, a single "true" motion underlying the trajectories of all visible elements was

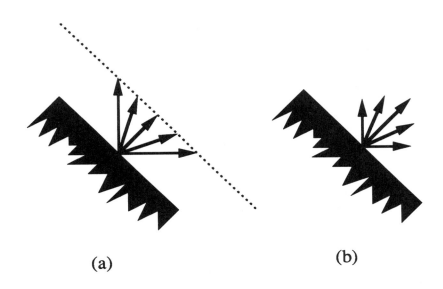

Figure 1: (a) This "velocity space" construction, adapted from Adelson and Movshon (1982), illustrates that the motion of a given point on a moving edge could be along any of the trajectories whose arrows end on the dashed line. (b) A complementary way to view this direction ambiguity is to consider that local motion signals from a straight edge are *strongest* in the perpendicular direction, while covering a range of possible directions.

defined, and the nature and juxtaposition of those elements was manipulated in order to investigate the nature of perceptual grouping of motion signals in a direction discrimination task. The apertures at times contained only straight lines extending to the circumference of the aperture, as in the classical description of the aperture problem, and at other times contained line ends or intersections that provided potential information concerning the "true" motion of the display. The results of the experiments revealed a strong tendency of the visual system to form some kind of average of information in the local motion signals derived from oriented straight contours, as displayed in Figure 1b, if that was the only kind of information present. The addition of identifiable features whose trajectories could be tracked over time affected direction discrimination in in complex way, however. At times featural trajectories were prepotent, completely determining a subject's perceived direction, while at other times a compromise between the trajectories supported by the featural information and those supported by the averaging process appeared to be struck.

The results of Mingolla, Todd, and Norman (1992) illustrate the dynamic nature of the processes whereby the visual system combines a diverse set of local motion signals that "belong" to the same object and separate them from similar sets of signals taken to come

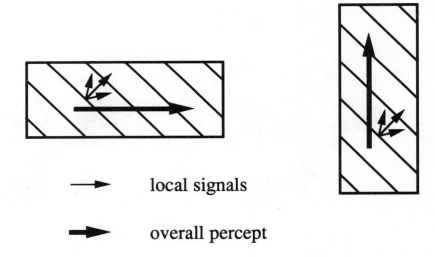

Figure 2: In the barberpole illusion a striped pattern is perceived as moving in the direction of elongation of a rectangular frame (or aperture). Local motion signals, indicated in by thin arrows at only one location within each rectangle, are generated throughout the interior lengths of each diagonal line. As indicated in Figure 1b, these local motion signals express both ambiguity and preference regarding direction of motion. Despite the large area covered by the ambiguous and diagonal-motion-preferring local signals, the resulting percept is horizontal or vertical, depending on the configuration of a visible frame (or aperture).

from different neighboring (segmentation) or overlapping (transparency) objects. Our theory introduces processes that clarify how such grouping effects are generated both during motion perception and the perception of static form.

Some of the subtle issues about grouping that occur during static form perception are illustrated in Figure 4. As pointed out by Prazdny (1984), reversing the contrast relative to a neutral background of one set of dots in a Glass pattern can weaken, or even annihilate, the grouping percept (Figures 4a and 4b). The paradoxical nature of this result can best be appreciated by juxtaposing it with the striking illusory groupings that can be sustained between the like directions-of-contrast (Figure 4c) of the Kanizsa square, as well as the opposite directions-of-contrast (Figure 4d) of the "reverse contrast Kanizsa square" (Cohen and Grossberg, 1984; Grossberg and Mingolla, 1985b; Prazdny, 1984; Shapley and Gordon, 1985). A similar pattern of sensitivity to direction-of-contrast at short-range and insensitivity to direction-of-contrast at long-range also occurs in motion perception. For "short-range" motion, reversal of dot contrast between frames of random dot cinematograms abolishes

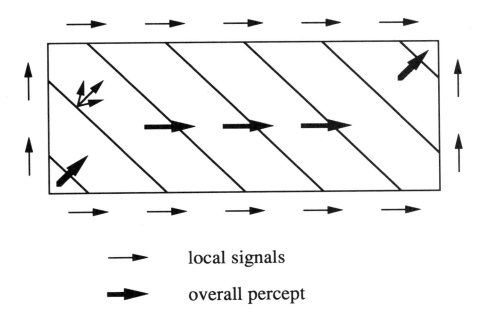

Figure 3: A closer look at the barberpole illusion reveals that for sufficiently large displays, whose length-to-width ratio is not too far from unity, the overall motion percept may vary across different areas of the display. The perceived motion near the lower left and upper right corners of the rectangle may be diagonal, while horizontal motion is seen through the bulk of the display. Unambiguous motion signals are generated in the region where each diagonal line meets the horizontal and vertical contours of the rectangle. (These unambiguous signals are diagramed *outside* the rectangle for clarity.) Evidently, such unambiguous signals exert an influence on the percept that is disproportionate to their areal extent, since the overall percept throughout a diagonal line tends to be a resultant (horizontal plus vertical to diagonal or horizontal plus horizontal to horizontal) of those signals.

a coherent motion percept, whereas long-range apparent motion can occur using displays whose contrast with respect to the background reverses between frames (Anstis and Mather, 1985). These different grouping effects are explained within the theory that is reviewed below.

2. The Static and Motion Boundary Contour Systems

The central theme of our modeling work is thus the investigation of mechanisms whereby an array of locally ambiguous motion signals can be globally reorganized into coherent object motion signals. Our investigation of the global reorganization of local motion signals is conducted within a theoretical framework which has already analysed analogous problems

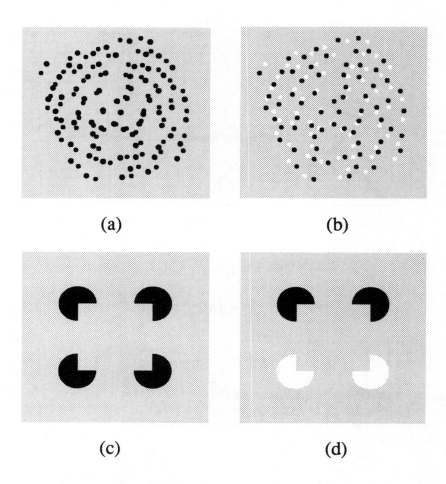

Figure 4: The subtleties of the interaction of spatial scale and direction-of-contrast are revealed by the juxtposition of two classical visual phenomena. Parts (a) and (b) illustrate how the formation of Glass patterns is destroyed by reversing the contrast of one field of dots (Prazdny, 1984), while the reversal of contrast of inducers in the Kanizsa square configuration from (c) to (d) does not significantly weaken illusory contour formation.

within the domain of static form perception. This latter theory has been called FACADE Theory, because its visual representations are predicted to multiplex together properties of Form-And-Color-And-DEpth in prestriate cortical area V4. FACADE Theory describes the neural architecture of two subsystems, the Boundary Contour System (BCS) and the Feature Contour System (FCS), whose properties are computationally complementary (Grossberg, Mingolla, and Todorović, 1989). The BCS generates an emergent 3-D boundary segmentation of edges, texture, shading, and stereo information at multiple spatial scales (Grossberg, 1987a, 1987b, 1990; Grossberg and Marshall, 1989; Grossberg and Mingolla, 1985a, 1985b,

1987). The FCS compensates for variable illumination conditions and fills-in surface properties of brightness, color, and depth among multiple spatial scales (Cohen and Grossberg, 1984; Grossberg, 1987a, 1987b; Grossberg and Mingolla, 1985a; Grossberg and Todorović, 1988).

The BCS provided a new computational rationale as well as a model of the neural circuits governing classical cortical cell types such as simple cells, complex cells, and hypercomplex cells in cortical areas V1 and V2. The theory also predicted a new cell type, the bipole cell (Cohen and Grossberg, 1984; Grossberg and Mingolla, 1985a) whose properties have been supported by neurophysiological experiments (von der Heydt, Peterhans, and Baumgartner, 1984; Peterhans and von der Heydt, 1989). Many heretofore unexplained phenomena about the perception of Form-And-Color-And-Depth have been clarified by FACADE Theory, and an ever increasing number of laboratories have been carrying out experiments suggested by its concepts and mechanisms (Eskew et al., 1991; Humphreys, Quinlan, and Riddoch, 1989; Kellman and Shipley, 1991; Meyer and Dougherty, 1987; Mikaelian, Linton, and Phillips, 1990; Paradiso and Nakayama, 1991; Peterhans and von der Heydt, 1989; Prinzmetal, 1990; Prinzmetal and Boaz, 1989; Sutter, Beck, and Graham, 1989; Todd and Akerstrom, 1987; Watanabe and Sato, 1989; Watanabe and Takeichi, 1990).

This BCS model, now called the *Static* BCS model, consists of several parallel copies, such that each copy is activated by a different range of receptive field sizes. Each Static BCS copy is further subdivided into two hierarchically organized systems (Figure 5): a *Static Oriented Contrast Filter*, or *SOC Filter*, for preprocessing quasi-static images (the eye never ceases to jiggle in its orbit); and a Cooperative-Competitive Feedback Loop, or CC Loop, for generating coherent emergent boundary segmentations of the filtered signals.

After the development of the Static BCS reached a certain level of clarity, it focused attention on the perplexing question: How do static and motion perception systems differ? Some regions of visual cortex are specialized for motion processing, notably region MT (Albright, Desimone, and Gross, 1984; Maunsell and van Essen, 1983; Newsome, Gizzi, and Movshon, 1983; Zeki, 1974a, 1974b). However, even the earliest stages of visual cortical processing, such as simple cells in V1, require stimuli that change through time for their maximal activation and are direction-sensitive (DeValois, Albrecht, and Thorell, 1982; Heggelund, 1981; Hubel and Wiesel, 1962, 1968, 1977; Tanaka, Lee, and Creutzfeldt, 1983). Why has evolution generated regions such as MT, when even V1 is change-sensitive and direction-sensitive? What computational properties are achieved by the $V1 \to MT$ cortical stream that cannot be achieved by the parallel $V1 \to V2 \to V4$ cortical stream?

Analysis of the SOC Filter design revealed that one of its basic properties made it unsuitable for motion processing. In particular, SOC Filter output signals cannot adequately discriminate the direction-of-motion of a moving figure. Further analysis of how this happens and how it can be overcome led to a new theory of biological motion perception that was outlined in Grossberg (1987b) and quantitatively specified and analysed in Grossberg (1990, 1991) and Grossberg and Rudd (1989a, 1989b, 1992). The results of this analysis suggested that the motion form perception system shares many design features with the static form perception system, but that it has incorporated the minimal changes needed to achieve sensitivity to both local and global properties of direction-of-motion. In fact, Grossberg (1990, 1991) has suggested that the two systems are parallel subsystems of a larger, symmetric system design, called Form-Motion (FM) Symmetry, that is predicted to govern the devel-

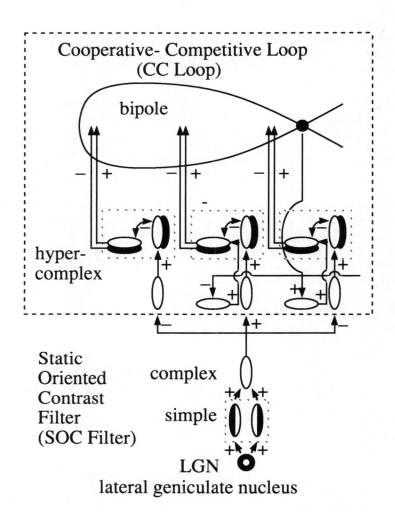

Figure 5: The static Boundary contour system consists of two main parts, the Static Oriented Contrast Filter (SOC Filter) and the Cooperative-Competitive Loop (CC Loop). The SOC Filter determines the locally preferred orientations of contrast differences in the input, while the CC Loop groups, selects, and sharpens contours, both "real" and "illusory."

opment of visual cortex. See Grossberg (this volume) for a discussion of FM Symmetry and the differences between the Static BCS and Motion BCS.

Correspondingly, this new theory of biological motion perception consists of a neural architecture called a *Motion Boundary Contour System*, or *Motion BCS*. The Motion BCS consists of several parallel copies, such that each copy is activated by a different range of receptive field sizes, as in the Static BCS. Also as in the Static BCS, each Motion BCS copy is further subdivided into hierarchically organized subsystems: a *Motion Oriented Contrast Filter*, or *MOC Filter*, for preprocessing moving images; and a Motion Cooperative-Competitive Feedback Loop, or MOCC Loop, for generating coherent emergent boundary segmentations of the filtered signals. The MOC Filter provides a model for the early stages of the $V1 \to MT$ processing stream in the visual cortex. Evidence for the MOC Filter includes its ability to explain many classical and recent data about short-range and long-range motion properties, and about cortical cell properties, that have not yet been explained by alternative models. Grossberg and Rudd (1989a) have also shown how properties of other motion perception models can be assimilated into different parts of the Motion BCS design.

The computer simulations of motion perception data that were reported by Grossberg and Rudd (1989a, 1989b, 1992) used a 1-dimensional version of the MOC Filter. The present article simulates data which require a 2-dimensional version of the MOC Filter. In addition, outputs from the 2-dimensional MOC filter input to a 2-dimensional version of the MOCC Loop. In order to motivate these results, we first discuss key motion segmentation and grouping phenomena, including the aperture problem, barberpole illusion, and motion capture. Then we discuss how these phenomena suggest the existence of a MOCC Loop which is analogous to the static CC Loop of Grossberg and Mingolla (1985a, 1985b, 1987) but is specialized to process moving images. The MOC Filter of Grossberg and Rudd (1989a, 1992) is reviewed in Grossberg (this volume). In this chapter, we show how the MOC Filter model can be extended to process 2-dimensional moving images. These extensions include hypotheses concerning the role of end-stopped simple cells, the spatial layout of simple cell receptive fields, and competition among signals sensitive to different directions-of-motion. We illustrate these concepts through computer simulations which study how the Motion BCS responds to changes in the bounding orientations, shapes, and motion directions of an object. These results are used to explain data about the aperture problem, barberpole illusion, and motion capture.

3. Global Segmentation and Grouping: From Locally Ambiguous Motion Signals to Coherent Object Motion Signals

As described in the previous section, the barberpole illusion (Figures 2 and 3) offers compelling evidence for long range cooperative processes in motion perception. Cooperativity among motion signals was studied by Lappin and Bell (1976) and more recently by Williams and Sekuler (1984). Random dot cinematograms were displayed in which successive displacements of dots over frames were not uniform across all dots, but rather were sampled from a rectangular distribution of possible directions. On any frame each dot's displacement was independent of both its own history of displacements and of the displacement of other dots in that frame. For appropriate parameter choices, observers reported perceiving a coherent global motion in the direction of the mean of sampled local dot motions, as well as perceiving the individual motions of each dot. Williams and Sekuler also reported hysteresis effects, whereby the percept of coherent motion persisted while motion distribution parameters were altered from a relatively narrow range, which easily supported the coherent percept, to a wider range, for which coherent motion was not ordinarily seen if presented initially. Thus cooperativity among motion signals involves averaging of directional information and hysteresis, as well as sharpening and choice.

Ramachandran (1981) and Ramachandran and Inada (1985) provided additional evi-

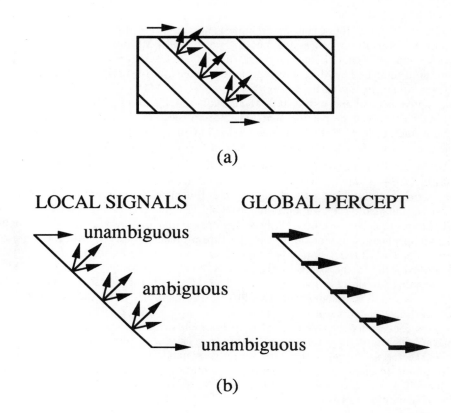

Figure 6: (a) By focusing on the distribution of motion signals on a single diagonal line, as indicated in the barberpole illusion, the phenomenon can be analysed as a form of motion capture. (b) The unambiguous signals from the line ends help to enhance signals of like direction (cooperation) and suppress signals of different directions (competition) within the interior of the line segment.

dence for cooperation among motion signals in a phenomena termed "motion capture," in which strong and unambiguous motion signals can actively reorganize motion signals in ambiguous regions where there are no locally preferred motion directions. From this perspective, the barberpole illusion can be analysed as displaying its own form of "motion capture," as indicated in Figure 6. That is, signals from the ambiguous interior region of the diagonal line segment are being "captured" by signals from the unambiguous ends. This enhancement of horizontal signals in the interior region is accompanied by the suppression of diagonal and vertical signals. Long-range cooperation is thus accompanied by short-range competition in order to enforce a clear choice of percept all along the line. As noted below, this cooperative-competitive interaction is analogous to that of the Static CC Loop.

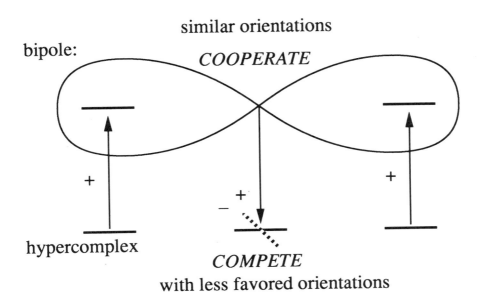

Figure 7: In the static cooperative-competitive feedback loop (CC Loop) signals of like orientation (here horizontal) initiated from "bottom up" input data arrive at two locations separated by some distance (indicated by the horizontal bar above the word "hypercomplex" and the corresponding bar at the lower right). Cooperative bipole cells, identified with projections of V2 cells, can, if they receive sufficient stimulation on *both* sides of their receptive field centers, send positive feedback to cells of like orientation in the hypercomplex layer. There cells of differing orientation and like position compete with each other to determine the most favorable orientation at each location. Because of the long-range feedback, local orientational preferences are modulated to be consistent with contextual data. While in this example, the completion is colinear, the "figure eight" shape of the bipole cell indicates that completion can be curvilinear, so long as a consistent co-occurrence of oriented signals favors interpolation at the bipole center. See Figure 28 and equations (A13) through (A21) of the Appendix of Grossberg and Mingolla (1985b) for a detailed explanation of static bipole cells.

4. Coherent and Resonant Completion: Static and Motion CC Loops

Static BCS mechanisms also combine long-range cooperation with short-range competition among orientationally tuned cells within a feedback network, namely the CC Loop (Figure 7). Since publication of the CC Loop equations in Grossberg and Mingolla (1985a, 1985b, 1987), increasing physiological evidence for the model has been reported. In particular, Peterhans and von der Heydt (1989) have recently reported evidence for cooperative linking interactions in cortical area V2, and Gray *et al.* (1989) and Eckhorn *et al.* (1988) have reported resonant interactions involving phase locking of spike trains of V2 cells with nonoverlapping receptive fields, which also occurs due to the long-range cooperative interactions of the CC Loop (Grossberg and Somers, 1991).

Given that the Static CC Loop can enhance consistent signals and suppress inconsistent or ambiguous signals from orientationally-tuned cells, and given that *in vivo* early orientationally tuned cells, such as simple cells, are sensitive to motion, the question naturally arose whether motion segmentation could be accomplished by the Static CC Loop. Our daily experiences with the dynamic geometry of form perception for static and moving contours suggests that this is not true. As illustrated in Figure 8, a *static* form system is concerned with *orientation* of contours, while a *motion* form system is concerned with *direction* of moving contours. Moreover, these systems must be different because the geometries governing static orientations and motion directions are different. For example, the opposite orientation of "vertical" is "horizontal," a difference of 90°, whereas the opposite direction of "up" is "down," a difference of 180°. Grossberg (1991, this volume) has suggested an explanation of this difference. Keeping track of direction as well as orientation requires an additional degree of freedom, since a segment of a given orientation may be moving in any of several directions, and a given direction of motion can be observed for segments of any of several orientations. This is just another way of stating the aperture problem.

The striking similarity of cooperative and competitive grouping requirements for both static and moving images, together with the existence of different geometries for static and motion perception, suggested that parallel versions of the CC Loop exist within a Static BCS and a Motion BCS. Additional analyses of short-range and long-range motion data suggested, moreover, that parallel versions of OC Filters exist within the Static BCS and the Motion BCS (Grossberg and Rudd, 1989a, 1992). This taxonomy of "static" and "motion" CC Loops does not, however, imply logical exclusivity of function. When a contour moves, the *static* CC Loop may operate to determine the best coherent *orientations* of the moving contour, at the same time that the *motion* CC Loop determines the best coherent *directions* of that contour. Moreover, the orientationally-based boundary segmentations that are generated by the Static BCS are suggested to help select directionally-based boundary segmentations of the Motion BCS at correctly calibrated depths (Grossberg, 1991, this volume). This interaction has been predicted to occur via the $V1 \to V2 \to MT$ pathway.

5. Many Orientations Can Move in the Same Direction

How then do the motion analogs of the cooperative bipole cells (Figure 7) of the static CC Loop function? In analysing the properties of static bipole cells, certain images by John Kennedy proved invaluable (Kennedy, 1979). As shown in Figure 9, static bipole cells are capable of choosing orientations which are not locally preferred for generating positive feedback to cooperatively link and complete static boundaries. Each bipole cell can generate an output signal only if *both* of its oriented receptive fields receive large enough inputs from cells with similarly oriented receptive fields. These input cells at the previous processing level consist of formal analogs of hypercomplex cells. The output signals from the bipole cells feed back to the hypercomplex cells (Figure 7), where they bias the competition among the hypercomplex cells towards the orientations and positions that are most favored by the bipole cells.

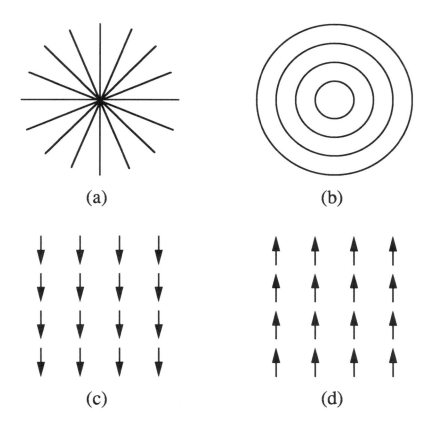

Figure 8: As indicated by such static rebound phenomena as the Mackay effect, in which prolonged fixation on a pattern of radial line segments (a) results in an aftereffect of perceived circular contours (b), the opposite of an static *orientation* signal is a signal of perpendicular orientation. The waterfall illusion, in which prolonged adaptation to motion signals as from a waterfall (c) is followed by the sensation of upward motion when one looks at a neutral scene (d), indicates that the opposite of a motion *direction* signal is not 90° but 180° from that signal.

The Motion CC, or MOCC, Loop must, however, cope with an additional degree of freedom, since it considers direction as well as orientation. Thus motion bipole cells are organized in a fashion that differs somewhat from that of their static analogs. As indicated in Figure 10, motion bipole cells are postulated to exist in families sensitive to motion signals of different directions but whose sources are arrayed in patterns of similar orientation. Expressed differently, while some motion bipole cells are assumed to favor the direction of motion perpendicular to the orientation induced by the elongated axis of their two lobes

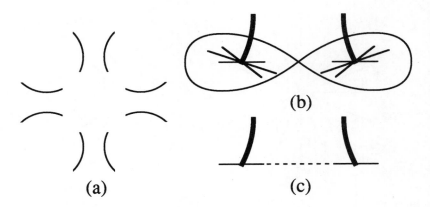

Figure 9: (a) This illusory contour display, adapted with permission from Kennedy (1979), indicates that boundary completion in the static CC Loop can choose orientations that are not locally preferred if global organizational factors are sufficiently powerful. (b) The locally preferred directions at the bottom ends of the two top curves of (a) are perpendicular to the ends of the curves and thus tilted off the horizontal. (c) Cooperation among signals that are horizontal enhances and completes horizontal signals (cooperation) while suppressing non-horizontal signals (competition) along the illusory contour.

(Figure 10b), others have a similar spatial layout but different preferred directions of motion (Figure 10a and 10c).

As in the Static CC Loop, each bipole cell of the MOCC Loop can generate an output signal only if *both* of its receptive fields receive large enough inputs from cells that are sensitive to a similar direction-of-motion. These input cells at the previous processing level consist of formal analogs of hypercomplex cells. The output signals from the bipole cells feed back to the hypercomplex cells, where they bias the competition among the hypercomplex cells towards the directions and positions that are most favored by the bipole cells.

It is intuitively clear that this is just the sort of cooperative feedback, propagated inward between pairs, or larger numbers, of flanking inducers, that is needed to explain phenomena like the barberpole illusion and motion capture. Our analysis elevates this intuition into a computationally precise theory. As a result, both orientationally-sensitive grouping during static form perception and directionally-sensitive grouping during motion form perception are predicted to utilize bipole cells. The explanatory power of this homology strengthens the case that both the Static BCS and Motion BCS architectures are variations of a common cortical design.

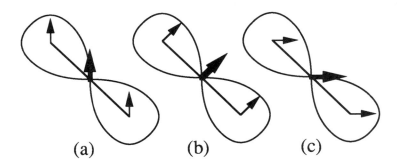

(a)   (b)   (c)

Figure 10: As with static bipole cells, motion bipoles are sensitive to bottom-up direction signals from motion hypercomplex cells (indicated by thin arrows), which send excitatory signals to each lobe of the bipole cell. If sufficient excitatory activity is sensed in *both* lobes, the bipole sends feedback signals of like direction (indicated by thick arrows) to the hypercomplex layer. While analogous to their static counterparts, motion bipole cells fundamentally different, insofar as they must cope with the additional degree of freedom imposed by the simultaneous determination of a globally consistent motion direction over many possible contour orientations. Thus families of bipole cells are presumed to exist, such that cells whose major axes are the same (diagonal in the present case) can be maximally sensitive to motion signals of different directions, ranging from (a) vertical to (b) diagonal (perpendicular to the major axis) to (c) horizontal for the three bipoles shown here.

6. Joining Sensitivity to Direction-of-Motion with Insensitivity to Direction-of-Contrast

In order to design a MOCC Loop, its hypercomplex cells need to be sensitive to a prescribed direction-of-motion. These cells may be excited by image contours with different orientations all moving in the same direction within a prescribed region of perceptual space. In order to synthesize directional sensitivity from several different orientations, the MOC Filter needs to have a somewhat different circuit design than the SOC Filter. As noted in Grossberg (this volume), the reason for this modification is that complex cells of the SOC Filter are insensitive to direction-of-motion as well as to direction-of-contrast. Consequently, the SOC Filter cannot be used to process the direction-of-motion of a moving figure. This deficiency arises from the way in which the SOC Filter becomes insensitive to direction-of-contrast at its complex cell level.

The main design problem leading to a MOC Filter is to make the minimal changes in the SOC Filter that are needed to model an oriented, contrast-sensitive filter whose outputs are insensitive to direction-of-contrast—a property that is just as important for moving images as for static images—yet is also *sensitive* to direction-of-motion—a property that is certainly essential in a motion perception system. Along the way, the MOC Filter introduces an extra degree of computational freedom which achieves several important properties in one stroke: sensitivity to direction-of-motion, long-range motion interactions, and binocularity (Grossberg and Rudd, 1989a, 1992).

7. Pooling Orientations and Directions-of-Contrast to Compute 2-D Directions-of-Motion

The Grossberg-Rudd model (Figure 11; see Grossberg, this volume) was used to explain and simulate motion data which exhibit a natural one-dimensional, or 1-D, symmetry; for example, apparent motion between colinear groups of flashes. The model needs further development to explain data concerning the motion of 2-D shapes, since such shapes may move in directions that may or may not be perpendicular to the orientations of their boundaries. The type of new issues that arise in the 2-D case are illustrated by the following example. Consider the lower right corner of a homogeneous rectangular form of relatively high luminance that is moving diagonally upward and to the right on a homogeneous background of relatively low luminance (Figure 12). Both the regions of horizontal and vertical contrast near the corner provide signals to the MOC Filter, provided that the sustained cells of Level 2 are spatially laid out as indicated in Figure 13. Here the direction of motion is diagonal to the orientational preference of the individual sustained cells. These vertically and horizontally oriented cells contribute to the total signal that codes movement in the diagonally upward direction. So too do cells whose orientation is perpendicular to the diagonal direction. On the other hand, cells whose orientation is colinear with the direction of motion should not be included.

Accordingly, each motion detector is assumed to receive weighted inputs from all sustained-transient cells whose orientations differ from the preferred direction-of-motion by no more than 90 degrees, and whose preferred positions lie within a prescribed distance from the preferred position of the motion detector. As indicated in Figure 14, the long range filter (Level 5, Figure 11) can simultaneously accept motion signals from both the horizontal and vertical edges of the moving corner, despite the gating of one set of signals by transient "luminance increasing" on-cells and gating of another set by "luminance decreasing" off-cells (Level 3, Figure 11). Thus while simultaneous increase and decrease of luminance is logically impossible in an infinitesimal area, and while a too rapid change from increase to decrease may be unresolvable by sustained cells at Level 2, the simultaneous increase and decrease of luminance at different orientations and different locations, but in the same direction, are pooled by the long-range filter. This sort of long-range filtering by MOC Filter complex cells is not the same as the still longer-range cooperative grouping by CC Loop bipole cells.

8. Sustained-Transient Gating before Short-Range Spatial Filtering

In attempting to simulate the 2-D MOC filter on the computer, a computational problem was noticed whose solution was not required in the 1-D simulations of Grossberg and Rudd (1989a, 1992). In particular, Figure 11 shows short-range spatial filtering of sustained cell outputs before gating the result with a transient on-cell or off-cell. The different spatial layouts of sustained cells in Figure 14 made it difficult to select a regular spatial arrangement of transient cells that could be used for all cases. This analysis suggested that each sustained cell is first gated by a transient on-cell and off-cell at its own location before each sustained-transient cell inputs to the several directionally-sensitive spatial filters to which it contributes. While these 2-D simulation problems were being confronted, Grossberg (1990, 1991, this

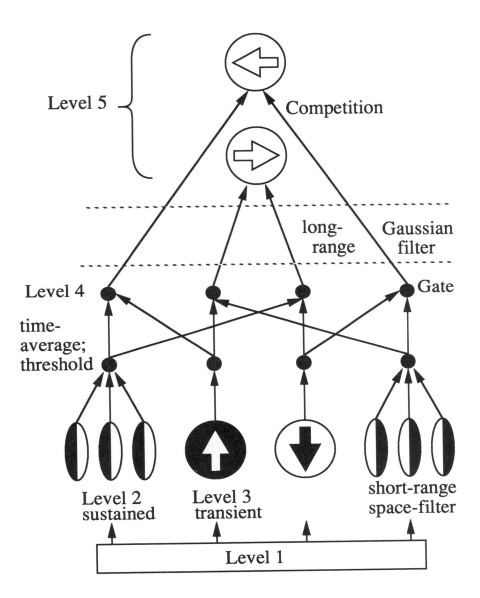

Figure 11: The Motion Oriented Contrast (MOC) Filter: Level 1 registers the input pattern. Level 2 consists of sustained response cells with oriented receptive fields that are sensitive to direction-of-contrast. Level 3 consists of transient response cells with unoriented receptive fields that are sensitive to direction-of-change in the total cell input. Level 4 cells combine sustained cell and transient cell signals to become sensitive to direction-of-motion and sensitive to direction-of-contrast. Level 5 cells combine Level 4 cells to become sensitive to direction-of-motion and insensitive to direction-of-contrast.

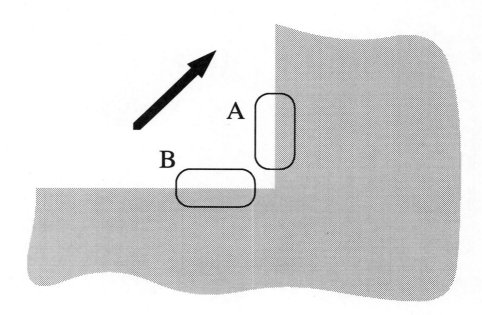

Figure 12: The lower right corner of a horizontally oriented rectangular region of homogeneous high luminance moves diagonally upward and to the right over a background of homogeneous low luminance. In region A a dark-to-light (luminance increasing over time) transition occurs at a vertical edge, while in region B a light-to-dark (luminance decreasing over time) transition occurs at a horizontal edge.

volume) observed that the FM Symmetry principle requires a similar spatial arrangement. We therefore modify the MOC Filter, as in Figure 15, by computing all sustained-transient combinations at each position before combining their outputs via directionally sensitive short-range spatial filters.

9. Endstopping: Generation of a Terminator or Corner Advantage in Motion Signals

Another 2-D MOC Filter refinement involves an *endstopping* operation. The need for this was illustrated in our discussion of the barberpole illusion in Section 1. There we noted that motion signals near terminators or corners tend to be better indicators of object motion than signals generated from a relatively straight interior of a contour. In order for a motion signal at a terminator or corner to be effective, however, it must somehow be translated into a relatively large signal strength, since the region of ambiguous interior motion signals is often larger than the region of unambiguous terminator or corner motion signals.

We suggest that one source of this enhancement involves endstopping the sustained cells and/or the transient cells of the MOC Filter. Many simple cells, identified with cells at

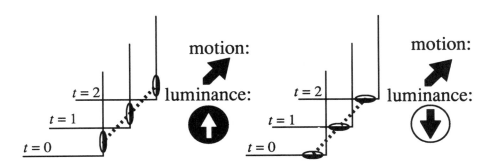

Figure 13: Over three successive time steps, the contours of the rectangle occupy the positions indicated, while luminance increases along the vertical edge and decreases along the horizontal edge. If certain of the sustained cell receptive fields sending inputs to Level 4 of the MOC Filter were arranged as indicated, a diagonal motion signal could be generated from both vertically and horizontally oriented cells, in conjunction with luminance gating signals of opposite signs.

Levels 2 and/or 4 of the MOC Filter, exhibit endstopping (Dreher, 1972). Endstopping is often informally described as consisting of inhibitory zones that occur along the major axis of orientationally tuned cells, beyond the contrast summing region, as indicated in Figure 16a. If this were all that was involved, then the end of a contrast edge that is oblique to a cell's preferred direction and partially overlaps the cell could escape the inhibitory zone and stimulate the cell more than a corresponding edge of the preferred direction (Figure 16b). This problem arises because the inhibition is anisotropic—that is, occurs only from the directions aligned with the cell's orientational axis.

The anatomical substrate for such anisotropy would be more difficult to implement than a scheme of isotropic inhibition among orientationally tuned cells. For such an isotropic inhibitory scheme (Figure 17), the *observed* inhibitory zones for stimulation with edge stimuli would still appear only at the ends of the cell's receptive field, owing to the interaction between isotropic inhibition and oriented receptive fields.

As illustrated in Figure 18a, strong endstopping can attenuate signals from all but the ends of a contour. Strong endstopping reduces the problem of determining motion direction to one of tracking an isolated region of activity in an upward diagonal direction, as in Figure 18b. Weak endstopping, as in Figure 18c, can partially attenuate interior signals, relative to signals at the ends. If endstopping is too weak, however, surviving signals indicating the "locally preferred" rightward horizontal direction can at least partially confound the computation of an upward diagonal object motion, as illustrated in Figure 18d. Marshall (1990) has also invoked end stopping in his explanation of the barberpole illusion, though in a manner different from that outlined here.

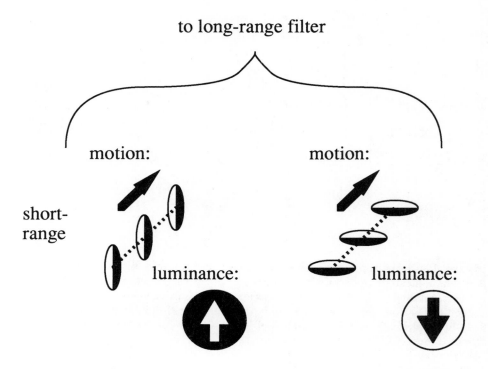

Figure 14: Signals arrive at the long-range filter (to Level 5 of the motion OC Filter) after several rectification and gating operations. Accordingly signals that were gated by both increases and decreases in luminance at (necessarily) different places can be combined into a coherent motion signal.

10. Directional Competition and Boundary Completion

The inward cooperative propagation by bipole cells of motion signals to locations between the strong end reactions can help to overcome this problem as it completes the motion signal along the entire contour (Figure 10). The cooperative feedback from the strong end reactions also leads to inhibition of other directional signals via short-range competition between directions at positions along the contour, as occurs during motion capture (see Section 2). Thus the type of CC Loop interactions among hypercomplex cells and bipole cells that help to select the globally preferred orientations in the Static BCS also help to select the globally preferred directions in the Motion BCS, with the caveat that "orientation" computations are replaced by "direction" computations.

11. 2-D MOC Filter Equations

We have found that several closely related computational realizations of the above heuristics can generate motion fields capable of distinguishing between contour orientation and con-

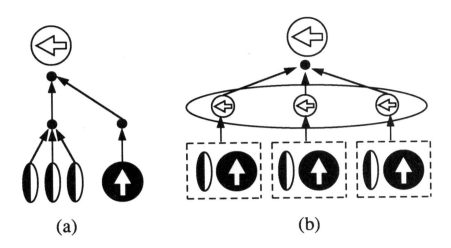

Figure 15: (a) The gating of sustained and transient cells within the MOC Filter of Grossberg and Rudd (1989a) was between several sustained cells (three shown) with aligned receptive field centers and a single transient cell. (b) The revised form of gating is one-to-one, between transient and sustained cells. Subsequently signals from spatially aligned gating cells (indicated by three small arrows) are pooled to form a single Level 4 signal.

tour direction-of-motion. We will here define a model that separates computations involving receptive fields with opposite directions-of-contrast until these are merged by the long-range Gaussian filter. This is the strategy followed by the Grossberg-Rudd model (Figure 11). Analogous properties have been simulated using a variant of this model in which receptive fields with opposite directions-of-contrast are merged at the short-range spatial filter stage (see Section 13). Other variations of the basic computational strategy also work, and might be used by different species *in vivo*. The description that follows is keyed to Figure 19.

Level 1: Stimulus Representation

Let $I_{pq}(t)$ denote the intensity of a time-varying image input at position (p,q) and time t.

Level 2: Oriented Sustained Receptive Fields

Let the output J_{ijk} of a receptive field centered at position (i,j) with orientation k be defined by

$$J_{ijk} = \sum_{pq} A_{ijpq}^{(k)} I_{pq} , \tag{1}$$

where $A_{ijpq}^{(k)}$ defines the value of a Gabor kernel at position (p,q) that is centered at position (i,j) with orientation k. The Gabor kernel is

$$A_{ijpq}^{(k)} = \alpha_A \ \exp\left[-\beta_A(\gamma_A\{B_{pqij}^{(k)}\}^2 + \{C_{pqij}^{(k)}\}^2)\right] \sin(\delta_A C_{pqij}^{(k)}), \tag{2}$$

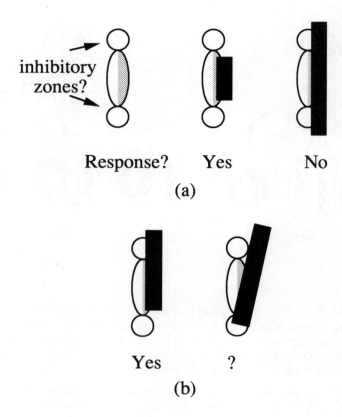

Figure 16: If the inhibitory end-zones of endstopped simple cells were laid out as caricatured in part (a), a cell would be unable to distinguish the termination of a line of its preferred orientation from the continuation of a line of an orientation slightly different from the cell's preferred orientation. In the resulting situation, diagrammed in (b), disinhibition of the (top) inhibitory zone occurs while the strength of response from the central region decreases.

where α_A is a constant that scales the Gabor amplitude, β_A scales the size of the kernel's Gaussian envelope, γ_A specifies the degree of elongation of the receptive field in the preferred orientation, and δ_A controls the frequency of the kernel's sinusoidal modulation. Terms

$$B^{(k)}_{ijpq} = (p-i)\cos(\frac{2\pi k}{K}) - (q-j)\sin(\frac{2\pi k}{K}) \qquad (3)$$

and

$$C^{(k)}_{ijpq} = (p-i)\sin(\frac{2\pi k}{K}) + (q-j)\cos(\frac{2\pi k}{K}) \qquad (4)$$

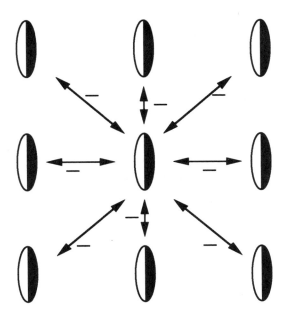

Figure 17: Spatially isotropic inhibition among like-oriented simple cells can generate functional enhancement of activity of oriented receptive fields near line ends or corners, often referred to as "endstopping."

describe the effect of shifting a receptive field centered at position $(0,0)$ to position (i,j), rotating it to orientation k, and evaluating it at position (p,q), given that K is the total number of orientations.

Level 3: Endstopped Sustained Cells

A competition of like-oriented Gabor receptive fields across neighboring positions and orientations leads to sustained cell activations x_{ijk} that are stronger at line ends and corners, as in equation

$$\frac{d}{dt}x_{ijk} = -\alpha_x x_{ijk} + (\beta_x - x_{ijk})J_{ijk} - (\gamma_x + x_{ijk})\sum_{pq\in D_{ij}} J_{pqk}, \tag{5}$$

where D_{ij} is the set of all positions within some radius E of (i,j); that is,

$$D_{ij} = \{(p,q) : (p-i)^2 + (q-j)^2 \leq E^2\}. \tag{6}$$

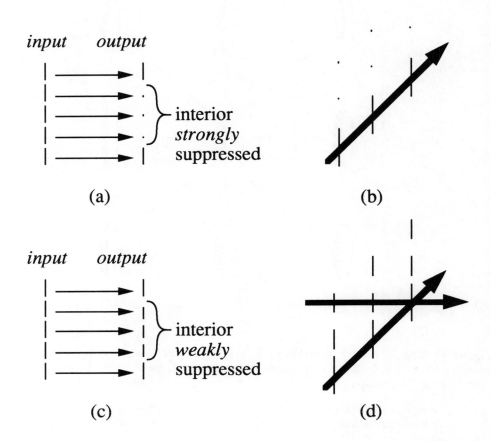

Figure 18: The enhancement of motion signals at line ends and corners can be strong or weak. (a) Strong inhibition can kill interior signals, making the problem of motion segmentation easy. (b) The surviving pools of activity at line ends can be directly tracked. (c) If endstopping is mild enough, however, locally preferred motion directions (perpendicular to edge contrast) survive, as shown in (d).

Level 4: Unoriented Transient Receptive Fields

The transient on-cell $y_{ij}^{(+)}$ responds to increments in the total input to a region F_{ij} surrounding (i, j), as in

Visual Motion Perception 317

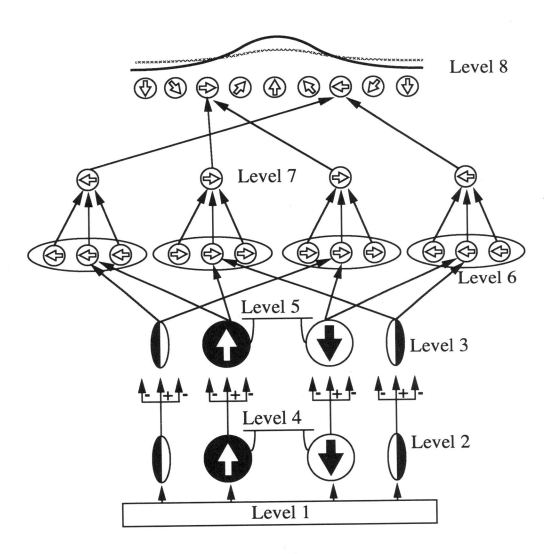

Figure 19: The 2-D MOC Filter embodies a number of variations and extensions of the architecture of the 1-D MOC Filter of Figure 11. See text for details.

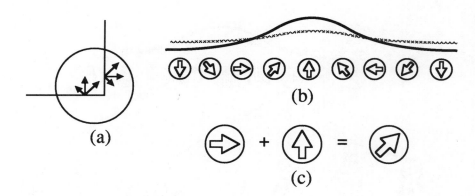

Figure 20: (a) At a corner the Gaussian filter combines signals of several directions. (The circle indicates the spatial domain of the Gaussian filter.) (b) The resolution of motion signals of many directions can be accomplished by a shunting "center-surround" competition with normalization among direction signals at a given location. Such a competition chooses the globally most consistent direction at a location. The solid line indicates the "on-center" and the dashed line the "off surround" of interaction weights across directions. (c) An input distribution with peaks at the rightward and upward directions is transformed into a single peaked distribution pointing upward and rightward by such a network.

$$y_{ij}^{(+)} = \left[\frac{d}{dt} \sum_{pq \in F_{ij}} I_{pq}\right]^+, \tag{7}$$

where F_{ij} is the set of all positions within some radius G of (i,j); that is,

$$F_{ij} = \{(p,q) : (p-i)^2 + (q-j)^2 \leq G^2\} \tag{8}$$

for some radius G. Likewise, transient off-cell $y_{ij}^{(-)}$ responds to decrement in the total input to a region surrounding (i,j), as in

$$y_{ij}^{(-)} = \left[-\frac{d}{dt} \sum_{pq \in F_{ij}} I_{pq}\right]^+ \tag{9}$$

where region F_{ij} is defined above.

Level 5: Center-Surround Transient Cells

In this version of the model, lateral inhibition among transient cells, analogous to the endstopping operation among simple cells, is implemented. The center-surround transient cell activations $Y_{ij}^{(+)}$ and $Y_{ij}^{(-)}$ obey the same type of equation (5) as the endstopped sustained cells; namely,

$$\frac{d}{dt}Y_{ij}^{(+)} = -\alpha_y Y_{ij}^{(+)} + (\beta_y - Y_{ij}^{(+)})y_{ij}^{(+)} - (\gamma_y + Y_{ij}^{(+)}) \sum_{pq \in H_{ij}} y_{pq}^{(+)} \tag{10}$$

and
$$\frac{d}{dt}Y_{ij}^{(-)} = -\alpha_y Y_{ij}^{(-)} + (\beta_y - Y_{ij}^{(-)})y_{ij}^{(-)} - (\gamma_y + Y_{ij}^{(-)}) \sum_{pq \in H_{ij}} y_{pq}^{(-)}, \quad (11)$$

where
$$H_{ij} = \{(p,q) : (p-i)^2 + (q-j)^2 \le L^2\}. \quad (12)$$

The transformation from Level 4 signals to Level 5 signals results in enhanced activity at line ends and corners.

Level 6: Sustained-Transient Simple Cells

As in Level 4 of the Grossberg-Rudd MOC Filter (Figure 11), multiplying outputs from sustained cells and transient cells starts to compute a local measure of motion direction. In particular, sustained cell activations that are sensitive to opposite directions-of-contrast give rise to activations that are sensitive to the same direction-of-motion by being multiplied with transient on-cells and off-cells, respectively. For example, the sustained-transient interaction

$$M_{ijm}^{(+)} = [x_{ijk}]^+ Y_{ij}^{(+)} \quad (13)$$

where
$$m = (k + \frac{K}{4}) \pmod{K}, \quad (14)$$

with K an integer multiple of 4, and the sustained-transient interaction

$$M_{ijm}^{(-)} = [x_{ij\hat{k}}]^+ Y_{ij}^{(-)} \quad (15)$$

where
$$\hat{k} = (k + \frac{K}{2}) \pmod{K}, \quad (16)$$

are both maximally sensitive to oriented contours of orientation k moving in direction m, even though the activations x_{ijk} and $x_{ij\hat{k}}$ sense opposite directions-of-contrast in an orientation that is perpendicular to m.

These quantities are time-averaged by the cells at which the sustained-transient interactions occur. Let $z_{ijm}^{(+)}$ and $z_{ijm}^{(-)}$ compute the activities of the sustained-transient cells which receive the inputs $M_{ijm}^{(+)}$ and $M_{ijm}^{(-)}$, respectively. Then

$$\frac{d}{dt}z_{ijm}^{(+)} = -\alpha_z z_{ijm}^{(+)} + M_{ijm}^{(+)} \quad (17)$$

and
$$\frac{d}{dt}z_{ijm}^{(-)} = -\alpha_z z_{ijm}^{(-)} + M_{ijm}^{(-)}. \quad (18)$$

For simplicity, the time averaging in (17) and (18) was computed in discrete time steps using equations

$$z_{ijm}^{(+)}(t) = \sum_{\tau=t-T}^{t} M_{ijm}^{(+)}(\tau)\zeta^{t-\tau}, \quad (19)$$

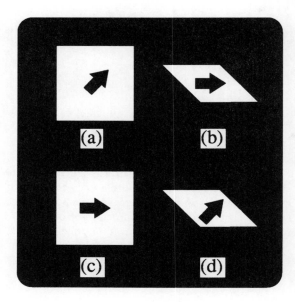

Figure 21: Simulations of the response of the 2-D MOC Filter to the motions of two simple figures are shown in subsequent figures. These figures are: (a) a square moving diagonally upward and to the right, (b) a parallelogram moving horizontally rightward, (c) a square moving horizontally rightward, and (d) a parallelogram moving diagonally up and to the right.

and

$$z_{ijm}^{(-)}(t) = \sum_{\tau=t-T}^{t} M_{ijm}^{(-)}(\tau)\zeta^{t-\tau}, \qquad (20)$$

where $0 < \zeta < 1$ and the number of time steps T was chosen large enough to provide a good approximation to (17) and (18), respectively. Comparison of the continuous and discrete time equations shows that $\zeta = \exp(-\alpha_z)$.

Level 7: Short-Range Spatial Filter: Pooling of Orientation Detectors into Local Direction Detectors

Each activation $z_{ijm}^{(+)}$ and $z_{ijm}^{(-)}$ is derived from an oriented receptive field whose orientation k is perpendicular to m. The next operation pools activations that are sensitive to the same direction-of-contrast using a short-range spatial filter. The spatial filter pools activations that lie along motion trajectories with different preferred directions-of-motion m. Activations $z_{ijn}^{(+)}$ or $z_{ijn}^{(-)}$ are accepted whose directions n are not too different from the trajectory direction m. This pooling operation exploits the relative advantage at line ends and corners achieved by endstopping to compute trajectory signals that begin to disambiguate a contour's direction-of-motion from its orientation.

Figure 22: A representation of a frame from an input sequence presented to the 2-D MOC Filter. A light square moves diagonally up and to the right against a dark background. The resolution of the image used in the simulation was 128 × 128 pixels; the distance between nearest receptive field centers for network activities depicted in Figures 23–30 is four pixel units.

The short-range spatial kernel $N_{ijpq}^{(m,n)}$ favors a motion trajectory with direction-of-motion m that is centered at position (i,j). It separately pools activations $z_{pqn}^{(+)}$ or $z_{pqn}^{(-)}$ with a weight that depends upon how close n is to m and (p,q) is to (i,j), as in

$$u_{ijm}^{(+)} = \sum_{pqn} N_{ijpq}^{(m,n)} z_{pqn}^{(+)} \tag{21}$$

and

$$u_{ijm}^{(-)} = \sum_{pqn} N_{ijpq}^{(m,n)} z_{pqn}^{(-)}, \tag{22}$$

where

$$N_{ijpq}^{(m,n)} = P_{ijpq}^{(m)} \left[\cos(2\pi(m-n)K^{-1})\right]^+ \tag{23}$$

and

$$P_{ijpq}^{(m)} = \alpha_p \exp\left[-\beta_p(\gamma_p\{Q_{ijpq}^{(m)}\}^2 + \{R_{ijpq}^{(m)}\}^2)\right] \cos(\delta_p R_{ijpq}^{(m)}). \tag{24}$$

In (24), just as in (2), α_p is a constant that scales the kernel amplitude, β_p determines the size of the kernel envelope, γ_p specifies the degree of kernel elongation in the preferred

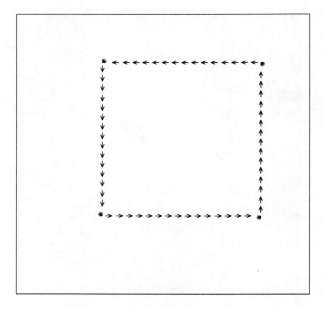

Figure 23: The output of the sustained cells of Level 2 in response to the moving square of Figure 22.

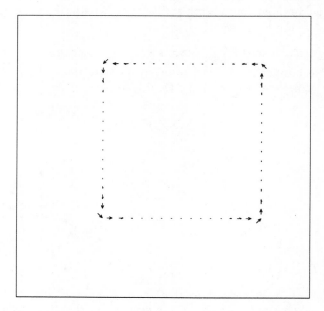

Figure 24: The output of the end-stopped sustained cells of Level 3 in response to the moving square of Figure 22.

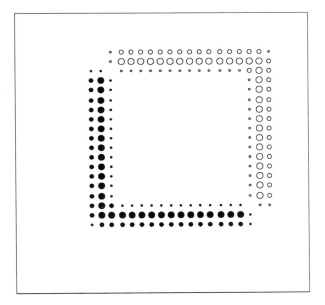

Figure 25: The output of the transient cells of Level 4 in response to the moving square of Figure 22.

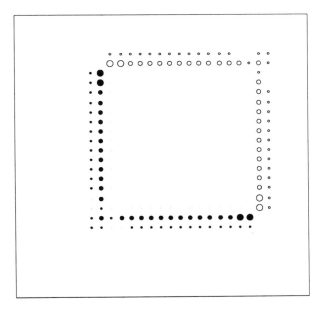

Figure 26: The output of the contrast-enhanced transient cells of Level 5 in response to the moving square of Figure 22.

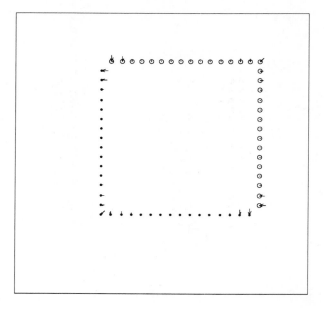

Figure 27: The output of the sustained-transient simple cells of Level 6 in response to the moving square of Figure 22.

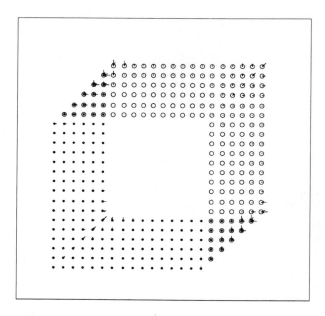

Figure 28: The pattern formed by the temporal smearing of the responses of sustained-transient simple cells of Level 6 to the moving square of Figure 22.

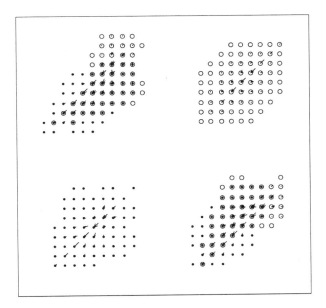

Figure 29: The output of the short-range spatial filter cells of Level 7 in response to the moving square of Figure 22.

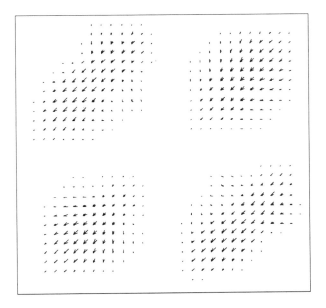

Figure 30: The output of the competitive cells of Level 8 in response to the moving square of Figure 22.

direction m, and δ_p specifies the frequency of the kernel's cosine modulation. The values

$$Q_{pqij}^{(m)} = (p-i)\cos(\frac{2\pi m}{K}) - (q-j)\sin(\frac{2\pi m}{K}) \tag{25}$$

and

$$R_{pqij}^{(m)} = (p-i)\sin(\frac{2\pi m}{K}) + (q-j)\cos(\frac{2\pi m}{K}) \tag{26}$$

describe the effect of shifting a receptive field centered at position $(0,0)$ to position (i,j), rotating it to point in direction m, and evaluating it at position (p,q).

The cell activations $u_{ijm}^{(+)}$ and $u_{ijm}^{(-)}$ are computed using algebraic equations (21) and (22) rather than the differential equations (17) and (18) used to compute $z_{ijm}^{(+)}$ and $z_{ijm}^{(-)}$, because $u_{ijm}^{(+)}$ and $u_{ijm}^{(-)}$ are assumed to respond much more quickly to their inputs than $z_{ijm}^{(+)}$ and $z_{ijm}^{(-)}$. In particular, the kernel $N_{ijpq}^{(m,n)}$ accumulates evidence for motion in direction m along a trajectory through position (i,j). The persistence of $z_{pqn}^{(+)}$ and $z_{pqn}^{(-)}$ activations as they temporally decay, together with the spatial anisotropy of the kernel, begin to overcome the uncertainties that arise from the aperture problem. Kernel anisotropy is hypothesized to arise, at least in part, in response to experiences with the trajectories of moving contours during an early phase of brain development. Such trajectories are, with high probability, approximately straight over sufficiently small regions.

The cell activations $u_{ijm}^{(+)}$ and $u_{ijm}^{(-)}$ both detect an estimate of direction-of-motion m at position (i,j), but they are sensitive to opposite directions-of-contrast. The long-range spatial filter combines signals that are sensitive to opposite directions-of-contrast, but the same direction-of-motion, to compute a more accurate estimate of motion direction, as indicated in Figure 17. The output signals $U_{ijm}^{(+)}$ and $U_{ijm}^{(-)}$ from Level 7 to this long-range filter are rectified versions of the Level 7 activations, as in

$$U_{ijm}^{(+)} = [u_{ijm}^{(+)} - \eta]^+ \tag{27}$$

and

$$U_{ijm}^{(-)} = [u_{ijm}^{(-)} - \eta]^+. \tag{28}$$

Level 8: Long-Range Spatial Filter and Directional Competition

The trajectory responses $U_{ijm}^{(+)}$ and $U_{ijm}^{(-)}$ are next combined across space and direction-of-contrast to generate a response ν_{ijm} that is sensitive to the consensus within a region surrounding position (i,j) of how much evidence there exists for motion in direction m. Thus we let

$$\nu_{ijm} = \sum_{pq \in W_{ij}} (U_{pqm}^{(+)} + U_{pqm}^{(-)})\exp\{-\alpha_\nu[(p-i)^2 + (q-j)^2]\}, \tag{29}$$

where

$$W_{ij} = \{(p,q) : (p-i)^2 + (q-j)^2 \le X^2\}. \tag{30}$$

Symbol	Figures	Level	Cell Type
↓	23, 32	2	oriented sustained
↓	24, 33	3	end-stopped oriented sustained
●	25, 34	4	unoriented transient (luminance decreasing)
○	25, 34	4	unoriented transient (luminance increasing)
●	26, 35	5	center-surround transient (luminance decreasing)
○	26, 35	5	center-surround transient (luminance increasing)
↓●	27, 36	6	sustained-transient simple (luminance decreasing)
↓○	27, 36	6	sustained-transient simple (luminance increasing)
↯●	28, 37	6	time-average of sustained-transient simple (luminance decreasing)
↯○	28, 37	6	time-average of sustained-transient simple (luminance increasing)
↯●	29, 38	7	short-range spatial filter (luminance decreasing)
↯○	29, 38	7	short-range spatial filter (luminance increasing)
↓	30, 39, 41, 43	8	motion hypercomplex

Table 1.

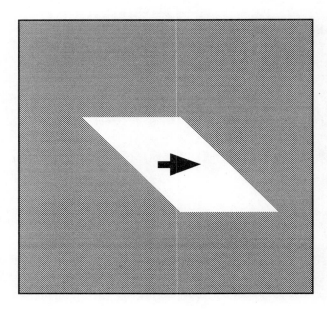

Figure 31: A representation of a frame from an input sequence presented to the 2-D MOC Filter. A light parallelogram moves rightward against a dark background. The resolution of the image used in the simulation was 128 × 128 pixels; the distance between nearest receptive field centers for network activities depicted in Figures 32–39 is four pixel units.

The output of the long-range filter undergoes a final competition to choose a consensual direction among all signals which have been grouped at a spatial location (Figure 20). This competition is in the form of a "center-surround" organization in "direction space," as given by

$$\frac{d}{dt}w_{ijm} = -\alpha_w w_{ijm} + (\beta_w - w_{ijm})\sum_n \nu_{ijn}\gamma_{nm} - (\delta_w + w_{ijm})\sum_n \nu_{ijn}\epsilon_{nm} \quad (31)$$

where

$$\gamma_{nm} = \gamma \exp[-\mu_w(m-n)^2] \quad (32)$$

and

$$\epsilon_{nm} = \epsilon \exp[-\nu_w(m-n)^2]. \quad (33)$$

The competition stage at Level 8 is not necessary to generate consistent direction signals for simple, noise-free inputs. Snowden (1989, 1990) has reported psychophysical and physiological evidence for such shunting inhibition among directional signals, for situations in which conflicting signals must be resolved. Williams and Phillips (1987) and Watson and Ahumada (1985) have proposed a similar direction-averaging mechanism.

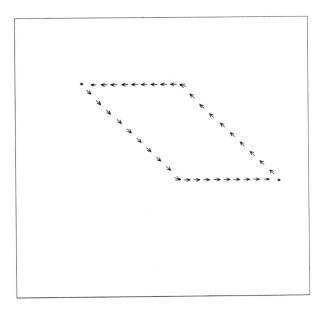

Figure 32: The output of the sustained cells of Level 2 in response to the moving parallelogram of Figure 31.

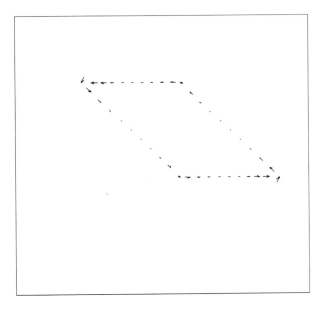

Figure 33: The output of the end-stopped sustained cells of Level 4 in response to the moving parallelogram of Figure 31.

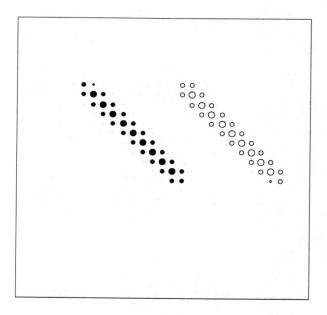

Figure 34: The output of the transient cells of Level 4 in response to the moving parallelogram of Figure 31.

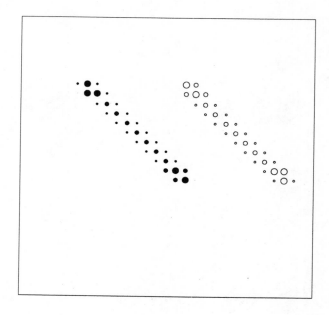

Figure 35: The output of the contrast-enhanced transient cells of Level 5 in response to the moving parallelogram of Figure 31.

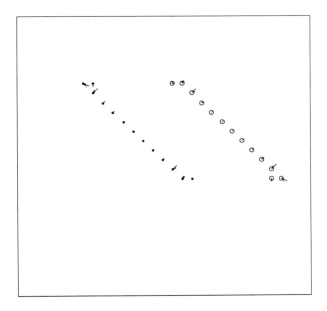

Figure 36: The output of the sustained-transient simple cells of Level 6 in response to the moving parallelogram of Figure 31.

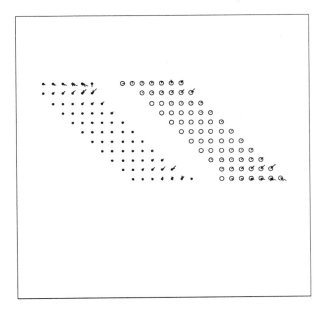

Figure 37: The pattern formed by the temporal smearing of the responses of sustained-transient simple cells of Level 6 to the moving parallelogram of Figure 31.

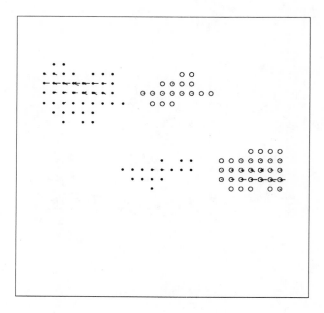

Figure 38: The output of the short-range spatial filter cells of Level 7 in response to the moving parallelogram of Figure 31.

Figure 39: The output of the competitive cells of Level 8 in response to the moving parallelogram of Figure 31.

Figure 40: A representation of a frame from an input sequence presented to the 2-D MOC Filter. A light square moves horizontally to the right against a dark background. The resolution of the image used in the simulation is 128 × 128 pixels; the distance between nearest receptive field centers for network activities depicted in Figure 41 is four pixel units.

12. Computer Simulations: Distinguishing Motion Direction from Boundary Orientation

Computer simulations were carried out on simple moving forms to illustrate how the MOC Filter can compute an accurate measure of motion direction even if the bounding contours of the form have orientations that are not perpendicular to the direction of motion. Four illustrative moving forms are analysed below: a rectangle moving diagonally upward; a rectangle moving to the right; a rhombus moving diagonally upward; and a parallelogram moving to the right (Figure 21). For the square moving diagonally, the resultant activation pattern at each stage of the MOC Filter is shown in order to aid the reader's intuition. The graphics conventions used are shown in Table 1.

Consider the case of a light square moving diagonally up and to the right on a dark background, as depicted in Figure 22. Figure 23 displays the response of the oriented sustained detectors of Level 2; the parameters controlling the scale of the detectors are such that only a single row or column is activated around the perimeter of the square, although several orientations are active at each position. Figure 24 displays the response of the end-stopped, oriented sustained detectors of Level 3. Note the attenuated response along the interiors of the segments bounding the square, and the enhanced response at corners. Figure 25 displays the response of increasing and decreasing transient detectors of Level 4. Note that the leading and trailing corners (upper right and lower left, respectively) have strong activity,

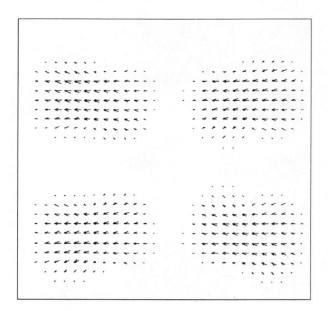

Figure 41: The output of the competitive cells of Level 8 in response to the moving square of Figure 40.

because the axis they define is in the direction of motion, while the other two corners have attenuated activity. Figure 26 displays the result of contrast enhancing competition of Level 5 transient detectors. Note that this competition occurs independently in the "increasing" and "decreasing" channels, permitting the greatest enhancement of activity at the corners where activity was weakest in Figure 25.

Figures 31–39 display a sequence of transformations corresponding to those of Figures 23–30, but this time for the case of a rhombus moving horizontally to the right, as depicted in Figure 31. The transformation from the output of oriented sustained cells of Level 2, shown in Figure 32, to the output of end-stopped cells of Level 3, shown in Figure 33, changes the relative distribution of active orientations at corners. Likewise, the transformation from Level 4 transient cells to Level 5 contrast-enhanced transient cells, shown in Figures 34 and 35 respectively, helps to create a stronger pool of activity at the ends of the diagonal lines, as was discussed with reference to the barberpole illusion in Section 2. The output of transient-sustained simple cells of Level 6, shown in Figure 36, contains an assortment of active local direction signals. The spatial trail formed by the smearing of those signals in time, shown in Figure 37, enables the short-range spatial filters of Level 7, shown in Figure 38, to more accurately register the prevailing direction of motion. The output of Level 8, shown in Figure 39, further smooths the signals over space, while sharpening their directional distribution. Note that the resulting pattern of activity is stronger at the 45° corners than at the 135° corners, and that the directional pattern at the latter is not symmetric about the horizontal. It may be that such effects underlie some of the anomalous motion percepts reported for

Figure 42: A representation of a frame from an input sequence presented to the 2-D MOC Filter. A light parallelogram moves diagonally up and to the right against a dark background. The resolution of the image used in the simulation was 128 × 128 pixels; the distance between nearest receptive field centers for network activities depicted in Figure 43 is four pixel units.

motions of curves such as those employed by Nakayama and Silverman (1988). Changing parameters in the simulation to include longer spatial filters at Level 7 would in any case reduce this effect and generate a more nearly horizontal pattern of responses.

For comparison, we have included the response at Level 8 to the cases where a horizontal square moves rightward (Figures 40 and 41) and where a parallelogram moves diagonally (Figures 42 and 43).

13. A Related MOC Filter

Similar computational properties are found if the mask responses in (1) are thresholded, as in

$$J_{ijk} = \left[\sum_{pq} A^{(k)}_{ijpq} I_{pq} - \epsilon_A\right]^+, \tag{31}$$

and the terms $M^{(+)}_{ijm}$ in (13) and $M^{(-)}_{ijm}$ in (15) are combined into a single time-averaged sustained-transient cell type z_{ijm}, instead of the separate streams (19) and (20). Then

$$z_{ijm}(t) = \sum_{\tau=t-T}^{t} (M^{(+)}_{ijm}(\tau) + M^{(-)}_{ijm}(\tau))\zeta^{t-\tau} \tag{32}$$

Figure 43: The output of the competitive cells of Level 8 in response to the moving parallelogram of Figure 42.

and

$$u_{ijm} = \sum_{pqn} \overline{N}^{(m,n)}_{ijpq} z_{pqn}, \qquad (33)$$

where

$$\overline{N}^{(m,n)}_{ijpq} = P^{(m)}_{ijpq} \{[\cos(2\pi(m-n)K^{-1})]^+\}^p, \qquad (34)$$

function P_{ijpq} is defined in (23), and parameter p in (34) enables the smoothing across directions n to be made more or less broad within a 180-degree span. Using this approach, the threshold η in (27) and (28) is not needed, because the masks (31) are already thresholded, and the functions $M^{(+)}_{ijm}$ and $M^{(-)}_{ijm}$ combine to generate a more central directional tendency at an early processing stage. Then

$$\nu_{ijm} = \sum_{pq \epsilon W} u_{pqm} \, \exp\{-\alpha_\nu[(p-i)^2 + (q-j)^2]\} \qquad (35)$$

suffices instead of (29).

14. Perception of Moving-Form-in-Depth: The V1 → V2 → MT Pathway

The Motion BCS circuitry suggests the types of processing stages, and their ordering, that are needed to effectively process moving images. As such, the circuit suggests experimental tests of the cell types and interactions in the $V1 \rightarrow MT$ cortical processing stream.

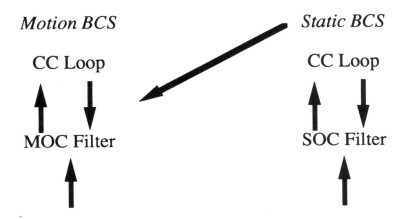

Figure 44: Model analog of $V1 \to V2 \to MT$ pathway: Stereo-sensitive emergent segmentations from the Static CC Loop help to select the depthfully correct combinations of motion signals in the MOC Filter.

The theory also helps to understand why an indirect, $V1 \to V2 \to MT$, cortical pathway from V1 to MT exists in addition to the direct $V1 \to MT$ pathway (DeYoe and van Essen, 1988) (Figure 44). As noted in Section 7, outputs from the MOC Filter sacrifice a measure of orientational specificity in order to effectively process direction-of-motion. However, precisely oriented binocular matches are important in the selection of cortical cells that are tuned to the correct binocular disparities (von der Heydt, Hänny, and Dürsteler, 1981). The Static BCS can carry out such precise oriented matches; the Motion BCS cannot. This fact suggests that a pathway from the Static BCS to the Motion BCS exists in order to help the Motion BCS to generate its motion segmentations at correctly calibrated depths.

Such a pathway needs to arise after the level of BCS processing at which cells capable of binocular fusion are chosen and binocularly rivalrous cells are suppressed. This occurs within the hypercomplex cells and bipole cells of the Static BCS (Grossberg, 1987b; Grossberg and Marshall, 1989), hence within the model analog of prestriate cortical area V2 (Figure 44). Thus the existence of a pathway from V2 and/or V4 to MT is consistent with the different functional roles of the Static BCS and the Motion BCS.

According to this reasoning, the $V2 \to MT$ pathway should occur at a processing stage prior to the one at which several orientations are pooled into a single direction-of-motion within each spatial scale. Thus, the pathway ends in the MOC Filter at a stage no later than Level 4 in Figure 11, or correspondingly, Level 7 in Figure 19. Such a pathway would join like orientations within like spatial scales between the Static BCS and the Motion BCS. It could thereby enhance the activation within the Motion BCS of those spatial scales and orientations that are binocularly fused within the Static BCS.

REFERENCES

Adelson, E.H. and Movshon, J.A. (1982). Phenomenal coherence of moving visual patterns. *Nature*, **300**, 523–525.

Albright, T.D., Desimone, R., and Gross, C.G. (1984). Columnar organization of directionally sensitive cells in visual area MT of the macaque. *Journal of Neurophysiology*, **51**, 16–31.

Anderson, S.J. and Burr, D.C. (1987). Receptive field size of human motion detection units. *Vision Research*, **27**, 621–635.

Anstis, S.M. and Mather, G. (1985). Effects of luminance and contrast on direction of ambiguous apparent motion. *Perception*, **14**, 167–179.

Anstis, S.M. and Ramachandran, V.S. (1987). Visual inertia in apparent motion. *Vision Research*, **27**, 755–764.

Braddick, O.J. (1974). A short range process in apparent motion. *Vision Research*, **14**, 519–527.

Burr, D.C., Ross, J., and Morrone, M.C. (1986). Smooth and sampled motion. *Vision Research*, **26**, 643–652.

Carpenter, G.A., Grossberg, S., and Mehanian, C. (1989). Invariant recognition of cluttered scenes by a self-organizing ART architecture: CORT-X boundary segmentation. *Neural Networks*, **2**, 169–181.

Cohen, M.A. and Grossberg, S. (1984). Neural dynamics of brightness perception: Features, boundaries, diffusion, and resonance. *Perception and Psychophysics*, **36**, 428–456.

Derrington, A. and Suero, M. (1991). Motion of complex patterns is computed from the perceived motions of their components. *Vision Research*, **31**(1), 139–150.

DeValois, R.L., Albrecht, D.G., and Thorell, L.G. (1982). Spatial frequency selectivity of cells in macaque visual cortex. *Vision Research*, **22**, 545–559.

DeYoe, E.A. and van Essen, D.C. (1988). Concurrent processing streams in monkey visual cortex. *Trends in Neuroscience*, **11**, 219–226.

Dreher, B. (1972). Hypercomplex cells in the cat's striate cortex. *Investigative Ophthalmology*, **11**, 355–356.

Eckhorn, R., Bauer, R., Jordan, W., Brosch, M., Kruse, W., Munk, M., and Reitboeck, H.J. (1988). Coherent oscillations: A mechanism of feature linking in the visual cortex? *Biological Cybernetics*, **60**, 121–130.

Eskew, R.T., Jr., Stromeyer, C.F., III, Picotte, C.J., and Kronauer, R.E. (1991). Detection uncertainty and the facilitation of chromatic detection by luminance contours. *Journal of the Optical Society of America A*, **8**, 394–403.

Foster, K.H., Gaska, J.P., Nagler, M., and Pollen, D.A. (1985). Spatial and temporal frequency selectivity of neurons in visual cortical areas V1 and V2 of the macaque monkey. *Journal of Physiology*, **365**, 331–363.

Gray, C.M., Konig, P., Engel, A.K., and Singer, W. (1989). Oscillatory responses in cat visual cortex exhibit inter-columnar synchronization which reflects global stimulus properties. *Nature*, **338**, 334–337.

Grossberg, S. (1973). Contour enhancement, short-term memory, and constancies in reverberating neural networks. *Studies in Applied Mathematics*, **52**, 217–257.

Grossberg, S. (1982). **Studies of mind and brain: Neural principles of learning, perception, development, cognition, and motor control.** Boston: Reidel Press.

Grossberg, S. (1987a). Cortical dynamics of three-dimensional form, color, and brightness perception, I: Monocular theory. *Perception and Psychophysics*, **41**, 87–116.

Grossberg, S. (1987b). Cortical dynamics of three-dimensional form, color, and brightness perception, II: Binocular theory. *Perception and Psychophysics*, **41**, 117–158.

Grossberg, S. (Ed.) (1987c). **The adaptive brain, II: Vision, speech, language, and motor control.** Amsterdam: Elsevier/North-Holland.

Grossberg, S. (1990). Neural facades: Visual representations of static and moving form-and-color-and-depth. *Mind and Language*, **5**, 411–456.

Grossberg, S. (1991). Why do parallel cortical systems exist for the perception of static form and moving form? *Perception and Psychophysics*, **49**, 117–141.

Grossberg, S. and Marshall, J. (1989). Stereo boundary fusion by cortical complex cells: A system of maps, filters, and feedback networks for multiplexing distributed data. *Neural Networks*, **2**, 29–51.

Grossberg, S. and Mingolla, E. (1985a). Neural dynamics of form perception: Boundary completion, illusory figures, and neon color spreading. *Psychological Review*, **92**, 173–211.

Grossberg, S. and Mingolla, E. (1985b). Neural dynamics of perceptual grouping: Textures, boundaries, and emergent segmentations. *Perception and Psychophysics*, **38**, 141–171.

Grossberg, S. and Mingolla, E. (1987). Neural dynamics of surface perception: Boundary webs, illuminants, and shape-from-shading. *Computer Vision, Graphics, and Image, Processing*, **37**, 116–165.

Grossberg, S., Mingolla, E., and Todorović, D. (1989). A neural network architecture for preattentive vision. *IEEE Transactions on Biomedical Engineering*, **36**, 65–84.

Grossberg, S. and Rudd, M.E. (1989a). A neural architecture for visual motion perception: Group and element apparent motion. *Neural Networks*, **2**, 421–450.

Grossberg, S. and Rudd, M.E. (1989b). Neural dynamics of visual motion perception: Group and element apparent motion. *Investigative Ophthalmology Supplement*, **30**, 73.

Grossberg, S. and Rudd, M.E. (1992). Cortical dynamics of visual motion perception: Short-range and long-range motion. *Psychological Review*, in press.

Grossberg, S. and Somers, D. (1991). Synchronized oscillations during cooperative feature linking in a cortical model of visual perception. *Neural Networks*, **4**, 453–466.

Grossberg, S. and Todorović, D. (1988). Neural dynamics of 1-D and 2-D brightness perception: A unified model of classical and recent phenomena. *Perception and Psychophysics*, **43**, 241–277.

Heggelund, P. (1981). Receptive field organization of simple cells in cat striate cortex. *Experimental Brain Research*, **42**, 89–107.

Hubel, D.H. and Wiesel, T.N. (1962). Receptive fields, binocular interaction and functional architecture in the cat's visual cortex. *Journal of Physiology*, **160**, 106–154.

Hubel, D.H. and Wiesel, T.N. (1968). Receptive fields and functional architecture of monkey striate cortex. *Journal of Physiology*, **195**, 215–243.

Hubel, D.H. and Wiesel, T.N. (1977). Functional architecture of macaque monkey visual cortex. *Proceedings of the Royal Society of London B*, **198**, 1–59.

Humphreys, G.W., Quinlan, P.T., and Riddoch, M.J. (1989). Grouping processes in visual search: Effects with single- and combined-feature targets. *Journal of Experimental Psychology: General*, **118**, 258–279.

Kellman, P.J. and Shipley, T.F. (1991). A theory of visual interpolation in object perception. *Cognitive Psychology*, **23**, 141–221.

Kennedy, J.M. (1979). Subjective contours, contrast, and assimilation. In C. F. Nodine and D. F. Fisher (Eds.) **Perception and pictorial representation**. Ney York: Praeger.

Lappin, J.S. and Bell, H.H. (1976). The detection of coherence in moving random-dot patterns. *Vision Research*, **16**, 161–168.

Livingstone, M.S. and Hubel, D.H. (1987). Psychophysical evidence for separate channels for the perception of form, color, movement, and depth. *Journal of Neuroscience*, **7**, 3416–3468.

Marr, D. and Ullman, S. (1981). Directional selectivity and its use in early visual processing. *Proceedings of the Royal Society of London B*, **211**, 151–180.

Marshall, J.A. (1990). Self-organizing neural networks for perception of visual motion. *Neural Networks*, **3**, 45–74.

Maunsell, J.H.R. and van Essen, D.C. (1983). Response properties of single units in middle temporal visual area of the macaque. *Journal of Neurophysiology*, **49**, 1127–1147.

Meyer, G.E. and Dougherty, T. (1987). Effects of flicker-induced depth on chromatic subjective contours. *Journal of Experimental Psychology: Human Perception and Performance*, **13**(3), 353–360.

Mikaelian, H.H., Linton, M.J., and Phillips, M. (1990). Orientation-specific luminance aftereffects. *Perception and Psychophysics*, **47**(6), 575–582.

Mingolla, E., Todd, J.T., and Norman, F. (1992). The perception of globally coherent motion. *Vision Research*, in press.

Movshon, J.A., Adelson, E.H., Gizzi, M.S., and Newsome, W.T. (1985). The analysis of moving visual patterns. In C. Chagas, R. Gattas, and C. Gross (Eds.) **Pattern recognition mechanisms**. New York: Springer-Verlag.

Nakayama, K. and Silverman, G.H. (1984). Temporal and spatial characteristics of the upper displacement limit for motion in random dots. *Vision Research*, **24**, 293–299.

Nakayama, K. and Silverman, G.H. (1985). Detection and discrimination of sinusoidal grating displacements. *Journal of the Optical Society of America*, **2**, 267–273.

Nakayama, K. and Silverman, G.H. (1988). The aperture problem, II: Spatial integration of velocity information along contours. *Vision Research*, **28**(6), 747–753.

Newsome, W.T., Gizzi, M.S., and Movshon, J.A. (1983). Spatial and temporal properties of neurons in macaque MT. *Investigative Ophthalmology and Visual Sciences*, **24**, 106.

Paradiso, M. and Nakayama, K. (1991). Brightness perception and filling-in. *Vision Research*, **31**(7/8), 1221–1236.

Peterhans, E. and von der Heydt, R. (1989). Mechanisms of contour perception in monkey visual cortex II. Contours bridging gaps. *Journal of Neuroscience,* **9**(5), 1749–1763.

Pollen, D.A., Gaska, J.P., and Jacobsen, L.D. (1989). Physiological constraints on models of visual cortical function. In R.M.J. Cotterill (Ed.), **Models of brain function**. Cambridge: Cambridge University Press, pp. 115–135.

Prazdny, K. (1984). On the perception of glass patterns. *Perception,* **13**, 469–478.

Prinzmetal, W. (1990). Neon colors illuminate reading units. *Journal of Experimental Psychology: Human Perception and Performance,* **16**(3), 584–597.

Prinzmetal, W. and Boaz, K. (1989). Functional theory of illusory conjunctions and neon colors. *Journal of Experimental Psychology: General,* **118**(2), 165–190.

Ramachandran, V.S. (1981). **Perception of apparent movement: Studies in cognitive sciences, Volume II**. Irvine: University of California School of Social Sciences.

Ramachandran, V.S. and Inada, V. (1985). Spatial phase and frequency in motion capture of random-dot patterns. *Spatial Vision,* **1**, 57–67.

Shapley, R. and Gordon, J. (1985). Nonlinearity in the perception of form. *Perception and Psychophysics,* **37**, 84–88.

Snowden, R.J. (1989). Motions in orthogonal directions are mutually suppressive. *Journal of the Optical Society of America,* **6**(7), 1096–1101.

Snowden, R.J. (1990). Suppressive interactions between moving patterns: Role of velocity. *Perception and Psychophysics,* **47**(1), 74–78.

Sutter, A., Beck, J., and Graham, M. (1989). Contrast and spatial variables in texture segregation: Testing a simple spatial-frequency channels model. *Perception and Psychophysics,* **46**(4), 312–332.

Tanaka, M., Lee, B.B., and Creutzfeldt, O.D. (1983). Spectral tuning and contour representation in area 17 of the awake monkey. In J. D. Mollon and L. T. Sharpe (Eds.), **Colour vision**. New York: Academic Press.

Todd, J.T. and Akerstrom, R.A. (1987). Perception of three-dimensional form from patterns of optical texture. *Journal of Experimental Psychology: Human Perception and Performance,* **13**(2), 242–255.

von der Heydt, R., Hänny, P., and Dürsteler, M.R. (1981). The role of orientation disparity in stereoscopic perception and the development of binocular correspondence. In E. Grastyán and P. Molnár (Eds.), **Advances in physiological sciences, volume 16: Sensory functions**. Elmsford, NY: Pergamon Press.

von der Heydt, R., Peterhans, E., and Baumgartner, G. (1984). Illusory contours and cortical neuron responses. *Science,* **224**, 1260–1262.

Wallach, H. (1976). **On perception**. New York: Quadrangle/The New York Times Book Company.

Watanabe, T. and Sato, T. (1989). Effects of luminance contrast on color spreading and illusory contour in the neon color spreading effect. *Perception and Psychophysics,* **45**(4), 427–430.

Watanabe, T. and Takeichi, H. (1990). The relation between color spreading and illusory contours. *Perception and Psychophysics,* **47**(5), 457–467.

Watson, A.B. and Ahumada, A.J. (1985). Model of human visual-motion sensing. *Journal of the Optical Society of America A*, **2**, 322–342.

Welch, L. (1989). The perception of moving plaids reveals two motion-processing stages. *Nature* **337**, 734–736.

Williams, D. and Phillips, G. (1987). Cooperative phenomena in the perception of motion direction. *Journal of the Optical Society of America A*, **4**(5), 878–885.

Williams, D. and Sekuler, R. (1984). Coherent global motion percepts from stochastic local motions. *Vision Research*, **24**, 55–62.

Wilson, H.R. (1991). A psychophysically motivated model for two-dimensional motion perception. *Investigative Ophthalmology and Visual Science*, **32**(4), 893. Abstract #1101.

Zeki, S.M. (1974a). Functional organization of a visual area in the posterior bank of the superior temporal sulcus of the rhesus monkey. *Journal of Physiology (London)*, **236**, 549–573.

Zeki, S.M. (1974b). Cells responding to changing image size and disparity in the cortex of the rhesus monkey. *Journal of Physiology (London)*, **242**, 827–841.

NEURAL CIRCUITS FOR VISUAL ATTENTION IN THE PRIMATE BRAIN

Robert Desimone

Abstract

The cortical mechanism for object recognition involves a series of visual areas along a pathway from the occipital to the temporal cortex. At some point along this pathway, most likely starting in visual area V4, spatially directed attention plays a major role in determining how cells process information. When more than one stimulus is located within a neuron's receptive field in either area V4 or the inferior temporal (IT) cortex, the neuron's response will be determined primarily by the stimulus at the locus of attention. The influence of the ignored stimulus will be filtered out. This filtering of unwanted information is under the control of a number of cortical and subcortical attentional control structures which appear to be closely associated with the oculomotor system. Within the attentional control system, there appears to be a competitive interaction among all locations within the visual field. Competition within the system is biased either by external instruction to attend to one location or by the intrinsic properties of the stimuli themselves. Stimuli located at the location that is the winner of the competition will be processed preferentially in extrastriate cortex. If one location is put at a competitive disadvantage to others as a result of focal dysfunction in one of the attention control structures, then that point will lose the competition to any other visually stimulated points outside the dysfunctional zone. Ultimately, signals from the attentional control system gate either selective inputs to extrastriate neurons or the neurons themselves.

Finally, an important additional source of inputs to the attentional control system may arise from structures involved in memory formation. The anterior portion of IT cortex contains neurons whose responses are attenuated according to the similarity of the current stimulus to memory traces. This attenuation occurs whether the memory trace is actively held in short-term working memory or is passively held in a longer term store. Thus, IT neurons appear to function as adaptive memory filters that are maximally activated by new, unexpected, or not recently seen stimuli. The activation level of IT cortex may serve as a trigger for attentional and orienting systems, causing the subject to attend or orient to the stimulus causing the activation.

I thank Leslie Ungerleider, Mortimer Mishkin, Ricardo Gattass, and Earl Miller for valuable comments on the manuscript.

1. Cortical Pathways for Object Recognition and Spatial Vision

Our work on the neural mechanisms of visual attention and memory focuses on the extrastriate visual cortex of monkeys. The extrastriate cortex is located beyond the primary visual, or striate, cortex and contains well over twenty distinct areas, each with visual functions [for reviews, see Weller and Kaas (1981); Maunsell and Newsome (1987), Desimone and Ungerleider (1989), and Felleman and Van Essen (1991)]. These cortical areas appear to be organized within separate processing systems. According to a model originally proposed by Ungerleider and Mishkin (1982) there are two major processing systems, or "pathways", both of which originate in the striate cortex. One of the pathways is directed ventrally into the temporal lobe and is crucial for object recognition. The other is directed dorsally into the parietal lobe and is crucial for spatial perception and visuomotor performance. Recently, evidence for a third integrative processing pathway within the cortex of the superior temporal cortex of monkeys has emerged, which may either mediate the perception of complex visual motion, or integrate the object and spatial systems, or even carry out both functions (Boussaoud, Ungerleider, and Desimone, 1990; Morel and Bullier, 1991; Baizer, Ungerleider, and Desimone, 1991). The three major pathways are shown diagramatically in Figure 1.

Within each of the major cortical pathways, visual areas appear to form processing hierarchies. Anatomical studies have shown that virtually all connections between successive pairs of areas in the occipitotemporal processing pathway are reciprocal, i.e. projections from one area to the next are reciprocated by projections from the second area back onto the first. Yet, in spite of this reciprocity of anatomical connections, physiological and behavioral studies indicate that much of the processing in the occipitotemporal pathway is actually hierarchical (Desimone, Schein, Moran, and Ungerleider, 1985; Desimone and Ungerleider, 1989; Felleman and Van Essen, 1991). The earliest neuronal response latencies recorded in physiological studies increase steadily as one proceeds from the retina towards IT cortex. Likewise, the average receptive field size of individual neurons (i.e. the portion of the visual field from which a neuron receives information) also increases as one moves along the pathway towards the temporal lobe, consistent with the notion that the large receptive fields of cells in later areas are built up from the smaller fields of cells in earlier areas. Complexity of neuronal processing also increases from one area to the next along the pathway. For example, whereas neurons in striate cortex perform local spatial filtering operations, neurons in V2 have been shown to respond to "virtual" or illusory contours in certain figures (von der Heydt and Peterhans, 1989) and neurons in IT cortex are sensitive to global or overall object features, such as shape, irrespective of position on the retina (Schwartz, Desimone, Albright, and Gross, 1983; Desimone, Albright, Gross, and Bruce, 1984; Tanaka, Saito, Fukada, and Moriya, 1991). Finally, studies in which striate

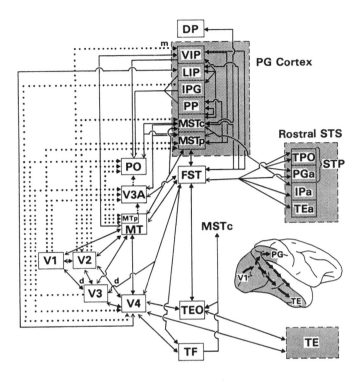

Figure 1. Some of the visual cortical areas and their connections in the macaque monkey. Solid lines indicate projections that originate from both central and peripheral field representations, and dotted lines indicate projections that arise exclusively from peripheral field representations. Solid arrowheads indicate "forward" connections, open arrowheads indicate "backward" connections, and lines connected with closed arrowheads at both ends indicate "intermediate" connections. The small 'd's and 'm's indicate projections limited to the dorsal or medial portions of certain visual areas. The diagram shows a dorsal pathway directed into the parietal cortex (cytoarchitectonic area PG) that is specialized for spatial vision, a ventral pathway directed into the inferior temporal cortex (cytoarchitectonic area TE) that is specialized for object vision, and a third pathway directed into the superior temporal sulcus (STS). Although the functions of this latter pathway are not yet clear, its connections suggest that it may either integrate information from the object vision and spatial visual systems or mediate the perception of complex object motion in space or carry out both functions. Adapted from Desimone and Ungerleider (1989) and Boussaoud, Ungerleider, and Desimone (1990).

cortex has been either reversibly deactivated (by cooling) or removed have found a complete loss of visually driven neuronal activity in certain visual areas (V2, IT) located along the occipitotemporal pathway (Rocha-Miranda et al., 1975; Schiller and Malpeli, 1977). Conversely, neurons in striate cortex appear to be largely unaffected by removal of area V2 (Sandell and Schiller, 1982) or, indeed, most of the non-visual association cortex (Nakamura, Schein, and Desimone, 1986).

In addition to the hierarchical aspects of visual processing, there are also important parallel aspects. Most extrastriate visual areas contain maps of the retina in which all retinal locations are analyzed in parallel by cortical cells with different receptive fields. Like spatial location, many other visual features are processed in parallel within individual areas, although there is some degree of specialization among the areas, particularly between the areas of the occipitotemporal and occipitoparietal systems. For example, area V4, which contributes to the occipitotemporal object recognition pathway, contains cells sensitive to orientation, length, width, spatial frequency, and color at every location in the visual field (Zeki, 1978; Desimone and Schein, 1987; Schein and Desimone, 1990). Area MT, which contributes to the occipitoparietal spatial vision pathway, contains cells sensitive to orientation, direction and speed of motion, and binocular disparity (Zeki, 1974; Maunsell and Van Essen, 1983a, 1983b; Albright, 1984). Within several extrastriate areas there is evidence that different features are preferentially analyzed within different anatomical substructures (see Desimone and Ungerleider, 1989).

2. Attention and Object Recognition

In spite of the massively parallel architecture of the visual cortex, our ability to process visual information in parallel is in some ways surprisingly limited. It is not usually possible, for example, to fully process or recognize within a single moment more than one or two faces in a crowded scene or more than a couple of words on a page of text, even though all of the unrecognized elements stimulate the retina. Thus, somehow the brain filters the incoming parallel stream of visual information such that only selected items have access to neural recognition or mnemonic systems. This filtering or selection process is commonly termed selective attention. The term attention is also applied to many other psychological phenomenon that we will not consider in this chapter, including alertness, arousal, and vigilance. Selective attention is not unlike the selection problem inherent in oculomotor control - we can move our eyes to only one of the many potential targets in the world at a time. We need a mechanism to select which of the many potential targets in a scene will gain control over our eye muscles at a given moment just as we need a mechanism to select which of many visual objects will gain access to our memory. In this sense, memory and muscle are very much alike. Indeed, attention and eye movements are typically linked, as shifts in the spatial locus of attention are commonly (but not always) followed by a shift in gaze.

The fact that the brain serializes the access of external stimuli to object recognition systems is presumably due to a limitation in a central processing resource, a limitation that has been explored at the computational, psychological, and physiological level (see Van Essen, Olshausen, Anderson, and Gallant, 1991). At the computational level, Tsotsos (1990) has recently shown on the basis of a complexity analysis that if one tried to recognize, in parallel, and at full resolution, the images of multiple independent objects randomly distributed across the retina, one would face a combinatorial explosion.

At the physiological level, one limited resource is the receptive field of an individual neuron. Receptive field size increases steadily along the occipitotemporal recognition pathway, such that by the inferior temporal cortex the receptive field of an individual neuron may encompass an entire scene (Gross, Rocha-Miranda, and Bender, 1972; Desimone, Albright, Gross, and Bruce, 1984). Because the images of several independent objects might be contained within the receptive field of such a neuron, the receptive-field resource would have to be shared by all of them. If the response of a neuron was influenced by all of the objects in the scene, without regard to spatial location, the response would presumably carry a very ambiguous message to the rest of the visual system.

A. Attentional Gating in Extrastriate Cortex

Several years ago, we discovered a possible solution to this puzzle in our neuronal recordings in areas V4 and IT (Moran and Desimone, 1985). We studied the responses of neurons in V4 and IT cortex to two stimuli placed simultaneously within the recorded neuron's receptive field. When we required the monkey to attend to one or another of the stimuli, we found that the neuron's response was gated by the locus of the animal's attention within the receptive field. That is, the neuron's response was determined primarily by the attended stimulus; the influence of the unattended stimulus was greatly attenuated.

If, for example, a neuron was color selective and responded to red but not to green stimuli, we placed both a red and a green stimulus simultaneously inside the neuron's receptive field. When the animal was instructed (as a result of task demands) to attend to the locus of the red stimulus, the neuron would give a good response, but if the animal was instructed to attend to the locus of the green stimulus, the neuron would give a poor response (Fig. 2). It is important to emphasize that there was no sensory difference between the two task conditions - there were always the same red and green stimuli inside the field - but there was a large difference in how the animal internally processed the stimuli in the two conditions. In each case, the neuron responded as if the receptive field had contracted around the attended stimulus (but see below), suggesting that attention serves not only to filter out unwanted information, but also to increase the spatial resolution of neurons with large receptive fields. Results

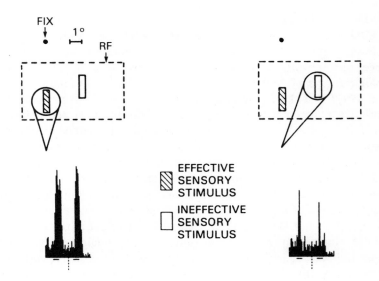

Figure 2. Effect of selective attention on the responses of a neuron in extrastriate area V4. A delayed matching to sample task was used to focus the monkey's attention on one of two simultaneously presented stimuli. The neuronal responses shown are from when the monkey attended to one location inside the receptive field (RF) and ignored another. At the attended location (circled), two stimuli (sample and test) were presented sequentially, and the monkey responded differently depending on whether they were the same or different. Irrelevant stimuli were presented simultaneously with the sample and test but at a separate location in the receptive field. In the initial mapping of the field, the cell responded well to red bars but not at all to green bars. A horizontal or a vertical red bar (effective sensory stimuli) was then placed at one location in the field and a horizontal or a vertical green bar (ineffective sensory stimuli) at another. When the animal attended to the location of the red bar at the time of presentation of either the sample or test, the cell gave a good response (left), but when the animal attended to the location of the green bar, the cell gave only a small response (right) even though the two stimuli on the retina were identical in the two conditions. Thus, the responses of the cell were determined predominantly by the attended stimulus. The horizontal bars under the histograms indicate the 200 msec period when the sample and test stimuli were on. Because of the random delay between the sample and test presentations, the histograms were synchronized separately at the onsets of the sample and test stimuli (indicated by the vertical dashed lines). Adapted from Moran and Desimone (1985).

consistent with these have been found in subsequent studies in IT cortex (Richmond and Sato, 1987; Sato, 1988; Spitzer and Richmond, 1991). The results are also consistent with those from studies of attention-modulated event-related potentials in humans (e.g., Mangun and Hillyard, 1990). Below we will consider whether receptive field contraction is a possible mechanism for attention gating.

Surprisingly, the attentional effects we observed in V4 depended on both the attended and ignored stimuli being located within the recorded neuron's receptive field. If one stimulus was located within the receptive field and one was outside, the locus of the animal's attention had no effect on the neuron's response. Thus, attention appears to work only locally in V4, possibly at the level of a cortical module. Receptive fields in IT cortex were too large to test the effects of attention outside of them. These results suggest that the filtering of unwanted information is at least a two-stage process, with the first stage working over a small spatial range in V4, and a second stage working over a much larger spatial range in IT cortex.

Since V4 cells provide IT cells with their visual input (directly or indirectly), why didn't the visual system implement the spatial attention mechanism in a single stage in V4? If the responses of all V4 cells to ignored stimuli were blocked regardless of whether the attended stimulus was in their receptive field or not, IT cells would receive information only about attended stimuli and there would not be any need to separately gate their responses with attention. Alternatively, why didn't the visual system implement the attentional mechanism in a single stage in IT cortex?

One reason for not implementing the entire attentional mechanism in one stage in V4 may be that V4 is too early a processing stage for such a mechanism. V4, like other retinotopically organized visual areas, is thought to be a site of operations that depend on parallel processing of information throughout the visual field, such as operations that underlie color constancy (Zeki, 1980; Allman, Miezin, and McGuinnes, 1985; Desimone, Schein, Moran, and Ungerleider, 1985; Schein and Desimone, 1990). If color constancy and other context effects depended on interactions among V4 cells with widely distributed receptive fields, these operations might be severely disrupted were all processing in V4 blocked other than to an attended stimulus. On the other hand, IT may be too late a processing stage for the entire attentional mechanism to be implemented there. Both because IT receptive fields are extremely large and because the V4 receptive fields providing inputs to IT cortex are already several times larger than those in the primary visual cortex, it might be impossible for the spatial attentional mechanism to achieve high spatial resolution by modulating IT cells alone. A two-step mechanism involving V4 and IT may be the best compromise among the various constraints.

Now that we know that visual processing in extrastriate cortex is under the control of selective attention, the next important task is to understand how this control comes about. Clearly, some of the neural signals that modulate extrastriate neuronal responses with spatial attention must come from outside the object recognition pathway itself. Possible candidates include, at a minimum, the posterior parietal cortex, parts of prefrontal cortex, the pulvinar nucleus of the thalamus, and the superior colliculus in the midbrain, all of which have been implicated in the control of spatial attention (for a recent review, see Goldberg and Colby, 1989). If we knew the anatomical substrate of the gating process within the extrastriate cortex itself, that could help narrow down the potential source.

B. Two Classes of Attentional Models

Broadly speaking, all potential anatomical models for attentional gating of extrastriate neuronal responses fall into either of two classes. In the first class of models, attentional mechanisms determine which specific subclass of *inputs* activate a given extrastriate neuron, whereas in the second class attentional mechanisms determine which specific subclass of extrastriate *cells* are activated. The notion of input gating readily lends itself to the metaphor of receptive fields expanding and contracting with attention. If, for example, each of the major dendritic branches of an extrastriate neuron received synaptic inputs carrying visual information from a different portion of the receptive field, then inhibition of one of the branches by an attentional mechanism would effectively cause the neuron's receptive field to shrink. The "shifter-circuit" model of Anderson and Van Essen (1987) is an example of an input gating model, and another example, a scheme of ours, is illustrated for area V4 in Figure 3A.

According to the input-gating model of Figure 3A, attentional mechanisms regulate which inputs from V2 activate a given neuron in V4. When attention is directed to a specific location in the visual field, distal attentional mechanisms, represented by the "flashlight", excite an intermediary neuron with a relatively small receptive field in a corresponding location. Only the selection of the intermediary neuron is under dynamic control - all other features of the model are fixed. When selected, the intermediary neuron blocks (possibly through shunting inhibition) just those V2 inputs to a V4 neuron that carry information from locations *outside* the intermediary neuron's own receptive field. Because the intermediary neuron's receptive field is smaller than that of the V4 neuron, only the responses of the V4 neuron to stimuli at nonattended locations within the receptive field will be attenuated. If the intermediary neuron were connected only with V4 neurons whose receptive fields overlapped those of its own (e.g. only the cells within a hypothetical V4 "hypercolumn") then attention to a given location within the visual field would affect visual processing only in nearby portions of V4's visual field representation. Processing of stimuli at more distant locations in V4 would be unaffected by the presence or absence of attention, which is what we found in our physiological study.

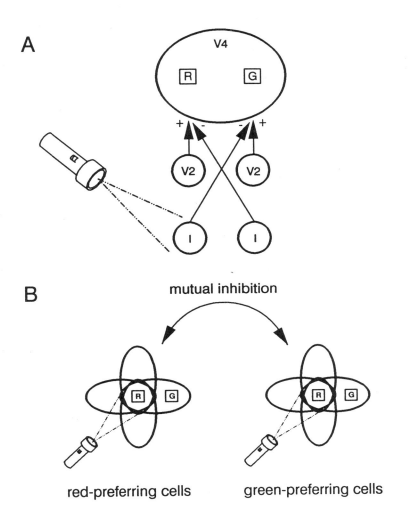

Figure 3. Two classes of models for attentional gating of extrastriate neurons. (A) Input-gating model, in which intermediary neurons gate the inputs from V2 cells to V4 cells. The circles and ellipses represent both the neurons and their receptive fields within areas V2 and V4 and a hypothetical intermediary attention control area. In the example shown, the V4 neuron's receptive field is stimulated by red and green bars. The source of instruction (e.g. "attend to the left") to the system is represented by the flashlight, which activates an intermediary neuron (I) whose receptive field is at the attended location. The intermediary neuron then blocks all inputs to the V4 neuron that arise from V2 neurons with receptive fields located outside the attended location. In this

The input-gating model of Figure 3A is fairly generic. The model does not specify the anatomical location of the intermediary neuron, nor does it even require that it be an individual neuron as opposed to an assembly of neurons with diverse properties that collectively accomplish the same function. The assembly might involve both excitatory and inhibitory neurons distributed across more than one anatomical structure. Furthermore, the small receptive field may be a "virtual" field in that it could be implemented by an ensemble of neurons each with a large receptive field but which collectively act on only a small spatial location.

In contrast to input-gating models, cell-gating models rely on inhibition of entire neurons rather than on inhibition of specific inputs to those neurons. The oscillation model of Crick and Koch (1990) is an example of a cell-gating model, and a different example, a scheme of ours, is illustrated in Figure 3B. Cell-gating models might initially seem implausible because our own physiological evidence shows that the responses of extrastriate cells appear to switch from one stimulus to another within the receptive field depending on which one is attended. When a red and a green stimulus are located within the receptive field of a red-preferring cell, for example, we found that the response was switched on or off depending on which stimulus was at the attended location. However, we are not able to physiologically record selective inputs to a cell being gated on or off. With extracellular recording, we can only record the final output of the neuron "as a whole". Furthermore, we can easily observe the effects of attention only when the receptive field contains stimuli with contrasting sensory properties. If a receptive field contains two red stimuli, for example, we are not able to measure the effects of attending to one or the other of them. Thus, it is logically possible that, rather than gating selective inputs on or off, the spatial attentional mechanism gates entire cells on or off depending on the sensory features of the stimulus at the attended location.

The cell-gating model of Figure 3B assumes some type of lateral inhibitory

---

Figure 3 (continued). case the V4 neuron would respond only to the red bar. (B) Cell-gating model, based on competitive interactions (e.g., lateral inhibition) between V4 cells with contrasting sensory preferences. In the example, red-preferring cells compete with green-preferring cells. The external source of instruction biases all cells that contain the attended location within their receptive field. Because a red stimulus is located at the attended location, there are more red-preferring V4 cells than green-preferring cells that are preferentially activated, and the green-preferring cells are therefore suppressed.

connection or competitive interaction between cells with both contrasting stimulus sensitivities and overlapping receptive fields in V4. As an example, consider red and green stimuli presented simultaneously and close together within the visual field. According to the model, the red stimulus will activate red-preferring cells in V4 and the green stimulus green-preferring cells, and both classes of cells will attempt to inhibit each other's response to their preferred stimulus. The inhibition might be mediated either directly by inhibitory interconnections (possibly shunting inhibition) or indirectly by means of a competitive inhibitory network. In the absence of attention to one or the other stimulus, neither the red nor green preferring cells have any advantage in the competition and all cells respond to their preferred stimulus. In the absence of attention, the magnitude of response to the preferred stimulus will depend on the specifics of the implementation. According to the model, the function of attention is to bias the competition so that one cell type wins. For the model to work, we assume that extrinsic attentional control mechanisms, again represented by a flashlight in the figure, potentiate the response of all V4 neurons whose receptive fields include an attended location in the visual field. Equivalently, the attentional input might simply potentiate the inhibitory action of cells in the affected zone. The cells in this zone should have scattered receptive fields that overlap in only a small region of the visual field. Whatever stimulus happens to be situated in the region of overlap will cause the competition in the zone to be biased in favor of cells preferentially activated by the features of that stimulus. For example, if a red stimulus happens to be at the attended location, then more red-preferring than green-preferring cells will have a potentiated response, and the red-preferring ones will win the competition. Consequently, the response of the green-preferring cells to the nearby but unattended green stimulus inside their receptive field will be attenuated. Conversely, the response of the red-preferring cells would be attenuated if the green stimulus was at the attended location. In this cell-gating model, V4 receptive fields do not actually shrink away from unattended stimuli; nonetheless, cells respond poorly to an ignored stimulus when there is an attended stimulus with different sensory qualities nearby, just as in the input-gating model. If attentional mechanisms gated cells in V2 in an analogous manner, this *would* have a modulating effect on downstream V4 receptive fields, but so far no evidence of such gating has been found in V2 (Moran and Desimone, unpublished data).

An advantage of the cell-gating model over the input-gating model is that the anatomical substrate of the attentional control mechanism in cortex is potentially much simpler. The input-gating model seems to require an elaborate anatomical infrastructure to enable inhibition of specific subsets of inputs to each cell in the cortex, whereas the cell gating model requires only a small excitatory drive, or bias, to a restricted region of cortex. This drive might originate from any or all of several parallel anatomical inputs to extrastriate cortex, including those from the pulvinar nucleus of the thalamus, the prefrontal cortex, or the posterior parietal cortex. On the other hand, one disadvantage of the cell-gating model is that it is not yet clear how it applies to attentional control of IT cells, whose receptive fields are often so large as to include all of the central visual field. The cell-gating mechanism depends on partial spatial overlap among receptive fields that are restricted in size. One possibility is that mechanism makes use of spatial nonuniformities in IT receptive fields. Although IT receptive fields are large, they often have a small zone of maximal sensitivity. Another possible answer might come from studying the role of attention in area TEO, a newly mapped area lying between V4 and IT with receptive fields that are intermediate in size between those in the two other areas (Boussaoud, Ungerleider, and Desimone, 1991). Other than theoretical considerations, there are currently no strong physiological grounds for choosing one class of attention-gating model over another.

The input-gating and cell-gating models of Figure 3 are so lacking in physiological detail it may be an overstatement to term them models. More accurately, they are starting points for the development of more specific models that will incorporate physiological and anatomical facts as they become known. Until more is known about the nature of attentional inputs entering the cortex, including where they come from, what layer they enter and which cells they contact, whether they are excitatory or inhibitory, and whether they gate cells or the inputs to cells, any specific model is likely to have a limited lifetime.

C. The Source of Attentional Control Signals

As already indicated, some of the structures that have been implicated in the control of spatial attention are the posterior parietal cortex, certain parts of the prefrontal cortex, the pulvinar nucleus of the thalamus, and the superior colliculus. Significantly, most or all of these structures have also been implicated in oculomotor control. The oculomotor and spatial attention systems are closely linked and may, in fact, be completely overlapping (Rizzolatti, 1983; Rizzolatti, Riggio, Dascola, and Umilta, 1987). Shifts of gaze typically follow shifts of attention, although, as we have seen, shifts of gaze are not imperative. It may be that the major difference between the two systems is simply whether or not the "go" signal is given for the eyes to actually move. Interestingly, the influence of the two systems on cortical visual processing may be largely the same regardless of whether the eyes move. When the eyes fixate a new target, the object recognition system in visual cortex is dominated by new input because of the large magnification of the foveal representation in the cortex. A similar shift in processing appears to occur when a new stimulus becomes the object of attention, even in the absence of an eye movement to it.

Our own work on potential subcortical sources of attentional control signals has focused on the pulvinar and the superior colliculus (Desimone, Wessinger, Thomas, and Schneider, 1990). The pulvinar has been implicated in attentional control on the basis of neuropsychological studies of humans with thalamic brain damage (Rafal and Posner, 1987), PET activation studies (LaBerge and Buchsbaum, 1990), as well as electrophysiological recording and chemical deactivation studies in monkeys (Petersen, Robinson, and Keys, 1985; Petersen, Robinson, and Morris, 1987; Robinson, Petersen, and Keys, 1986). This structure contains several different nuclei, each of which in turn probably contains several functionally distinct regions (Benevento and Rezak, 1976; Bender, 1981; Ungerleider, Galkin, and Mishkin, 1983). Robinson and his colleagues have shown in monkeys that reversible unilateral deactivation of a part of the pulvinar they term Pdm leads to a reversible impairment in the ability to disengage attention from the contralesional field (Petersen, Robinson, and Morris, 1987). Pdm has anatomical connections with the posterior parietal cortex, lesions of which cause a similar attentional impairment (Posner, Walker, Friedrich, and Rafal, 1984). This behavioral result led us to ask whether the lateral nucleus of the pulvinar (Pl), the part connected with areas V4 and IT, might have a corresponding role in the ability to focus attention on one stimulus in the visual field and ignore distractors.

We deactivated the lateral pulvinar on one side of the brain with the GABA agonist muscimol and tested the ability of monkeys to perform a color discrimination on a target in the opposite (contralesional) visual field, with or without a distractor in the ipsilesional (unaffected) field. The injection was large enough to affect a complete hemifield representation in the lateral pulvinar of one hemisphere. The results showed that the pulvinar impaired the discrimination of the target color in the presence of the distractor. That is, with the pulvinar deactivated the animal found it difficult to ignore the distractor in the ipsilesional (unaffected) hemifield and focus attention on the target in the contralesional field. However, if there was no distractor, the deactivation had little or no behavioral effect. These are just the results we would expect if the pulvinar deactivation caused the loss of an attentional control signal to extrastriate cortex. In fact, we hypothesize that it was the loss of a control signal to IT cortex that contributed most to the behavioral deficit, because IT cells have bilateral receptive fields which presumably encompassed both the target and the distracting stimulus. Without such a signal, IT cells would respond to both the attended stimulus in the contralesional field as well as to the distracting stimulus in the ipsilesional field. When there was no distractor, no gating of IT cell responses was necessary, which would explain why there was no behavioral effect of the pulvinar deactivation in this condition.

Surprisingly, we found that if the distractor was located within the same hemifield (i.e., the contralesional field) as the target, the behavioral impairment resulting from the pulvinar deactivation was much less than when the distractor was located in the opposite hemifield. This result is reminiscent of the extinction phenomenon reported after certain unilateral lesions in human patients (De Renzi, 1982). In extinction, stimuli located within the contralesional field are perceived normally if the ipsilesional field is blank, but may "disappear" if a distractor is presented in the ipsilesional field. Such a result could be explained if, within the attentional control system, there is a competitive interaction between stimuli across the two visual hemifields. A unilateral lesion would presumably bias the competition in favor of stimuli within the ipsilesional visual field, so that attention would preferentially be directed to stimuli within this field. When target and distractor are both located within the same hemifield, a unilateral lesion would not bias the competition and, presumably, the remaining portions of the attentional control system would still send the appropriate control signal to IT cortex. One conclusion from this line of reasoning is that the pulvinar cannot be the only source of attentional control signals.

Preliminary results from a second behavioral study suggest that the competition within the attentional control system is not limited to stimuli in opposite hemifields, but occurs among stimuli at all spatial locations. Unlike the previous study in which we deactivated the representation in the pulvinar of the entire hemifield, in this second study we deactivated in the superior colliculus the representation of just a small portion of the hemifield. In this case, we found that the monkey had difficulty in attending to a target within the affected portion of the visual field when there was a distractor located anywhere within the unaffected parts of the visual field, including within the same hemifield. Consistent with these behavioral results, we have also recently found that, in a difficult discrimination task with a target and distractor in the visual field, some neurons in the superficial layers of the superior colliculus give better responses to attended target stimuli than to physically identical distractors (Gattass and Desimone, 1991).

Although we have not tested the effects of restricted deactivation of just a portion of the hemifield representation in the pulvinar, the results so far are consistent with a competition model of attentional control, in which stimuli at all points within the visual field are in competition with one another. Competition within the system is biased either by external instruction to attend to one location or by the intrinsic properties of the stimuli themselves [see Duncan and Humphreys (1989), and Bundesen (1990)]. This idea is similar to the notion of a 'saliency map' underlying attention, as suggested by Koch and Ullman (1985) and Triesman and her colleagues (Treisman and Souther, 1985; Triesman, 1986, 1988). In the Koch and Ullman model, the saliency map contains a "winner-take-all" competitive network.

Based on our work, we believe that there is no single saliency map in the attentional control system, but rather a series of maps (or possibly a distributed map) that are closely associated with structures involved in oculomotor control. As in the oculomotor system, the different components of the attentional control system each likely makes unique contributions to attention (Posner and Petersen, 1990); however, these unique contributions may be difficult to isolate with lesion techniques. Because some functions must be shared across structures, no single discrete lesion will be sufficient to totally eliminate attentional control; rather, a discrete lesion will throw the remaining portions of the system out of "balance". A stimulus located at the point in space that is the winner of the competition within the attentional control system will be processed preferentially in extrastriate cortex. If a point is put at a competitive disadvantage as a result of dysfunction in one of the attentional control structures, then that point will lose the competition to any other visually stimulated points (i.e. distractors) outside the dysfunctional zone.

3. Closing the Loop

At the beginning of this chapter we pointed out the role that attention plays in controlling access of external stimuli to the neural systems underlying storage of visual memories. Some recent studies of our own and others suggest a way in which mnemonic systems might, in turn, influence which stimuli activate attentional systems. In our study, we examined the responses of neurons in the most anterior ventral portion of IT cortex in monkeys performing tasks that required them to hold items in memory while concurrently viewing other items (Miller, Li, and Desimone, 1991). We used two variants of the task, one which tapped working memory, i.e. memory for the recency of occurrence of a familiar item, and another one which tapped recognition memory, i.e. memory underlying judgements of novelty or familiarity.

In the working memory variant of the task, we found that the neuronal response to an incoming visual stimulus was attenuated if it matched a stimulus actively held in memory, even when several other stimuli intervened in the retention interval. Other studies of working memory in IT have also reported memory matching effects (Gross et al., 1979; Mikamai and Kubota, 1980; Baylis and Rolls, 1987; Vogels and Orban, 1989; Eskandar, Richmond, and Optican, 1991), as well as high maintained activity of IT cells during retention intervals (Fuster and Jervey, 1982; Miyashita and Chang, 1988) but this is the first evidence for mnemonic effects spanning intervening items within a trial. It is not clear if IT cortex is the earliest visual cortical area showing such effects, for there is some evidence for short-term memory matching phenomena in V4 (Haenny, Maunsell, and Schiller, 1988; Haenny and Schiller, 1988).

In the recognition memory variant of the task, we found that for about one-third of the neurons in this area, neuronal responses to a set of initially novel stimuli systematically declined as the stimuli gradually became familiar to

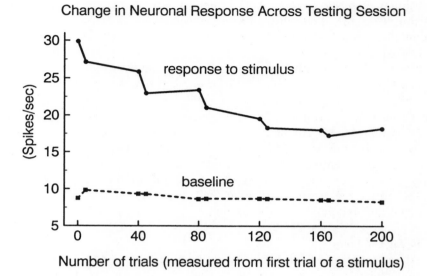

Figure 4. Average response of 25 IT neurons that showed a significant attenuation of response across the test session to repeated presentations of 20 stimuli that the monkey had never seen before. Each stimulus was presented ten times during the session, for a total of 200 trials per session. Responses were averaged across all stimuli used with each cell, and the trial number for a given stimulus was measured from the first trial on which that stimulus appeared. The staircase appearance of the curve results from different numbers of intervening trials between presentations of a given stimulus. Fewer intervening trials for a given stimulus caused a sharper drop in response than did larger numbers of intervening trials. The baseline is the average spontaneous firing rate of the cells, measured before each trial. Adapted from Miller, Li, and Desimone (1991).

the animal over the course of a recording session lasting about an hour. The latter result was obtained even when a large number of experimental trials (and many minutes) intervened between repetitions of the same stimulus.

Together, our results show that neuronal responses in IT cortex are attenuated according to the similarity of the current stimulus to memory traces, whether the trace is actively held in short-term working memory or is passively held in a longer term store. Thus, IT neurons appear to function as adaptive memory filters that are maximally activated by new, unexpected, or not recently seen stimuli. Contact with new stimuli will eventually cause an adjustment of synaptic weights in IT so that the "pool" of activated neurons will shrink and the overall level of cortical activation will be reduced. This process would not merely precede memory storage, but would be a critical component of the storage mechanism. This is similar to Mishkin's (1991) notion that memory formation depends, in part, on the development of a small subset of IT neurons that are activated by a recognized stimulus. The results are consistent with recent physiological findings of Riches, Wilson, and Brown (1991), as well as a recent PET activation study by Squire and colleagues (Squire, Ojemann, Miezin, Petersen, Videen, and Raichle, 1991). In the latter study, blood flow in response to visually presented words in occipitotemporal cortex was reduced if the words had been recently seen before.

The fact that IT cortex has a reduced level of activation for familiar or expected stimuli suggests that a high level of cortical activation may itself serve as a trigger for attentional and orienting systems, causing the subject to orient to the stimulus causing the activation. As the subject gains more experience with the stimulus, the activation level of cortex should be reduced, ultimately reducing the drive on the orienting systems. This link between the mnemonic and attentional systems would "close the loop" between the two systems, resulting in orienting behavior that is influenced by both current stimuli and prior memories. Such a mechanism has a number of similarities to the adaptive resonance theory (ART) of Carpenter and Grossberg (1987a, 1987b, 1990). In this multi-layer network model, a mismatch between the current stimulus and stored memory representations causes an attentional facilitation of the cells in a layer holding an internal representation of the stimulus. The attentional facilitation then leads to an updating of memory traces by the current stimulus.

REFERENCES

Allman, J., Miezin, F., and McGuinness, E. (1985). Stimulus specific responses from beyond the classical receptive field: Neurophysiological mechanisms for local-global comparisons in visual neurons. *Annual Review of Neuroscience*, **8**, 407-430.

Albright, T.D. (1984). Direction and orientation selectivity of neurons in visual area MT of the macaque. *Journal of Neurophysiology,* **52,** 1106-1130.

Anderson, C.H. and Van Essen, D.C. (1987). Shifter circuits: A computational strategy for dynamic aspects of visual processing. *Proceedings of the National Academy of Science, USA,* **84,** 6297-6301.

Baizer, J., Ungerleider, L., and Desimone, R. (1991). Organization of visual inputs to posterior parietal and inferior temporal cortex in the macaque. *Journal of Neuroscience,* **11,** 168-190.

Baylis, G.C. and Rolls, E.T. (1987). Responses of neurons in the inferior temporal cortex in short term and serial recognition memory tasks. *Experimental Brain Research,* **65,** 614-622.

Bender, D.B. (1981). Retinotopic organization of macaque pulvinar. *Journal of Neurophysiology,* **46,** 672-693.

Benevento, L.A. and Rezak, M. (1976). The cortical projections of the inferior pulvinar and adjacent lateral pulvinar in the rhesus monkey *(Macaca mulatta)*: An autoradiographic study. *Brain Research,* **108,** 1-24.

Boussaoud, D., Ungerleider, L., and Desimone, R. (1990) Pathways for motion analyses: Cortical connections of the medial superior temporal and fundus of the superior temporal visual areas in the macaque. *Journal of Comparative Neurology,* **296,** 462-495.

Boussaoud, D., Desimone, R., Ungerleider, L. (1991). Visual topography of area TEO in the macaque. *Journal of Comparative Neurology,* **306,** 554-575.

Bundesen, C. (1990). A theory of visual attention. *Psychological Review,* **97,** 523-547.

Carpenter, G.A. and Grossberg, S. (1987a). A massively parallel architecture for a self-organizing neural pattern recognition machine. *Computer Vision, Graphics, and Image Processing,* **37,** 54-115.

Carpenter, G.A. and Grossberg, S. (1987b). ART 2: Self-organization of stable category recognition codes for analog input patterns. *Applied Optics,* **26,** 4919-4930.

Carpenter, G.A. and Grossberg, S. (1990). ART 3: Hierarchical search using chemical transmitters in self-organizing pattern recognition architectures. *Neural Networks,* **3,** 129-152.

Crick, F. and Koch, C. (1990). Some reflections on visual awareness. *Cold Spring Harbor Symposium on Quantitative Biology,* **55,** 953-962.

De Renzi, E. (1982). **Disorders of space exploration and cognition.** Wiley, Chichester.

Desimone, R., Albright, T.D., Gross, C.G., and Bruce, C. (1984). Stimulus selective properties of inferior temporal neurons in the macaque. *Journal of Neuroscience,* **4,** 2051-2062.

Desimone, R. and Schein, S.J., (1987). Visual properties of neurons in area V4 of the macaque: sensitivity to stimulus form. *Journal of Neurophysiology,* **57,** 835-868.

Desimone, R., Schein, S.J., Moran, J., and Ungerleider, L.G. (1985). Contour, color and shape analysis beyond the striate cortex. *Vision Research,* **25,** 441-452.

Desimone, R. and Ungerleider, L.G. (1989). Neural mechanisms of visual processing in monkeys. In F. Boller and J. Grafman (Eds.), **Handbook of Neuropsychology, Vol. II,** Elsevier, Amsterdam.

Desimone, R., Wessinger, M., Thomas, L., and Schneider, W. (1990). Attentional control of visual perception: Cortical and subcortical mechanisms. *Cold Spring Harbor Symposium on Quantitative Biology,* **55,** 963-971.

Duncan, J. and Humphreys, G.W. (1989). Visual search and stimulus similarity. *Psychological Review,* **96,** 433-458.

Eskandar, E.N., Richmond, B.J., and Optican, L.M. (1991). Inferior temporal neurons convey information about stimulus patterns and their behavioral significance. *Society for Neuroscience Abstracts,* **17,** 443.

Felleman, D.J. and Van Essen, D.C. (1991). Distributed hierarchical processing in the primate cerebral cortex. *Cerebral Cortex,* **1,** 1-47.

Gattass, R. and Desimone, R. (1991). Attention-related responses in the superior colliculus of the macaque. *Society for Neuroscience Abstracts,* **17,** 545.

Goldberg, M.E. and Colby, C.L. (1989). The neurophysiology of spatial vision. In F. Boller and J. Grafman (Eds.), **Handbook of Neuropsychology, Vol. II,** Elsevier, Amsterdam.

Gross, C.G., Bender, D.B., and Gerstein, G.L. (1979). Activity of inferior temporal neurons in behaving monkeys. *Neuropsychologia,* **17,** 215-229.

Gross, C.G., Rocha-Miranda, C.E., and Bender, D.B. (1972) Visual properties of neurons in inferotemporal cortex of the macaque. *Journal of Neurophysiology,* **35,** 96-111.

Haenny, P.E., Maunsell, J.H.R., and Schiller, P.H. (1988). State dependent activity in monkey visual cortex. II. Extraretinal factors in V4. *Experimental Brain Research,* **69,** 245-259.

Haenny, P.E. and Schiller, P.H. (1988). State dependent activity in monkey visual cortex. I. Single-cell activity in V1 and V4 on visual tasks. *Experimental Brain Research,* **69,** 225-244.

Koch, C. and Ullman, S. (1985). Shifts in selective visual attention: towards the underlying neural circuitry. *Human Neurobiology,* **4,** 219-227.

LaBerge, D. and Buchsbaum, M.S. (1990). Positron emission tomographic measurements of pulvinar activity during an attention task. *Journal of Neuroscience,* **10,** 613-619.

Mangun, G.R. and Hillyard, S.A. (1990). Allocation of visual attention to spatial locations: Tradeoff functions for event-related brain potentials and detection performance. *Perception and Psychophysics,* **47,** 532-550.

Maunsell, J.H.R. and Newsome, W.T. (1989). Visual processing in monkey extrastriate cortex. *Annual Review of Neuroscience,* **10,** 363-401.

Maunsell, J.H.R. and Van Essen, D.C. (1983a). Functional properties of neurons in middle temporal visual area of the macaque monkey. I. Selectivity for stimulus direction, speed, and orientation. *Journal of Neurophysiology,* **49,** 1127-1147.

Maunsell, J.H.R. and Van Essen, D.C. (1983b). Functional properties of neurons in middle temporal visual area of the macaque monkey. II. Binocular interactions and sensitivity to binocular disparity. *Journal of Neurophysiology,* **49,** 1148-1167.

Mikami, A. and Kubota, K. (1980). Inferotemporal neuron activities and color discrimination with delay. *Brain Research,* **182,** 65-78.

Miller, E., Li, L., and Desimone, R. (1991). A neural mechanism for working and recognition memory in inferior temporal cortex. *Science,* **254,** 1377-1379.

Mishkin, M. (1991). Cerebral memory circuits. In: **Yakult International Symposium on Perception, Cognition, and Brain.** Yakult Honsha Co., Ltd.

Miyashita, Y., and Chang, H.S. (1988). Neuronal correlate of pictorial short-term memory in the primate temporal cortex. *Nature,* **331,** 68-70.

Moran, J., and Desimone, R. (1985). Selective attention gates visual processing in the extrastriate cortex. *Science,* **229,** 782-784.

Morel, A. and Bullier, J. (1991). Anatomical segregation of two cortical visual pathways in macaque monkey. *Visual Neuroscience,* **4,** 555-578.

Nakamura, R.K., Schein, S.J., and Desimone, R. (1986). Visual responses from cells in striate cortex of monkeys rendered chronically 'blind' by lesions of nonvisual cortex. *Experimental Brain Research,* **63,** 185-190.

Petersen, S.E., Robinson, D.L., and Keys, W. (1985). Pulvinar nuclei of the behaving rhesus monkey: Visual responses and their modulation. *Journal of Neurophysiology,* **54,** 867-886.

Petersen, S., Robinson, D.L., and Morris, J.D. (1987). Contributions of the pulvinar to visual spatial attention. *Neuropsychologia,* **25,** 97-105.

Posner, M.I. and Petersen, S.E. (1990). The attention system of the human brain. *Annual Review of Neuroscience,* **13,** 25-42.

Posner, M.I., Walker, J.A., Friedrich, F.J., and Rafal, R.D. (1984). Effects of parietal lobe injury on covert orienting of visual attention. *Journal of Neuroscience,* **4,** 1863-1874.

Rafal, R.D. and Posner, M.I. (1987). Deficits in human visual spatial attention following thalamic lesions. *Proceedings of the National Academy of Science USA,* **84,** 7349-7353.

Riches, I.P., Wilson, F.A.W., and Brown, M.W. (1991). The effects of visual stimulation and memory on neurons of the hippocampal formation and the neighboring parahippocampal gyrus and inferior temporal cortex of the primate. *Journal of Neuroscience,* **11,** 1763-1778.

Richmond, B.J. and Sato, T. (1987). Enhancement of inferior temporal neurons during visual discrimination. *Journal of Neurophysiology,* **6,** 1292-1306.

Rizzolatti, G. (1983). Mechanisms of selective attention in mammals. In J.P. Ewert (Ed.) **Advances in Vertibrate Neuroethology,** Plenum Press, London.

Rizzolatti, G., Riggio, L., Isabella, D., and Umilta, C. (1987). Reorienting attention across the horizontal and vertical meridians: Evidence in favor of a premotor theory of attention. *Neuropsychologia,* **25,** 31-40.

Robinson, D.L., Petersen, S.E., and Keys, W. (1986). Saccade-related and visual activities in the pulvinar nuclei of the behaving rhesus monkey. *Experimental Brain Research,* **62,** 625-634.

Rocha-Miranda, C.E., Bender, D.B., Gross, C.G., and Mishkin, M. (1975). Visual activation of neurons in inferotemporal cortex depends on striate cortex and forebrain commissures. *Journal of Neurophysiology,* **38,** 475-491.

Sandell, J.H. and Schiller, P.H. (1982). Effects of cooling area 18 on striate cortex cells in the squirrel monkey. Journal of Neurophysiology, 48, 38-48.

Sato, T. (1988). Effects of attention and stimulus interaction on visual responses of inferior temporal neurons in macaque. *Journal of Neurophysiology,* **60,** 344-364.

Schein, S. and Desimone, R. (1990). Spectral properties of V4 neurons in the macaque. *Journal of Neuroscience,* **10,** 3369-3389.

Schiller, P.H. and Malpelli, J.G. (1977). The effect of striate cortex cooling on area 18 cells in the monkey. Brain Research, 126, 366-369.

Schwartz, E.L., Desimone, R., Albright, T.D., and Gross, C.G. (1983) Shape recognition and inferior temporal neurons. *Proceedings of the National Academy of Sciences USA,* **80,** 5776-5778.

Spitzer, H., Desimone, R., and Moran, J. (1988). Increased attention enhances both behavioral and neuronal performance. *Science,* **240,** 338-340.

Spitzer, H. and Richmond, B.J. (1991). Task difficulty: ignoring, attending to, and discriminating a visual stimulus yield progressively more activity in inferior temporal cortex. *Experimental Brain Research,* **83,** 340-348.

Squire, L., Ojemann, J., Miezin, F., Petersen, S., Videen, T., and Raichle, M. (1991). A functional anatomical study of human memory. *Society for Neuroscience Abstracts,* **17,** 4.

Tanaka, K., Saito, H., Fukada, Y., and Moriya, M. (1991). Coding visual images of objects in the inferotemporal cortex of the macaque monkey. *Journal of Neurophysiology,* **66,** 170-189.

Treisman, A. (1986) Features and objects in visual processing. *Scientific American,* **255,** 114-125.

Treisman, A. (1988). Features and objects: The Fourteenth Bartlett Memorial Lecture. *Quarterly Journal of Experiental Psychollgy,* **40A,** 201-237.

Treisman, A. and Souther, J. (1985). Search asymmetry: A diagnostic for preattentive processing of separable features. *Journal of Experimental Psychology,* **114,** 285-310.

Tsotsos, J.K. (1990). Analyzing vision at the complexity level. *Behavioral and Brain Sciences,* **13,** 423-469.

Ungerleider, L.G., Galkin, T.W., and Mishkin, M. (1983). Visuotopic organization of projections from striate cortex to inferior and lateral pulvinar in rhesus monkey. *Journal of Comparative Neurology,* **217,** 137-157.

Ungerleider, L.G. and Mishkin, M. (1982). Two cortical visual systems. In D.J. Ingle, M.A. Goodale, R.J.W. Mansfield (Eds), **Analysis of Visual Behavior,** MIT Press, Cambridge.

Van Essen, D.C., Olshausen, B., Anderson, C.H., and Gallant, J.L. (1991). Pattern recognition, attention, and information bottlenecks in the primate visual system. *Proceedings of SPIE Conference on Information Processing: From Neurons to Chips,* **1473,** 17-28.

Vogels, R. and Orban, G.A. (1989). Effects of task related stimulus attributes on infero-temporal neurons studied in the discriminating monkey. *Society for Neuroscience Abstracts,* **16,** 621.

von der Heydt, R. and Peterhans, E. (1989). Mechanisms of contour perception in monkey visual cortex. I. Lines of Pattern Discontinuity. *Journal of Neuroscience,* **9,** 1731-1748.

Weller, R.E. and Kaas, J.H. (1981). Cortical and subcortical connections of visual cortex in primates. In C.N. Woolsey (Ed.), **Cortical Sensory Organization, Vol. 2, Multiple Visual Areas,** Humana Press, Clifton NJ.

Zeki, S.M. (1974). Functional organization of a visual area in the posterior bank of the superior temporal sulcus of the rhesus monkey. *Journal of Physiology (London),* **236,** 549-573.

Zeki, S.M. (1978). Uniformity and diversity of structure and function in rhesus monkey prestriate visual cortex. *Journal of Physiology (London),* **277,** 273-290.

Zeki, S.M. (1980). The representation of colours in the cerebral cortex. *Nature,* **284,** 412-418.

Zeki, S.M. (1983). Color coding in the cerebral cortex: The reaction of cells in monkey visual cortex to wavelengths and colours. *Neuroscience* **9,** 741-765.

ATTENTIVE SUPERVISED LEARNING AND RECOGNITION BY AN ADAPTIVE RESONANCE SYSTEM

Gail A. Carpenter[†], Stephen Grossberg[‡], Natalya Markuzon[§],
John H. Reynolds[¶], and David B. Rosen[¶]

Center for Adaptive Systems
and
Department of Cognitive and Neural Systems
Boston University
111 Cummington Street
Boston, Massachusetts 02215 USA

1. Introduction

ARTMAP is a class of neural network architectures that employ attentional mechanisms to perform incremental supervised learning of recognition categories and multidimensional maps. The first ARTMAP system (Carpenter, Grossberg, and Reynolds, 1991) was used to classify binary vectors. This article describes a more general ARTMAP system that learns to classify analog as well as binary vectors (Carpenter, Grossberg, Markuzon, Reynolds, and Rosen, 1992). This generalization is accomplished by replacing the ART 1 modules (Carpenter and Grossberg, 1987a) of the binary ARTMAP system with Fuzzy ART modules (Carpenter, Grossberg, and Rosen, 1991a). Where ART 1 dynamics are described in terms of set-theoretic operations, Fuzzy ART dynamics are described in terms of fuzzy set-theoretic operations (Zadeh, 1965). Hence the new system is called Fuzzy ARTMAP. Also described is an ARTMAP *voting strategy*. This voting strategy is based on the observation that ARTMAP fast learning typically leads to different adaptive weights and recognition categories for different orderings of a given training set, even when overall predictive accuracy of all simulations is similar. The different category structures cause the set of test set items where errors occur to vary from one simulation to the next. The voting strategy uses an ARTMAP system that is trained several times on input sets with different orderings. The final prediction for a given test set item is the one made by the largest number of simulations. Since the set of items making erroneous predictions varies from one simulation to the next, voting cancels many of the errors. Further, the voting strategy can be used to assign confidence estimates to competing predictions given small, noisy, or incomplete training sets.

[†] Supported in part by British Petroleum (89-A-1204), DARPA (AFOSR 90-0083), the National Science Foundation (NSF IRI 90-00530) and the Office of Naval Research (ONR N00014-91-J-4100).

[‡] Supported in part by the Air Force Office of Scientific Research (AFOSR 90-0175), DARPA (AFOSR 90-0083) and the Office of Naval Research (ONR N00014-91-J-4100).

[§] Supported in part by National Science Foundation (NSF IRI 90-00530) and British Petroleum (89-A-1204).

[¶] Supported in part by DARPA (AFOSR 90-0083).

Acknowledgements: The authors wish to thank Cynthia E. Bradford for her valuable assistance in the preparation of the manuscript.

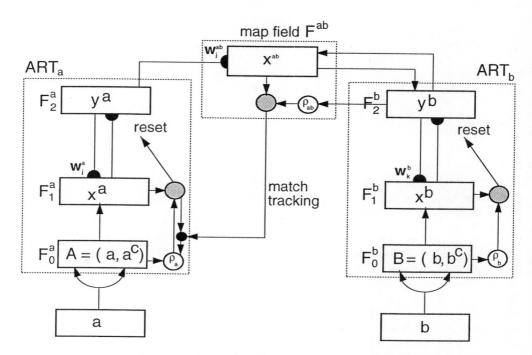

Figure 1. Fuzzy ARTMAP architecture. The ART_a complement coding preprocessor transforms the M_a-vector **a** into the $2M_a$-vector $\mathbf{A} = (\mathbf{a}, \mathbf{a}^c)$ at the ART_a field F_0^a. **A** is the input vector to the ART_a field F_1^a. Similarly, the input to F_1^b is the $2M_b$-vector $(\mathbf{b}, \mathbf{b}^c)$. When a prediction by ART_a is disconfirmed at ART_b, inhibition of map field activation induces the match tracking process. Match tracking raises the ART_a vigilance (ρ_a) to just above the F_1^a to F_0^a match ratio $|\mathbf{x}^a|/|\mathbf{A}|$. This triggers an ART_a search which leads to activation of either an ART_a category that correctly predicts **b** or to a previously uncommitted ART_a category node.

Simulations illustrate Fuzzy ARTMAP performance as compared to benchmark back propagation and genetic algorithm systems. In all cases, Fuzzy ARTMAP simulations lead to favorable levels of learned predictive accuracy, speed, and code compression in both on-line and off-line settings. Fuzzy ARTMAP is also easy to use. It has a small number of parameters, requires no problem-specific system crafting or choice of initial weight values, and does not get trapped in local minima.

Each ARTMAP system includes a pair of Adaptive Resonance Theory modules (ART_a and ART_b) that create stable recognition categories in response to arbitrary sequences of input patterns (Figure 1). During supervised learning, ART_a receives a stream $\{\mathbf{a}^{(p)}\}$ of input patterns and ART_b receives a stream $\{\mathbf{b}^{(p)}\}$ of input patterns, where $\mathbf{b}^{(p)}$ is the correct prediction given $\mathbf{a}^{(p)}$. These modules are linked by an associative learning network and an internal controller that ensures autonomous system operation in real time. The controller is designed to create the minimal number of ART_a recognition categories, or "hidden units," needed to meet accuracy criteria. It does this by realizing a Minimax Learning Rule that enables an ARTMAP system to learn quickly, efficiently, and accurately as it conjointly *minimizes* predictive error and *maximizes* predictive generalization. This scheme automatically links predictive success to category size on a trial-by-trial basis using only local operations. It works by increasing the vigilance parameter ρ_a of ART_a by the minimal amount needed to correct a predictive error at ART_b.

Parameter ρ_a calibrates the minimum confidence that ART_a must have in a recognition category, or hypothesis, activated by an input $\mathbf{a}^{(p)}$ in order for ART_a to accept that category, rather than search for a better one through an automatically controlled process of hypothesis testing. Lower values of ρ_a enable larger categories to form. These lower ρ_a values lead to broader generalization and higher code compression. A predictive failure at ART_b increases ρ_a by the minimum amount needed to trigger hypothesis testing at ART_a, using a mechanism called *match tracking* (Carpenter, Grossberg, and Reynolds, 1991). Match tracking sacrifices the minimum amount of generalization necessary to correct a predictive error. Hypothesis testing leads to the selection of a new ART_a category, which focuses attention on a new cluster of $\mathbf{a}^{(p)}$ input features that is better able to predict $\mathbf{b}^{(p)}$. Due to the combination of match tracking and fast learning, a single ARTMAP system can learn a different prediction for a rare event than for a cloud of similar frequent events in which it is embedded.

Whereas binary ARTMAP employs ART 1 systems for the ART_a and ART_b modules, Fuzzy ARTMAP substitutes Fuzzy ART systems for these modules. Fuzzy ART shows how computations from fuzzy set theory can be incorporated naturally into ART systems. For example, the intersection ($\cap$) operator that describes ART 1 dynamics is replaced by the AND operator ($\wedge$) of fuzzy set theory (Zadeh, 1965) in the choice, search, and learning laws of ART 1 (Figure 2). Especially noteworthy is the close relationship between the computation that defines fuzzy subsethood (Kosko, 1986) and the computation that defines category choice in ART 1. Replacing operation $\cap$ by operation $\wedge$ leads to a more powerful version of ART 1. Whereas ART 1 can learn stable categories only in response to binary input vectors, Fuzzy ART can learn stable categories in response to either analog or binary input vectors. Moreover, Fuzzy ART reduces to ART 1 in response to binary input vectors.

In Fuzzy ART, learning always converges because all adaptive weights are monotone nonincreasing. Without additional processing, this useful stability property could lead to the unattractive property of category proliferation as too many adaptive weights converge to zero. A preprocessing step, called complement coding, uses on-cell and off-cell responses to prevent category proliferation. Complement coding normalizes input vectors while preserving the amplitudes of individual feature activations. Without complement coding, an ART category memory encodes the degree to which critical features are consistently present in the training exemplars of that category. With complement coding, both the degree of absence and the degree of presence of features are represented by the category weight vector. The corresponding computations employ fuzzy OR ($\vee$, maximum) operators, as well as fuzzy AND ($\wedge$, minimum) operators.

This article includes summaries of the ART, Fuzzy ART, and Fuzzy ARTMAP systems. Section 2 describes the main characteristics of ART models, and Section 3 describes Fuzzy ART. Section 4 shows how two Fuzzy ART unsupervised learning modules are linked to form the Fuzzy ARTMAP supervised learning system. Sections 5 and 6 present two classes of benchmark simulation results. Section 5 describes a simulation task of learning to identify which points lie inside and which lie outside a given circle. Fuzzy ARTMAP on-line learning

ART 1 (BINARY)	FUZZY ART (ANALOG)

CATEGORY CHOICE

$$T_j = \frac{|\mathbf{I} \cap \mathbf{w}_j|}{\alpha + |\mathbf{w}_j|} \qquad T_j = \frac{|\mathbf{I} \wedge \mathbf{w}_j|}{\alpha + |\mathbf{w}_j|}$$

MATCH CRITERION

$$\frac{|\mathbf{I} \cap \mathbf{w}|}{|\mathbf{I}|} \geq \rho \qquad \frac{|\mathbf{I} \wedge \mathbf{w}|}{|\mathbf{I}|} \geq \rho$$

FAST LEARNING

$$\mathbf{w}_j^{(\text{new})} = \mathbf{I} \cap \mathbf{w}_j^{(\text{old})} \qquad \mathbf{w}_j^{(\text{new})} = \mathbf{I} \wedge \mathbf{w}_j^{(\text{old})}$$

$\cap$ = logical AND intersection $\qquad$ $\wedge$ = fuzzy AND minimum

Figure 2. Comparison of ART 1 and Fuzzy ART.

(also called incremental learning) is demonstrated, with test set accuracy increasing from 88.6% to 98.0% as the training set increased in size from 100 to 100,000 randomly chosen points. With off-line learning, the system needed from 2 to 13 epochs to learn all training set exemplars to 100% accuracy, where an epoch is defined as one cycle of training on an entire set of input exemplars. Test set accuracy then increased from 89.0% to 99.5% as the training set size increased from 100 to 100,000 points. Application of the voting strategy improved an average single-run accuracy of 90.5% on five runs to a voting accuracy of 93.9%, where each run trained on a fixed 1,000-item set for one epoch. These simulations are compared with studies by Wilensky (1990) of back propagation systems. These systems used at least 5,000 epochs to reach 90% accuracy on training and testing sets.

Section 6 describes Fuzzy ARTMAP performance on a benchmark letter recognition task developed by Frey and Slate (1991). Each database training exemplar represents a capital letter, in one of a variety of fonts and with significant random distortions, as a

16-dimensional feature vector. Each feature is assigned a value from 0 to 15. A number from 0 to 25 identifies the letters A–Z. Frey and Slate used this database to train a variety of classifiers that incorporate Holland-style genetic algorithms (Holland, 1980). Trained on 16,000 exemplars and tested on 4,000 exemplars, the best performing classifier had a test-set error rate of about 17.3%, more than three times the minimal error rate of an individual Fuzzy ARTMAP system (5.3%) and more than four times the error rate of a Fuzzy ARTMAP voting system (4.0%). In fact, application of the voting strategy improved an average accuracy of 93.9% on five separate runs to a voting accuracy of 96.0%. Moreover, this improved ARTMAP performance did not require greater memory resources: Fuzzy ART-MAP created fewer than 1,070 ART_a recognition categories in all simulations, compared to 1,040–1,302 rules created by the most accurate genetic algorithms.

2. ART Systems and Fuzzy Logic

Adaptive Resonance Theory, or ART, was introduced as a theory of human cognitive information processing (Grossberg, 1976, 1980). The theory has since led to an evolving series of real-time neural network models for unsupervised category learning and pattern recognition. These models are capable of learning stable recognition categories in response to arbitrary input sequences with either fast or slow learning. Model families include ART 1 (Carpenter and Grossberg, 1987a), which can stably learn to categorize binary input patterns presented in an arbitrary order; ART 2 (Carpenter and Grossberg, 1987b), which can stably learn to categorize either analog or binary input patterns presented in an arbitrary order; and ART 3 (Carpenter and Grossberg, 1990), which can carry out parallel search, or hypothesis testing, of distributed recognition codes in a multi-level network hierarchy. Variations of these models adapted to the demands of individual applications have been developed by a number of authors.

Figure 3 illustrates one example from the family of ART 1 models, and Figure 4 illustrates a typical ART search cycle. As shown in Figure 4a, an input vector **I** registers itself as a pattern **X** of activity across level F_1. The F_1 output vector **S** is then transmitted through the multiple converging and diverging adaptive filter pathways emanating from F_1. This transmission event multiplies the vector **S** by a matrix of adaptive weights, or long term memory (LTM) traces, to generate a net input vector **T** to level F_2. The internal competitive dynamics of F_2 contrast-enhance vector **T**. A compressed activity vector **Y** is thereby generated across F_2. In ART 1, the competition is tuned so that the F_2 node that receives the maximal $F_1 \to F_2$ input is selected. Only one component of **Y** is nonzero after this choice takes place. Activation of such a winner-take-all node defines the category, or symbol, of the input pattern **I**. Such a category represents all the inputs **I** that maximally activate the corresponding node.

Activation of an F_2 node may be interpreted as "making a hypothesis" about an input **I**. When **Y** is activated, it generates a signal vector **U** that is sent top-down through the second adaptive filter. After multiplication by the adaptive weight matrix of the top-down filter, a net vector **V** inputs to F_1 (Figure 4b). Vector **V** plays the role of a learned top-down expectation. Activation of **V** by **Y** may be interpreted as "testing the hypothesis" **Y**, or "reading out the category prototype" **V**. The ART 1 network is designed to match the "expected prototype" **V** of the category against the active input pattern, or exemplar, **I**.

This matching process may change the F_1 activity pattern **X** by suppressing activation

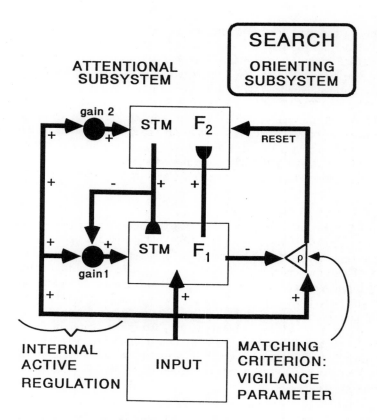

Figure 3. Typical ART 1 neural network (Carpenter and Grossberg, 1987a).

of all the feature detectors in **I** that are not confirmed by **V**. The resultant pattern **X*** encodes the pattern of features to which the network "pays attention". If the expectation **V** is close enough to the input **I**, then a state of *resonance* occurs as the attentional focus takes hold. The resonant state persists long enough for learning to occur; hence the term *adaptive resonance* theory. ART 1 learns prototypes, rather than exemplars, because the attended feature vector **X***, rather than the input **I** itself, is learned.

The criterion of an acceptable match is defined by a dimensionless parameter called *vigilance*. Vigilance weighs how close the input exemplar **I** must be to the top-down prototype **V** in order for resonance to occur. Because vigilance can vary across learning trials, recognition categories capable of encoding widely differing degrees of generalization, or morphological variability, can be learned by a single ART system. Low vigilance leads to broad generalization and abstract prototypes. High vigilance leads to narrow generalization and to prototypes that represent fewer input exemplars. In the limit of very high vigilance, prototype learning reduces to exemplar learning. Thus a single ART system may be used, say, to recognize abstract categories of faces and dogs, as well as individual faces and dogs.

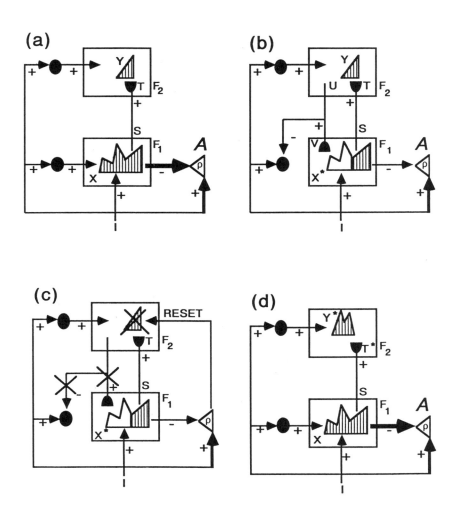

Figure 4. ART search for an F_2 code: (a) The input pattern **I** generates the specific STM activity pattern **X** at F_1 as it nonspecifically activates the orienting subsystem A. Pattern **X** both inhibits A and generates the output signal pattern **S**. Signal pattern **S** is transformed into the input pattern **T**, which activates the STM pattern **Y** across F_2. (b) Pattern **Y** generates the top-down signal pattern **U** which is transformed into the prototype pattern **V**. If **V** mismatches **I** at F_1, then a new STM activity pattern **X*** is generated at F_1. The reduction in total STM activity which occurs when **X** is transformed into **X*** causes a decrease in the total inhibition from F_1 to A. (c) If the matching criterion fails to be met, A releases a nonspecific arousal wave to F_2, which resets the STM pattern **Y** at F_2. (d) After **Y** is inhibited, its top-down prototype signal is eliminated, and **X** can be reinstated at F_1. Enduring traces of the prior reset lead **X** to activate a different STM pattern **Y*** at F_2. If the top-down prototype due to **Y*** also mismatches **I** at F_1, then the search for an appropriate F_2 code continues.

If the top-down expectation **V** and the bottom-up input **I** are too novel, or unexpected, to satisfy the vigilance criterion, then a bout of hypothesis testing, or memory search, is triggered. Search leads to selection of a better recognition code, symbol, category, or hypothesis to represent input **I** at level F_2. An *orienting subsystem* mediates the search process (Figure 3). The orienting subsystem interacts with the attentional subsystem, as in Figures 4c and 4d, to enable the attentional subsystem to learn about novel inputs without risking unselective forgetting of its previous knowledge.

The search process prevents associations from forming between **Y** and **X*** if **X*** is too different from **I** to satisfy the vigilance criterion. The search process resets **Y** before such an association can form. A familiar category may be selected by the search if its prototype is similar enough to the input **I** to satisfy the vigilance criterion. The prototype may then be refined in light of new information carried by **I**. If **I** is too different from any of the previously learned prototypes, then an uncommitted F_2 node is selected and learning of a new category is initiated.

A network parameter controls how deeply the search proceeds before an uncommitted node is chosen. As learning of a particular category self-stabilizes, all inputs coded by that category access it directly in a one-pass fashion, and search is automatically disengaged. The category selected is, then, the one whose prototype provides the globally best match to the input pattern. Learning can proceed on-line, and in a stable fashion, with familiar inputs directly activating their categories, while novel inputs continue to trigger adaptive searches for better categories, until the network's memory capacity is fully utilized.

The read-out of the top-down expectation **V** may be interpreted as a type of hypothesis-driven query. The matching process at F_1 and the hypothesis testing process at F_2 may be interpreted as query-driven symbolic substitutions. From this perspective, ART systems provide examples of new types of self-organizing production systems (Laird, Newell, and Rosenbloom, 1987). By incorporating predictive feedback into their control of the hypothesis testing cycle, ARTMAP systems embody self-organizing production systems that are also goal-oriented. ARTMAP systems are thus a new type of self-organizing expert system which is capable of stable autonomous fast learning about nonstationary environments that may contain a great deal of morphological variability. The fact that fuzzy logic may also be usefully incorporated into ARTMAP systems blurs even further the traditional boundaries between artificial intelligence and neural networks.

The Fuzzy ART model incorporates the design features of other ART models due to the close formal homolog between ART 1 and Fuzzy ART operations. Figure 2 summarizes how the ART 1 operations of category choice, matching, search, and learning translate into Fuzzy ART operations by replacing the set theory intersection operator ($\cap$) of ART 1 by the fuzzy set theory conjunction, or MIN operator ($\wedge$). Despite this close formal homology, Fuzzy ART is described as an algorithm, rather than a locally defined neural model. A neural network realization of Fuzzy ART is described elsewhere (Carpenter, Grossberg, and Rosen, 1991b). For the special case of binary inputs and fast learning, the computations of Fuzzy ART are identical to those of the ART 1 neural network. The Fuzzy ART algorithm also includes two optional features, one concerning learning and the other input preprocessing, as described in Section 3.

3. Summary of the Fuzzy ART Algorithm

ART field activity vectors: Each ART system includes a field F_0 of nodes that represent a current input vector; a field F_1 that receives both bottom-up input from F_0 and top-down input from a field F_2 that represents the active code, or category (Figure 3). The F_0 activity vector is denoted $\mathbf{I} = (I_1, \ldots, I_M)$, with each component I_i in the interval $[0,1]$, $i = 1, \ldots, M$. The F_1 activity vector is denoted $\mathbf{x} = (x_1, \ldots, x_M)$ and the F_2 activity vector is denoted $\mathbf{y} = (y_1, \ldots, y_N)$. The number of nodes in each field is arbitrary.

Weight vector: Associated with each F_2 category node $j (j = 1, \ldots, N)$ is a vector $\mathbf{w}_j \equiv (w_{j1}, \ldots, w_{jM})$ of adaptive weights, or LTM traces. Initially

$$w_{j1}(0) = \ldots = w_{jM}(0) = 1; \tag{1}$$

then each category is said to be *uncommitted*. After a category is selected for coding it becomes *committed*. As shown below, each LTM trace w_{ji} is monotone nonincreasing through time and hence converges to a limit. The Fuzzy ART weight vector $\mathbf{w}_j$ subsumes both the bottom-up and top-down weight vectors of ART 1.

Parameters: Fuzzy ART dynamics are determined by a choice parameter $\alpha > 0$; a learning rate parameter $\beta \in [0,1]$; and a vigilance parameter $\rho \in [0,1]$.

Category choice: For each input $\mathbf{I}$ and F_2 node j, the *choice function* T_j is defined by

$$T_j(\mathbf{I}) = \frac{|\mathbf{I} \wedge \mathbf{w}_j|}{\alpha + |\mathbf{w}_j|}, \tag{2}$$

where the fuzzy AND (Zadeh, 1965) operator $\wedge$ is defined by

$$(\mathbf{p} \wedge \mathbf{q})_i \equiv \min(p_i, q_i) \tag{3}$$

and where the norm $|\cdot|$ is defined by

$$|\mathbf{p}| \equiv \sum_{i=1}^{M} |p_i|. \tag{4}$$

for any M-dimensional vectors $\mathbf{p}$ and $\mathbf{q}$. For notational simplicity, $T_j(\mathbf{I})$ in (2) is often written as T_j when the input $\mathbf{I}$ is fixed.

The system is said to make a *category choice* when at most one F_2 node can become active at a given time. The category choice is indexed by J, where

$$T_J = \max\{T_j : j = 1 \ldots N\}. \tag{5}$$

If more than one T_j is maximal, the category j with the smallest index is chosen. In particular, nodes become committed in order $j = 1, 2, 3, \ldots$. When the Jth category is chosen, $y_J = 1$; and $y_j = 0$ for $j \neq J$. In a choice system, the F_1 activity vector $\mathbf{x}$ obeys the equation

$$\mathbf{x} = \begin{cases} \mathbf{I} & \text{if } F_2 \text{ is inactive} \\ \mathbf{I} \wedge \mathbf{w}_J & \text{if the } J^{th} F_2 \text{ node is chosen.} \end{cases} \tag{6}$$

Resonance or reset: *Resonance* occurs if the *match function* $|\mathbf{I} \wedge \mathbf{w}_J|/|\mathbf{I}|$ of the chosen category meets the vigilance criterion:

$$\frac{|\mathbf{I} \wedge \mathbf{w}_J|}{|\mathbf{I}|} \geq \rho; \tag{7}$$

that is, by (6), when the Jth category is chosen, resonance occurs if

$$|\mathbf{x}| = |\mathbf{I} \wedge \mathbf{w}_J| \geq \rho|\mathbf{I}|. \tag{8}$$

Learning then ensues, as defined below. *Mismatch reset* occurs if

$$\frac{|\mathbf{I} \wedge \mathbf{w}_J|}{|\mathbf{I}|} < \rho; \tag{9}$$

that is, if

$$|\mathbf{x}| = |\mathbf{I} \wedge \mathbf{w}_J| < \rho|\mathbf{I}|. \tag{10}$$

Then the value of the choice function T_J is set to 0 for the duration of the input presentation to prevent the persistent selection of the same category during search. A new index J is then chosen, by (5). The search process continues until the chosen J satisfies (7).

Learning: Once search ends, the weight vector $\mathbf{w}_J$ is updated according to the equation

$$\mathbf{w}_J^{(\text{new})} = \beta(\mathbf{I} \wedge \mathbf{w}_J^{(\text{old})}) + (1-\beta)\mathbf{w}_J^{(\text{old})}. \tag{11}$$

Fast learning corresponds to setting $\beta = 1$. The learning law used in the EACH system of Salzberg (1990) is equivalent to equation (11) in the fast-learn limit with the complement coding option described below.

Fast-commit slow-recode option: For efficient coding of noisy input sets, it is useful to set $\beta = 1$ when J is an uncommitted node, and then to take $\beta < 1$ after the category is committed. Then $\mathbf{w}_J^{(\text{new})} = \mathbf{I}$ the first time category J becomes active. Moore (1989) introduced the learning law (11), with fast commitment and slow recoding, to investigate a variety of generalized ART 1 models. Some of these models are similar to Fuzzy ART, but none includes the complement coding option. Moore described a category proliferation problem that can occur in some analog ART systems when a large number of inputs erode the norm of weight vectors. Complement coding solves this problem.

Input normalization/complement coding option: Proliferation of categories is avoided in Fuzzy ART if inputs are normalized. *Complement coding* is a normalization rule that preserves amplitude information. Complement coding represents both the on-response and the off-response to an input vector **a** (Figure 1). To define this operation in its simplest form, let **a** itself represent the on-response. The complement of **a**, denoted by $\mathbf{a}^c$, represents the off-response, where

$$a_i^c \equiv 1 - a_i. \tag{12}$$

The complement coded input **I** to the field F_1 is the 2M-dimensional vector

$$\mathbf{I} = (\mathbf{a}, \mathbf{a}^c) \equiv (a_1, \ldots, a_M, a_1^c, \ldots, a_M^c). \tag{13}$$

Note that

$$|\mathbf{I}| = |(\mathbf{a}, \mathbf{a}^c)|$$
$$= \sum_{i=1}^{M} a_i + (M - \sum_{i=1}^{M} a_i) \qquad (14)$$
$$= M,$$

so inputs preprocessed into complement coding form are automatically normalized. Where complement coding is used, the initial condition (1) is replaced by

$$w_{j1}(0) = \ldots = w_{j,2M}(0) = 1. \qquad (15)$$

4. Fuzzy ARTMAP Algorithm

The Fuzzy ARTMAP system incorporates two Fuzzy ART modules ART_a and ART_b that are linked together via an inter-ART module F^{ab} called a *map field*. The map field is used to form predictive associations between categories and to realize the *match tracking rule* whereby the vigilance parameter of ART_a increases in response to a predictive mismatch at ART_b. Match tracking reorganizes category structure so the predictive error is not repeated on subsequent presentations of the input. A circuit realization of the match tracking rule that uses only local real-time operations is provided in Carpenter, Grossberg, and Reynolds, (1991). The interactions mediated by the map field F^{ab} may be operationally characterized as follows.

ART_a and ART_b

Inputs to ART_a and ART_b are in the complement code form: for ART_a, $\mathbf{I} = \mathbf{A} = (\mathbf{a}, \mathbf{a}^c)$; for ART_b, $\mathbf{I} = \mathbf{B} = (\mathbf{b}, \mathbf{b}^c)$ (Figure 1). Variables in ART_a or ART_b are designated by subscripts or superscripts "a" or "b". For ART_a, let $\mathbf{x}^a \equiv (x_1^a \ldots x_{2M_a}^a)$ denote the F_1^a output vector; let $\mathbf{y}^a \equiv (y_1^a \ldots y_{N_a}^a)$ denote the F_2^a output vector; and let $\mathbf{w}_j^a \equiv (w_{j1}^a, w_{j2}^a, \ldots, w_{j,2M_a})$ denote the jth ART_a weight vector. For ART_b, let $\mathbf{x}^b \equiv (x_1^b \ldots x_{2M_b}^b)$ denote the F_1^b output vector; let $\mathbf{y}^b \equiv (y_1^b \ldots y_{N_b}^b)$ denote the F_2^b output vector; and let $\mathbf{w}_k^b \equiv (w_{k1}^b, w_{k2}^b, \ldots, w_{k,2M_b}^b)$ denote the kth ART_b weight vector. For the map field, let $\mathbf{x}^{ab} \equiv (x_1^{ab}, \ldots, x_{N_b}^{ab})$ denote the F^{ab} output vector, and let $\mathbf{w}_j^{ab} \equiv (w_{j1}^{ab}, \ldots, w_{jN_b}^{ab})$ denote the weight vector from the jth F_2^a node to F^{ab}. Vectors $\mathbf{x}^a, \mathbf{y}^a, \mathbf{x}^b, \mathbf{y}^b$, and $\mathbf{x}^{ab}$ are set to $\mathbf{0}$ between input presentations.

Map field activation

The map field F^{ab} is activated whenever one of the ART_a or ART_b categories is active. If node J of F_2^a is chosen, then its weights $\mathbf{w}_J^{ab}$ activate F^{ab}. If node K in F_2^b is active, then the node K in F^{ab} is activated by 1–to–1 pathways between F_2^b and F^{ab}. If both ART_a and ART_b are active, then F^{ab} becomes active only if ART_a predicts the same category as ART_b via the weights $\mathbf{w}_J^{ab}$. The F^{ab} output vector $\mathbf{x}^{ab}$ obeys

$$\mathbf{x}^{ab} = \begin{cases} \mathbf{y}^b \wedge \mathbf{w}_J^{ab} & \text{if the Jth } F_2^a \text{ node is active and } F_2^b \text{ is active} \\ \mathbf{w}_J^{ab} & \text{if the Jth } F_2^a \text{ node is active and } F_2^b \text{ is inactive} \\ \mathbf{y}^b & \text{if } F_2^a \text{ is inactive and } F_2^b \text{ is active} \\ \mathbf{0} & \text{if } F_2^a \text{ is inactive and } F_2^b \text{ is inactive.} \end{cases} \qquad (16)$$

By (16), $\mathbf{x}^{ab} = \mathbf{0}$ if the prediction $\mathbf{w}_J^{ab}$ is disconfirmed by $\mathbf{y}^b$. Such a mismatch event triggers an ART_a search for a better category, as follows.

Match tracking

At the start of each input presentation the ART_a vigilance parameter ρ_a equals a baseline vigilance $\overline{\rho_a}$. The map field vigilance parameter is ρ_{ab}. If

$$|\mathbf{x}^{ab}| < \rho_{ab}|\mathbf{y}^b|, \tag{17}$$

then ρ_a is increased until it is slightly larger than $|\mathbf{A} \wedge \mathbf{w}_J^a||\mathbf{A}|^{-1}$, where $\mathbf{A}$ is the input to F_1^a, in complement coding form. Then

$$|\mathbf{x}^a| = |\mathbf{A} \wedge \mathbf{w}_J^a| < \rho_a|\mathbf{A}|, \tag{18}$$

where J is the index of the active F_2^a node, as in (10). When this occurs, ART_a search leads either to activation of another F_2^a node J with

$$|\mathbf{x}^a| = |\mathbf{A} \wedge \mathbf{w}_J^a| \geq \rho_a|\mathbf{A}| \tag{19}$$

and

$$|\mathbf{x}^{ab}| = |\mathbf{y}^b \wedge \mathbf{w}_J^{ab}| \geq \rho_{ab}|\mathbf{y}^b|; \tag{20}$$

or, if no such node exists, to the shut-down of F_2^a for the remainder of the input presentation.

Map field learning

Learning rules determine how the map field weights w_{jk}^{ab} change through time, as follows. Weights w_{jk}^{ab} in $F_2^a \to F^{ab}$ paths initially satisfy

$$w_{jk}^{ab}(0) = 1. \tag{21}$$

During resonance with the ART_a category J active, $\mathbf{w}_J^{ab}$ approaches the map field vector $\mathbf{x}^{ab}$. With fast learning, once J learns to predict the ART_b category K, that association is permanent; i.e., $w_{JK}^{ab} = 1$ for all time.

5. Simulation: Circle-in-the-Square

The circle-in-the-square problem requires a system to identify which points of a square lie inside and which lie outside a circle whose area equals half that of the square. This task was specified as a benchmark problem for system performance evaluation in the DARPA Artificial Neural Network Technology (ANNT) Program (Wilensky, 1990). Wilensky examined the performance of 2-n-1 back propagation systems on this problem. He studied systems where the number (n) of hidden units ranged from 5 to 100, and the corresponding number of weights ranged from 21 to 401. Training sets ranged in size from 150 to 14,000. To avoid over-fitting, training was stopped when accuracy on the training set reached 90%. This criterion level was reached most quickly (5,000 epochs) in systems with 20 to 40 hidden units. In this condition, approximately 90% of test set points, as well as training set points, were correctly classified.

Fuzzy ARTMAP performance on this task after one training epoch is illustrated in Figures 5 and 6. As training set size increased from 100 exemplars (Figure 5a) to 100,000 exemplars (Figure 5d) the rate of correct test set predictions increased from 88.6% to 98.0% while the number of ART_a category nodes increased from 12 to 121. Each category node j required four learned weights $\mathbf{w}_j^a$ in ART_a plus one map field weight $\mathbf{w}_j$ to record whether category j predicts that a point lies inside or outside the circle. Thus, for example, 1-epoch training on 100 exemplars used 60 weights to achieve 88.6% test set accuracy. The map can be made arbitrarily accurate provided the number of ART_a nodes is allowed to increase as needed.

Figure 5 shows how a test set error rate is reduced from 11.4% to 2.0% as training set size increases from 100 to 100,000 in 1-epoch simulations. Test set error rate can be further reduced if exemplars are presented for as many epochs as necessary to reach 100% accuracy on the training set. The ARTMAP voting strategy provides a third way to eliminate test set errors. Recall that the voting strategy assumes a fixed set of training exemplars. Before each individual simulation the input ordering is randomly assembled. After each simulation the prediction of each test set item is recorded. Voting selects the outcome predicted by the largest number of individual simulations. In case of a tie, one outcome is selected at random. The number of votes cast for a given outcome provides a measure of predictive confidence at each test set point. Given a limited training set, voting across a few simulations can improve predictive accuracy by a factor that is comparable to the improvement that could be attained by an order of magnitude more training set inputs, as shown in the following example.

A fixed set of 1,000 randomly chosen exemplars was presented to a Fuzzy ARTMAP system on five independent 1-epoch circle-in-the-square simulations. After each simulation, inside/outside predictions were recorded on a 1,000-item test set. Accuracy on individual simulations ranged from 85.9% to 92.3%, averaging 90.5%; and the system used from 15 to 23 ART_a nodes. Voting by the five simulations improved test set accuracy to 93.9% (Figure 6c). In other words, test set errors were reduced from an average individual rate of 9.5% to a voting rate of 6.1%. Figure 6d indicates the number of votes cast for each test set point, and hence reflects variations in predictive confidence across different regions. Voting by more than five simulations maintained an error rate between 5.8% and 6.1%. This limit on further improvement by voting appears to be due to random gaps in the fixed 1,000-item training set. By comparison, a ten-fold increase in the size of the training set reduced the error by an amount similar to that achieved by five-simulation voting. For example, in Figure 5b, 1-epoch training on 1,000 items yielded a test set error rate of 7.5%; while increasing the size of the training set to 10,000 reduced the test set error rate to 3.3% (Figure 5c).

6. Simulation: Letter Image Recognition

Frey and Slate (1991) recently developed a benchmark machine learning task that they describe as a "difficult categorization problem" (p. 161). The task requires a system to identify an input exemplar as one of 26 capital letters A–Z. The database was derived from 20,000 unique black-and-white pixel images. The difficulty of the task is due to the wide variety of letter types represented: the twenty "fonts represent five different stroke styles

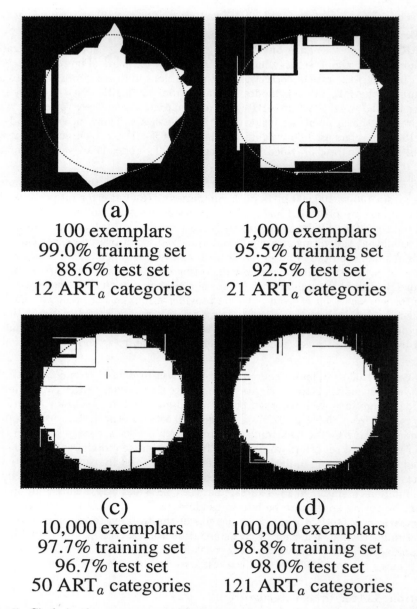

Figure 5. Circle-in-the-square test set response patterns after 1 epoch of Fuzzy ARTMAP training on (a) 100, (b) 1,000, (c) 10,000, and (d) 100,000 randomly chosen training set points. Test set points in white areas are predicted to lie inside the circle and points in black areas are predicted to lie outside the circle. The test set error rate decreases, approximately inversely to the number of ART_a categories, as the training set size increases.

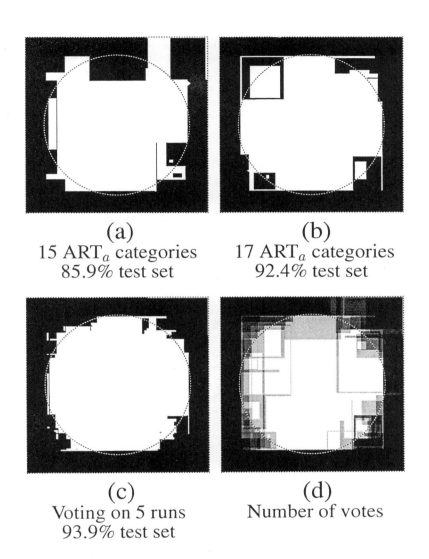

(a)
15 ART$_a$ categories
85.9% test set

(b)
17 ART$_a$ categories
92.4% test set

(c)
Voting on 5 runs
93.9% test set

(d)
Number of votes

Figure 6. Circle-in-the-square response patterns for a fixed 1,000-item training set. (a) Test set responses after training on inputs presented in random order. After 1 epoch that used 15 ART$_a$ nodes, test set prediction rate was 85.9%, the worst of 5 runs. (b) Test set responses after training on inputs presented in a different random order. After 1 epoch that used 23 ART$_a$ nodes, test set prediction rate was 92.3%, the best of 5 runs. (c) Voting strategy applied to five individual simulations. Test set prediction rate was 93.9%. (d) Cumulative test set response pattern of five 1-epoch simulations. Gray scale intensity increases with the number of votes cast for a point's being outside the circle.

(simplex, duplex, complex, and Gothic) and six different letter styles (block, script, italic, English, Italian, and German)" (p. 162). In addition each image was randomly distorted, leaving many of the characters misshapen. Sixteen numerical feature attributes were then obtained from each character image, and each attribute value was scaled to a range of 0 to 15. The resulting Letter Image Recognition file is archived in the UCI Repository of Machine Learning Databases and Domain Theories, maintained by David Aha and Patrick Murphy (ml_repository@ics.uci.edu).

Frey and Slate used this database to test performance of a family of classifiers based on Holland's genetic algorithms (Holland, 1980). The training set consisted of 16,000 exemplars, with the remaining 4,000 exemplars used for testing. Genetic algorithm classifiers having different input representations, weight update and rule creation schemes, and system parameters were systematically compared. Training was carried out for 5 epochs, plus a sixth "verification" pass during which no new rules were created but a large number of unsatisfactory rules were discarded. In Frey and Slate's comparative study, these systems had correct prediction rates that ranged from 24.5% to 80.8% on the 4,000-item test set. The best performance (80.8%) was obtained using an integer input representation, a reward sharing weight update, an exemplar method of rule creation, and a parameter setting that allowed an unused or erroneous rule to stay in the system for a long time before being discarded. After training, the optimal case, that had 80.8% performance rate, ended with 1,302 rules and 8 attributes per rule, plus over 35,000 more rules that were discarded during verification. (For purposes of comparison, a rule is somewhat analogous to an ART_a category in ARTMAP, and the number of attributes per rule is analogous to the size $|w_j^a|$ of ART_a category weight vectors.) Building on the results of their comparative study, Frey and Slate investigated two types of alternative algorithms, namely an accuracy-utility bidding system, that had slightly improved performance (81.6%) in the best case; and an exemplar/hybrid rule creation scheme that further improved performance, to a maximum of 82.7%, but that required the creation of over 100,000 rules prior to the verification step.

Fuzzy ARTMAP had an error rate on the letter recognition task that was consistently less than one third that of the three best Frey-Slate genetic algorithm classifiers described above. Moreover Fuzzy ARTMAP simulations each created fewer than 1,070 ART_a categories, compared to the 1,040–1,302 final rules of the three genetic classifiers with the best performance rates. With voting, Fuzzy ARTMAP reduced the error rate to 4.0% (Table 1). Most Fuzzy ARTMAP learning occurred on the first epoch, with test set performance on systems trained for one epoch typically over 97% that of systems exposed to inputs for the five epochs.

Table 1 shows how voting consistently improves performance. In each group, with $\alpha = 0.1$ or $\alpha = 1.0$ and with 1 or 5 training epochs, Fuzzy ARTMAP was run for 3 or 5 independent simulations, each with a different input order. In all cases voting performance was significantly better than performance of any of the individual simulations in a given group. In Table 1a, for example, voting caused the error rate to drop to 8.8%, from a 3-simulation average of 12.5%. With 1 training epoch, 3-simulation voting eliminated about 30–35% of the test set errors (Table 1a and 1c), and 5-simulation voting eliminated about 43% of the test set errors (Table 1e). In the 5-epoch simulations, where individual training set performance was close to 100%, 3-simulation voting still reduced the test set error rate by about 25% (Table 1b and 1d) and 5-simulation voting reduced the error rate by about 34% (Table 1f). The

TABLE 1

	% Correct Test Set Predictions	No. ART_a Categories	No. Epochs
(a) $\alpha = 0.1$ 3 simulations			
Average	87.5%	637	1
Range	87.0%-88.0%	619-661	1
Voting	91.2%		
(b) $\alpha = 0.1$ 3 simulations			
Average	89.7%	741	5
Range	89.3%-90.3%	726-757	5
Voting	92.2%		
(c) $\alpha = 1.0$ 3 simulations			
Average	92.1%	788	1
Range	91.8%-92.3%	762-807	1
Voting	94.8%		
(d) $\alpha = 1.0$ 3 simulations			
Average	94.0%	1,016	5
Range	93.8%-94.3%	988-1,055	5
Voting	95.5%		
(e) $\alpha = 1.0$ 5 simulations			
Average	91.8%	786	1
Range	91.2%-92.6%	763-805	1
Voting	95.3%		
(f) $\alpha = 1.0$ 5 simulations			
Average	93.9%	1,021	5
Range	93.4%-94.6%	990-1,070	5
Voting	96.0%		

Table 1. Voting strategy applied to sets of 3(a–d) or 5(e–f) Fuzzy ARTMAP simulations of the Frey-Slate character recognition task, with choice parameter $\alpha = 0.1$ (a,b) or $\alpha = 1.0$ (c–f); and with training on 1 epoch (a,c,e) or 5 epochs (b,d,f). (a) Voting eliminated 30% of the individual simulation test set errors, which dropped from a 3-simulation average rate of 12.5% to a voting rate of 8.8%. (b) Voting eliminated 24% of the errors, which dropped from 10.3% to 7.8%. (c) Voting eliminated 34% of the errors, which dropped from 7.9% to 5.2%. (d) Voting eliminated 25% of the errors, which dropped from 6.0% to 4.5%. (e) Voting eliminated 43% of the errors, which dropped from 8.2% to 4.7%. (f) Voting eliminated 34% of the errors, which dropped from 6.1% to 4.0%.

best overall results were obtained with $\alpha = 1.0$ and 5-epoch training, where voting reduced the 5-simulation average error rate of 6.1% to a voting error rate of 4.0% (Table 1f).

In summary, single-simulation fast-learn Fuzzy ARTMAP systems, with baseline vigilance $\overline{\rho_a} = 0$ and with choice parameters α ranging from 0.001 to 1.0, were trained on the 16,000-item input set of the Frey-Slate letter recognition task. After 1 to 5 epochs, individual Fuzzy ARTMAP systems had a robust prediction rate of 90% to 94% on the 4,000-item test set, with best performance obtained from the highest values of α. By pooling information across individual simulations, voting consistently eliminated 25%–43% of the errors giving a robust prediction rate of 92%–96%.

REFERENCES

Carpenter, G.A. and Grossberg, S. (1987a). A massively parallel architecture for a self-organizing neural pattern recognition machine. *Computer Vision, Graphics, and Image Processing*, **37**, 54–115.

Carpenter, G.A. and Grossberg, S. (1987b). ART 2: Stable self-organization of pattern recognition codes for analog input patterns. *Applied Optics*, **26**, 4919–4930.

Carpenter, G.A. and Grossberg, S. (1990). ART 3: Hierarchical search using chemical transmitters in self-organizing pattern recognition architectures. *Neural Networks*, **3**, 129–152.

Carpenter, G.A., Grossberg, S., Markuzon, N., Reynolds, J.H., and Rosen, D.B. (1992). Fuzzy ARTMAP: A neural network architecture for incremental supervised learning of analog multidimensional maps. *IEEE Transactions on Neural Networks*, in press.

Carpenter, G.A., Grossberg, S. and Reynolds, J.H. (1991). ARTMAP: Supervised real-time learning and classification of nonstationary data by a self-organizing neural network. *Neural Networks*, **4**, 565–588.

Carpenter, G.A., Grossberg, S., and Rosen, D.B. (1991a). Fuzzy ART: Fast stable learning and categorization of analog patterns by an adaptive resonance system. *Neural Networks*, **4**, 759–771.

Carpenter, G.A., Grossberg, S., and Rosen, D.B. (1991b). A neural network realization of Fuzzy ART. Technical Report CAS/CNS TR-91-021. Boston, MA: Boston University.

Frey, P.W. and Slate, D.J. (1991). Letter recognition using Holland-style adaptive classifiers. *Machine Learning*, **6**, 161–182.

Grossberg, S. (1976). Adaptive pattern classification and universal recoding, II: Feedback, expectation, olfaction, and illusions. *Biological Cybernetics*, **23**, 187–202.

Grossberg, S. (1980). How does a brain build a cognitive code? *Psychological Review*, **1**, 1–51.

Holland, J.H. (1980). Adaptive algorithms for discovering and using general patterns in growing knowledge bases. *International Journal of Policy Analysis and Information Systems*, **4**, 217–240.

Kosko, B. (1986). Fuzzy entropy and conditioning. *Information Sciences*, **40**, 165–174.

Laird, J.E., Newell, A., and Rosenbloom, P.S. (1987). SOAR: An architecture for general intelligence. *Artificial Intelligence*, **33**, 1–64.

Moore, B. (1989). ART 1 and pattern clustering. In D. Touretzky, G. Hinton, and T. Sejnowski (Eds.), **Proceedings of the 1988 connectionist models summer school**, pp. 174–185. San Mateo, CA: Morgan Kaufmann Publishers.

Rumelhart, D.E., Hinton, G., and Williams, R. (1986). Learning internal representations by error propagation. In D.E. Rumelhart and J.L. McClelland (Eds.). **Parallel distributed processing**. Cambridge, MA: MIT Press.

Salzberg, S.L. (1990). **Learning with nested generalized exemplars**. Hingham, MA: Kluwer Academic Publishers.

Wilensky, G. (1990). Analysis of neural network issues: Scaling, enhanced nodal processing, comparison with standard classification. DARPA Neural Network Program Review, October 29–30, 1990.

Zadeh, L. (1965). Fuzzy sets. *Information Control*, **8**, 338–353.

SYNCHRONIZED OSCILLATIONS FOR BINDING SPATIALLY DISTRIBUTED FEATURE CODES INTO COHERENT SPATIAL PATTERNS

Stephen Grossberg[†] and David Somers[‡]
Center for Adaptive Systems
and
Department of Cognitive and Neural Systems
Boston University
111 Cummington Street
Boston, MA 02215

Abstract

Neural network models are described for binding out-of-phase activations of spatially distributed cells into synchronized oscillations within a single processing cycle. These results suggest how the brain may overcome the temporal "jitter" inherent in multi-level processing of spatially distributed data. Coherent synchronous patterns of spatially distributed features are formed to represent and learn about external objects and events. Temporal jitter thus does not typically cause scenic parts to be combined into the wrong visual objects. During preattentive vision, such patterns may represent emergent boundary segmentations, including illusory contours. During attentive visual object recognition, such patterns may occur during an attentive resonant state that triggers new learning. Different properties of preattentive and attentive oscillations are predicted, and compared with neurophysiological data concerning rapid synchronization of cell activations in visual cortex.

[†] Supported in part by the Air Force Office of Scientific Research (AFOSR 90-0175), DARPA (AFOSR 90-0083), and the Office of Naval Research (N00014-91-J-4100).
[‡] Supported in part by NASA (NGT-50497).
Acknowledgements: The authors wish to thank Cynthia E. Bradford, Carol Y. Jefferson, and Kelly Dumont for their valuable assistance in the preparation of the manuscript.

1. Introduction: The Role of Synchrony in Spatial Pattern Coding

This chapter discusses neural network models that are capable of rapidly synchronizing the activities of spatially distributed cells within a single processing cycle. The chapter also reviews electrophysiological data describing rapid synchronization of spatially distributed cell activations in the visual cortex, and suggests a theoretical explanation for the functional role of such synchronization *in vivo*.

A key idea in this theoretical analysis concerns the hypothesis that the functional units of short term memory (STM) and long term memory (LTM) in the brain are *spatial patterns*. This hypothesis was originally derived from a mathematical analysis of the interactive dynamics of STM and LTM during associative learning by a neural network (Grossberg, 1968, 1969, 1971, 1982b). For definiteness, denote the input to the ith cell (or cell population) v_i by I_i. It was proved, under rather general conditions, that neural networks tend to learn the *spatial pattern* of relative input weights

$$\theta_i = \frac{I_i}{\sum_{k=1}^{n} I_k}. \tag{1}$$

The values θ_i measure the *relative* importance of the features coded by the cells v_i in the input pattern $(I_1, I_2, ..., I_n)$. It was also proved that the total input strength

$$I = \sum_{k=1} I_k \tag{2}$$

is used as energy to drive the learning process. When the relative importance of features remains fixed during a given time interval, the weights θ_i remain approximately constant, even if the total input I varies wildly through time. This may be expected to occur, for example, during sustained inspection of a fixed visual image. Thus the network tends to learn about the "temporally coherent part" of an input pattern. It tends to *factorize* the temporally coherent information $(\theta_1, \theta_2, ..., \theta_n)$ from the fluctuating energy I that mobilizes the learning process.

The mathematical analyses which led to the conclusion that the functional units of short term memory and long term memory are spatial patterns also provided a definition that could support this conclusion; namely, spatial patterns are temporally synchronized, spatially distributed activations that preserve their relative ordering through time. These results led to a series of investigations, some of which are reviewed in Grossberg (1982a, 1982b, 1988) and in Section 3 below.

Given these results, a potentially serious problem for brain processing can be described using Figure 1 as a guide. Figure 1 considers the processing of a visual image whose various parts are registered approximately synchronously at the eyes. Subsequent events at multiple processing stages may, however, destroy this synchrony. For example, spatially inhomogeneous noise may enable some cells to fire faster than others. Prior processing of previous images may also inhomogeneously habituate some neural pathways more than others, thereby slowing the processing rate in some pathways more than others. In all, even if an external source of spatially distributed information is synchronous, it may be represented with phase leads and lags at subsequent neural processing stages. If these various parts of the processed image are not ultimately recombined into synchronized spatial patterns, then they cannot be correctly learned and recognized by the brain as part of the same event.

Put another way, if the temporal "jitter" in the neural processing of distributed inputs is not corrected, then parts of different images that occur at different times could be scrambled together, leading to a maladaptive reorganization of scenic parts into the "wrong" visual objects. Although such scrambling can be made to occur *in vivo* using sufficient experimental

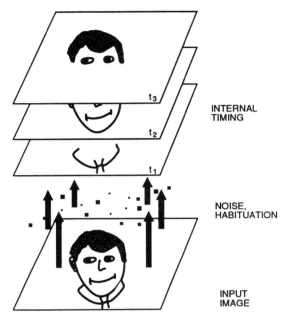

Figure 1. Multilevel processing of an image in noise may lead to phase leads and lags in the processing of different image regions. These temporally out-of-phase activations may be rapidly recombined into coherent spatial patterns by synchronized oscillations.

ingenuity, as in the case of "illusory conjunctions" (Treisman and Gelade, 1980), it would be all too frequent, with devastating results, in the absence of compensatory neural mechanisms. Synchronous binding of features into coherent spatial pattern codes has also been described as part of the process whereby complex acoustic signals are grouped into unitary auditory events (Bregman, 1990).

We suggest one such compensatory mechanism that reorganizes temporally out-of-phase spatial data into a synchronous spatial pattern, just so long as the phase lags and leads of these data lie within a critical time interval. In order to be effective during real-time perception, during which a series of rapidly changing images must be correctly processed, the synchronizer must be able to act quickly, within one or two processing cycles. This property holds in our model, as well as in the data.

2. Synchrony in Visual Cortical Cells

Using simultaneous, multielectrode, extracellular recordings, two labs (Eckhorn, Bauer, Jordan, Brosch, Kruse, Munk, and Reitbock, 1988; Gray and Singer, 1989; Gray, Konig, Engel, and Singer, 1989) have reported stimulus-evoked synchronized oscillations of 40–60 Hz in the primary visual cortex (Areas 17 and 18) of the cat. Coherence or synchrony of firing activity was found between cells within a cortical column (Eckhorn et al., 1988; Gray and Singer, 1989), in neighboring hypercolumns (Eckhorn et al., 1988; Gray et al., 1989), in distant hypercolumns (Gray et al., 1989), and lying in two different cortical areas (Eckhorn et al, 1988). Stimulus position, orientation, movement direction, and velocity were among the stimulus properties that yielded stimulus-evoked resonances. Synchronized oscillatory responses were frequently found at distant cortical positions when at least one of the primary coding properties was similar.

Using a single long moving bar as a stimulus, Gray et al. (1989) recorded simultaneously from cells which were separated by large cortical distances. They found that for cortical separations great enough that the receptive fields of the cells did not overlap (> 2 mm), coherent oscillations occurred only between cells with similar orientation preferences. Even at these large separations, the cross-correlations of the firing patterns of the cells indicated a tight synchrony, on average having a 0 ms phase lag. Nearly all phase-locked cells showed activity peaks within 3 ms of each other. Thus assuming a 25 ms period of oscillation, phase differences were typically less than 12% of the period.

Gray et al. repeated this procedure using two short disconnected bars as the stimuli. Using large interbar distances, only one bar would pass through the receptive field of one cell, and the second bar would pass through the receptive field of a second cell. When the two bars were moved colinearly in the same direction at the same speed, these two distant cells frequently synchronized their oscillations even though there was a gap between the stimuli. When the bars were moved in opposite directions, no synchrony resulted.

These stimulus-evoked resonances have been interpreted as reflecting the global properties of the stimulus, properties which might be expected to cohere in a single object. For instance, in the two bar stimulus paradigm, colinear but disconnected bars moving in the same direction at the same speed may be perceptually interpreted as a single continuous contour that is occluded in the middle, whereas two bars moving in opposite directions (without rotation) are likely to be perceptually interpreted as different contours. Segregation across occluding regions is a common problem that the visual system must solve regularly. In nature, occlusion may arise due to internal sources such as retinal veins in front of the photoreceptors or external sources such as a tree branch between the observer and the object of interest.

Before the present work was carried out, attempts to explain these oscillatory phenomena have typically been restricted to formal equations for the phase relations among abstractly defined oscillators (Atiya and Baldi, 1989; Baldi and Meir, 1990; Kammen, Holmes, and Koch, 1989). Herein we explain how suitably designed neural networks can give rise to such oscillations as an emergent property of their real-time dynamics. Moreover, we use neural networks which have previously been derived to explain and predict behavioral and neural data, other than the oscillatory phenomena themselves.

3. Different Predictions About Preattentive Visual Coding and Attentive Recognition Learning

Based upon the analyses of spatial pattern coding such as those reviewed in Section 1, Grossberg (1976b, 1978c) predicted that cortical codes would be expressed by resonant standing waves in which cooperatively linked cells oscillate in phase with one another. The mathematical analysis of such synchronous oscillations was begun in Ellias and Grossberg, (1975). It was also noted that these standing waves could be replaced by approach to an equilibrium point, or attractor, if no "slow" variables, such as inhibitory interneurons or chemical modulators, exist in the network. Both standing waves and equilibria can, in principle, support a feature-based cortical code. The standing waves were called "order-preserving limit cycles" to emphasize that the ordering, or relative importance, of feature detector activations should persist during each coding cycle, even if their absolute activations vary through time as the oscillation unfolds.

Mathematical analyses of both the standing wave and equilibrium point models were initiated in the 1970's. Studies of equilibrium point models led to a series of mathematical theorems, including a general theory for globally analysing equilibria and oscillations in competitive neural networks (Grossberg, 1978a, 1978b, 1980), and the Cohen-Grossberg model and theorem for content addressable memory (Cohen and Grossberg, 1983; Grossberg, 1982a). The Cohen-Grossberg model was designed to include the additive model, subsequently studied by Hopfield (1984), as well as the shunting model that describes interactions between cells that obey a membrane equation; see Grossberg (1988) for an historical overview. The present article continues the analysis of standing waves that was initiated in Ellias and Grossberg (1975).

The standing wave prediction was made in the context of a theory, called Adaptive Resonance Theory, or ART, which analyses the role of reciprocal top-down and bottom-up cortico-cortical and thalamo-cortical adaptive filters in the development of cortical feature detectors, recognition learning, attentional processing, and memory search (Grossberg, 1976a, 1976b, 1978c, 1982b). Within ART, a resonant standing wave can occur when bottom-up and top-down signals fuse into an attentional focus. Such an attentional focus can support new learning as it gives rise to a conscious perceptual experience. The predicted linkage between standing waves, attention, and conscious experience has recently begun to attract the interest of a large number of investigators.

Mathematical investigations of complete ART architectures have heretofore tended to analyse equilibrium point models (Figure 2), wherein slow variables that could sustain oscillations are eliminated for simplicity (Carpenter and Grossberg, 1987a, 1987b, 1990; Carpenter, Grossberg, and Reynolds, 1991; Carpenter, Grossberg, Markuzon, Reynolds, and Rosen, 1991). The present results illustrate how the ART standing waves predicted in Grossberg (1976b, 1978c) can be generated by the type of bottom-up and top-down feedback interactions among adaptive filters that are used in ART circuits.

After ART was introduced in order to analyse attentive learning and recognition, subsequent research by Grossberg and Mingolla focused on processes of preattentive vision. In this work, a new type of cortical cell, called a *bipole cell*, was predicted to exist (Grossberg, 1984, 1987a, 1987b; Grossberg and Mingolla, 1985a, 1985b, 1987). Bipole cells cooperatively link perceptual features into emergent boundary segmentations via cooperative-competitive feedback signals in a network called the CC Loop. The CC Loop is part of a larger neural model, called the Boundary Contour System (BCS), which suggested new perceptual roles for cortical simple cells, complex cells, hypercomplex cells, and bipole cells (Figure 3). In addition, the BCS was used to explain and predict a variety of psychophysical and perceptual data, notably data about texture segregation, subjective contours, filling-in of brightness and color, and 3-D surface perception.

The bipole cell receptive field was predicted to consist of two long, thin independent flanks which non-linearly sum inputs from cells with orientation preferences similar to the orientation of the long axis of the bipole (Figure 4). Bipole cell output signals can be activated if and only if both flanks are sufficiently activated. These signals feed excitatory input back to model hypercomplex cells in a lower network layer which have the same orientation preference and are positioned near the middle of the bipole cell. Through this cooperative feedback cycle, two disconnected but colinear contours can induce a boundary completion between them.

Although the bipole cell was predicted on perceptual grounds, its existence was soon supported by neurophysiological data. Recording from Area 18 of alert monkeys, von der Heydt, Peterhans, and Baumgartner (1984) found cells that responded to moving illusory contours. That is, with no local luminance information to suggest a contour and only with colinear inducing lines lying beyond the receptive fields of the cells, cells responded strongly when the global percept of the stimulus suggested an illusory contour. When the stimulus was altered so that an illusory contour was no longer perceived, the cells did not respond. Inducing lines on both sides of the site were necessary for the perception of the illusory contours by both the cells and human observers. These data indicate that some cells in visual cortex respond well to subjective contours, and that cells which are not directly activated by bottom-up signals require input from at least two colinear flanking regions in order to be activated.

In addition to this electrophysiological data, Gilbert and Wiesel (1989) provided ana-

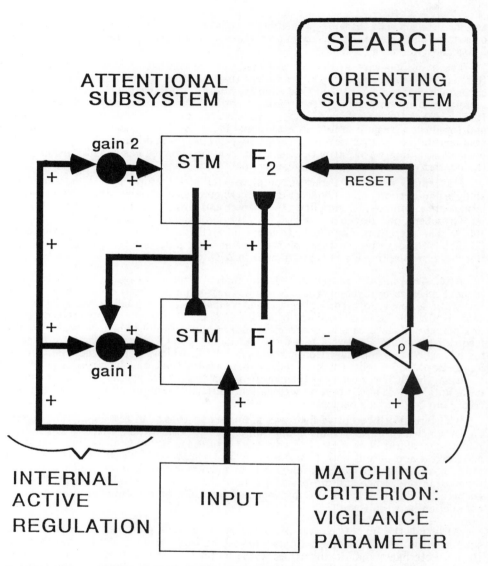

Figure 2: ART 1 system: Two successive stages, F_1 and F_2, of the attentional subsystem encode patterns of activation in short term memory (STM). Bottom-up and top-down adaptive filter pathways between F_1 and F_2 contain trainable long term memory (LTM) traces which multiply the signals in these pathways. The remainder of the circuit modulates these STM and LTM processes. Modulation by gain control enables F_1 to distinguish between bottom-up input patterns and top-down priming, or expectation, patterns, as well as to match these bottom-up and top-down patterns by the 2/3 Rule. Gain control signals also enable F_2 to react supraliminally to signals from F_1 while an input pattern is on. The orienting subsystem generates a reset wave to F_2 when sufficiently large mismatches between bottom-up and top-down patterns occur at F_1. This reset wave selectively and enduringly inhibits previously active F_2 cells until the input is shut off, and triggers a bout of hypothesis testing or memory search for a better F_2 representation. (Reprinted with permission, Carpenter and Grossberg, 1990.)

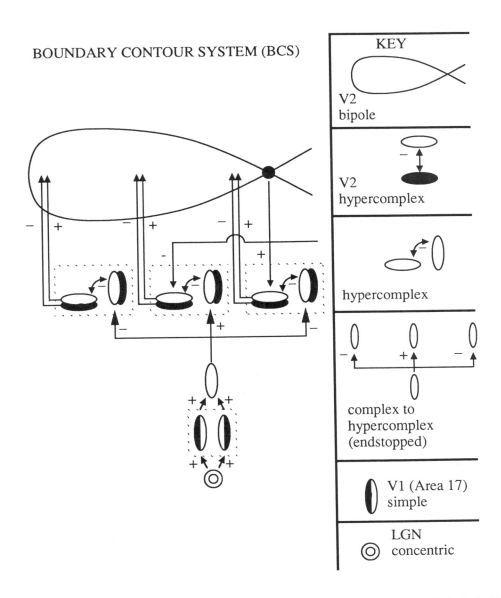

Figure 3. The static Boundary Contour System circuit described by Grossberg and Mingolla (1985a). The circuit is divided into an oriented contrast-sensitive filter (SOC Filter) followed by a cooperative-competitive feedback network (CC Loop). Multiple copies of this circuit are used, each corresponding to a different range of receptive field sizes of the SOC Filter. The depicted circuit has been used to analyse data about monocular vision. A binocular generalization of the circuit has also been described (Grossberg, 1987b; Grossberg and Marshall, 1989).

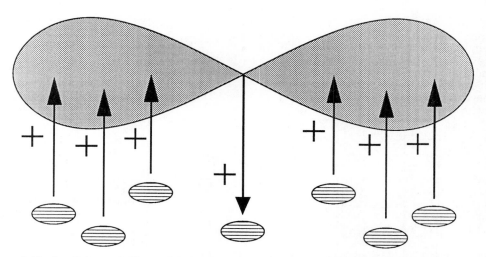

Figure 4. A bipole cell fires only if it is activated by sufficient input of similar orientation and/or direction to both of its receptive fields. It then delivers positive feedback signals to the layers of cells by which it was activated.

tomical evidence from Areas 17 and 18 of the cat. They remarked that "a prominent and stereotypical feature of cortical circuitry in the striate cortex is a plexus of long-range horizontal connections running for 6–8 mm parallel to the cortical surface". Using retrograde labeling, they found that these connections run between cortical columns of similar orientation preferences. Individual cells tended to have long and narrow receptive fields and greater than 90% of the connections appeared to be excitatory. Gilbert and Wiesel noted that while like orientation was necessary to achieve labelling, it was not sufficient. They speculated that there were "subthreshold contextual influences" at work. However, from these labeling techniques they were unable to determine a consistent relation between the orientation of the axis of the axonal fields and the orientation preferences of the columns to which they connected. While such evidence is not conclusive proof of the existence of long-range cooperative bipole cells in visual cortex, it does seem to strongly support the biological plausibility of the bipole mechanism.

The present article demonstrates that both the CC Loop and ART circuits can cooperatively link cells into stimulus-specific standing waves wherein cell activities oscillate in phase with each other. More generally, we show that Cooperative Bipole Coupling, Adaptive Filter Coupling, Nearest Neighbor Coupling, and Random Connection Coupling can all generate synchronous oscillations, but typically not within one or two processing cycles.

Finer differences between the global structure of these oscillations may be used to infer their different functional roles, while also testing predictions of the preattentive BCS theory and the attentive ART theory. In particular, a preattentive BCS resonance is predicted to complete across gaps in two stimulus inducers, as during the Gray et al (1989) two bar experiments and the perception of illusory contours (Grossberg and Mingolla, 1985a, 1985b). In contrast, an attentive ART resonance is predicted not to complete across gaps in stimulus inputs. It can "confirm the hypothesis" that input features are present and can bind them into coherent recognition codes, but it does not activate new features that are not already represented in the input data (Carpenter and Grossberg, 1991; Grossberg, 1987c; Grossberg and Stone, 1986). Synchronized oscillations may thus be generated in different parts of the brain by circuits that carry out different functional tasks, such as preattentive vision and attentive visual object recognition. The existence of synchronized oscillations in two different parts of the brain does not, imply that they carry out similar functions. Rather, they are suggested to share a basic computational format for the efficient processing of spatial patterns.

4. Specification of the Model

The source of the 40–60 Hz oscillations that have been reported has yet to be identified. With an average period of 16–25 ms, such oscillations may arise from local network effects, such as a feedback loop between an inhibitory interneuron and an excitatory cell, or the oscillations may be due to single cell membrane effects, such as the influence of a slow channel or second messenger. In the present simulations, we investigated how neural circuits that have already been shown to have strong behavioral and neural predictive value could act to synchronize their cell activations in a stimulus-specific manner.

The starting point for our work is the analysis by Ellias and Grossberg (1975) of oscillations within a neural network of excitatory potentials X_i and inhibitory interneuronal potentials Y_i which obey the equations

$$\frac{d}{dt}X_i = -AX_i + (B - X_i)[\sum_{k=1}^{n} f(X_k)C_{ki} + I_i] - X_i \sum_{k=1}^{n} f(Y_k)D_{ki} \tag{3}$$

and

$$\frac{d}{dt}Y_i = -EY_i + \sum_{k=1}^{n} X_k F_{ki}. \tag{4}$$

Each excitatory potential X_i in (3) obeys a membrane, or shunting, equation (Grossberg, 1973; Rall, 1955a, 1955b, 1956; Sperling and Sondhi, 1968), whereas each inhibitory interneuronal potential Y_i is approximated by an additive equation in (4). In equation (3), parameter A is the passive decay rate, B is the excitatory saturation point, C_{ki} is the excitatory path strength from cell k to cell i, I_i is an external input, and D_{ki} is the inhibitory path strength from cell k to cell i. In (4), parameter E is the passive decay rate, and F_{ki} is the excitatory path strength from cell k to cell i. A case of particular interest is the 2-dimensional E-G network

$$\frac{d}{dt}X = -AX + (B - X)(C[X - \Gamma]^+ + I) - DX[Y - \Gamma]^+ \tag{5}$$

and

$$\frac{dY}{dt} = E(X - Y), \tag{6}$$

where $[w]^+ = \max(w, 0)$, which was shown (Ellias and Grossberg, 1975) to undergo a series of Hopf bifurcations from equilibrium to frequency-dependent oscillations as the arousal level I is parametrically increased. This input-dependent oscillatory behavior is representative of visual cortical neurons that fire repetitively only when stimulated.

The parameters used in the present simulations were chosen as follows: $A = 1$, $B = 1$, $C = 20$, $D = 33.3$, $\Gamma = 0.4$, $E = F = 0.025$. These values were prescribed in Ellias and Grossberg's original work. The choice $E = 0.025$ was made to give the oscillator strong relaxation properties due to the relative time scale differences between the passive decay rates A and E of the excitatory and inhibitory cells. These parameters also produce a desirable "spike-like" waveform. For these parameter choices, the (X, Y) unit in (5) and (6) exhibited stable limit cycle oscillations for inputs between $I = 0.7$ and $I = 2.25$. The present results are consistent with the hypothesis that relaxation oscillators couple more rapidly and more reliably than sinusoidal oscillators for a variety of architectures (Somers and Kopell, 1991).

The simulations reported here utilized a one-dimensional array of 64 (X,Y) units organized, as in Ellias and Grossberg (1975), in a ring to avoid boundary effects. In order to connect these oscillatory units, a cooperative feedback loop among the potentials X_i was implemented. Thus each excitatory-inhibitory unit (X_i, Y_i) in the array obeys the equations:

$$\frac{d}{dt}X_i = -AX_i + (B - X_i)(C[X_i - \Gamma]^+ + \alpha C[Z_i - \Gamma]^+ + I_i) - DX_i[Y_i - \Gamma]^+ \tag{7}$$

and

$$\frac{d}{dt}Y_i = -EY_i + FX_i. \tag{8}$$

In (7), Z_i is the activity of the ith coupling unit. This term will be specified below. Parameter α calibrates the size of the excitatory coupling strength relative to that provided by the self-excitatory term $C[X_i - \Gamma]^+$. In these simulations, α was parametrically increased from 0—the no-coupling case—to test the effects of excitatory interneuronal coupling on the coherence of the oscillations.

5. Cooperative Coupling Mechanisms

Several coupling architectures were investigated; namely, Cooperative Bipole Coupling, Adaptive Filter Coupling, Nearest Neighbor Coupling, and Random Connectivity Coupling (Figure 5). This analysis illustrates the robust nature of the synchrony phenomenon. Each coupling unit Z_i could be interpreted biologically as either the output signal from the dendritic tree of an X_i cell, or as another cell that sends an excitatory connection to the X_i cell. For example, the Cooperative Bipole Coupling (Figure 5a) could be interpreted as a bipolar dendritic tree in which both compartments must be sufficiently activated to provide input to the X_i cell. Alternately, this coupling unit could be interpreted as a distant cell, perhaps lying in Area 18 (Grossberg and Mingolla, 1985a; von der Heydt, Peterhans, and Baumgartner, 1984), having the same dendritic properties and making a monosynaptic connection with the X_i cell. The coupling units are made explicit in the following equations:

Cooperative Bipole Coupling

$$Z_i = \left[\frac{P(\text{Right}_i)^n}{Q^n + (\text{Right}_i)^n} + \frac{P(\text{Left}_i)^n}{Q^n + (\text{Left}_i)^n} - \Gamma_{cpl}\right]^+ \tag{9}$$

where

$$\text{Right}_i = \frac{1}{\text{width}} \sum_{j=1}^{\text{width}} [X_{i+j} - \Gamma]^+ \tag{10}$$

and

$$\text{Left}_i = \frac{1}{\text{width}} \sum_{j=-1}^{-\text{width}} [X_{i+j} - \Gamma]^+. \tag{11}$$

Parameters P, Q, and n in (9) characterize a sigmoidal, or S-shaped, response curve that indicates a typical nonlinear summation within each cell compartment (Grossberg, 1973, 1982b; Rall, 1955a, 1955b, 1956). For these simulations, the values $P = 1$, $Q = 0.10$, and $n = 5$ were chosen. Parameter Γ_{cpl} is the coupling threshold. The choice $\Gamma_{cpl} = P$ was made in order to guarantee that both compartments or flanks needed to be sufficiently activated

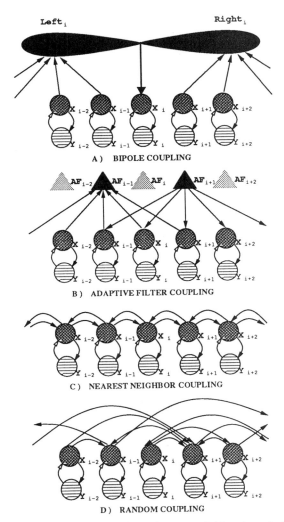

Figure 5: Coupling architectures. Figures A–D show the connectivities characteristic of the four coupling architectures implemented in the simulations. Each (X, Y) pair represents a single oscillator and the lines indicate typical connections. In (A) the bowtie-shaped symbol represents a single bipole cell. Coupling inputs from the right of oscillator i feed exclusively into the right half of the bipole. Coupling inputs from the left feed exclusively into the left half of the bipole. The two halves of the bipole are combined and the rectified, thresholded coupling signal that results feeds to oscillator i. There is a single bipole cell corresponding to each (X, Y) oscillator pair. In (B) the triangle-shaped symbols represent adaptive filter (AF) elements. Each adaptive filter element may receive coupling input from many oscillator units, and in turn may feed coupling output back to many oscillator units. In these simulations there are as many adaptive filter elements as there are oscillator units. In (C) each oscillator unit is connected reciprocally and equally with its immediate neighbor on either side and with no other oscillator units. In (D) each oscillator is connected with a fixed number of randomly chosen oscillators. All oscillators send out the same number of coupling outputs, but typically receive different numbers of coupling inputs. The filled arrowheads indicate excitatory connections, while unfilled arrowheads indicate inhibitory connections. The only inhibitory connections are from the Y cells to the X cells.

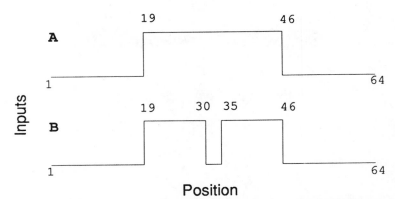

Figure 6: Single bar and double bar system inputs. Simulations were run on two different input images: a single, long bar (A); or two short bars separated by a slit (B). The values of I_i are displayed for each position i along the 64-position ring. Each high value ($I_i = 1.0$) represents a position on a bar in the image and is sufficient to drive an uncoupled (X, Y) pair into its oscillatory regime. Each low value $I_i = 0.05$ represents part of the background of the image and is not sufficient to drive oscillations in an uncoupled (X, Y) pair; rather, an equilibrium value is approached. Within the course of a simulation run, all input values were constant and did not themselves oscillate.

before Z_i could be activated, and thus before coupling feedback from Z_i to X_i could be generated. The width of the flanks was parametrically varied in the simulations.

Adaptive Filter Coupling

$$Z_i = \left[\frac{1}{\text{fan out}} \sum_{j=-\text{fan out}/2}^{+\text{fan out}/2} AF_{i+j} - \Gamma_{cpl}\right]^+ \qquad (12)$$

where

$$AF_i = \frac{1}{\text{fan in}} \sum_{j=-\text{fan in}/2}^{+\text{fan in}/2} [X_{i+j} - \Gamma]^+. \qquad (13)$$

The Adaptive Filter coupling (Figure 5b) assumes that many inputs fan-in, bottom-up, to each coupling compartment AF_i, and that these AF_i's fan-out, top-down, to many compartments Z_i. In the general case (fan in > 1, fan out > 1), this coupling can be realized, for example, by letting each Z_i collect signals in the excitatory dendritic tree that feeds X_i. In the case fan out = 1, AF_i and Z_i may be collapsed into the same dendritic structure, and it is not necessary to postulate intervening cells.

Nearest Neighbor Coupling

$$Z_i = 1/2[X_{i-1} - \Gamma]^+ + 1/2[X_{i+1} - \Gamma]^+ \qquad (14)$$

Nearest Neighbor coupling (Figure 5c) is defined by excitatory signalling between each cell and its two immediate neighbors.

Random Connection Coupling

$$Z_i = \left[\frac{1}{\text{num conn}} \sum_{\text{random } j}^{\text{num conn}} [X_{i+j} - \Gamma]^+ - \Gamma_{cpl}\right]^+. \qquad (15)$$

Random coupling (Figure 5d) sums the active inputs of a number of random connections. If the total input is sufficiently great, an excitatory signal to X_i is activated. Note that while the coupling is chosen randomly, the same random connections remain in effect throughout the course of a simulation run.

6. Methods

The typical paradigm for the simulations is as follows. At the start of each simulation, the (X, Y) units were given random initial conditions and the coupling variables Z were set equal to zero. The test stimuli were either long single bar images or short disconnected double bar images (Figure 6). For the single bar stimulus, the central region ($i = 19$ to $i = 46$) was set to the target value. The double bar stimulus differed in that a central slit region ($i = 31$ to $i = 34$) was set to the background value. Input array positions corresponding to part of a bar were given values sufficient to generate stable limit cycle oscillations in an uncoupled unit [equations (5) and (6)]. At positions not on a bar, the input value was insufficient to sustain oscillations in its unit. For a given simulation run, the inputs and parameters were chosen and held fixed. Within a series of runs, a parameter or pair of parameters, such as the synaptic coupling strength α and/or the widths of coupling interaction, were varied in order to determine their effects.

7. Results

The primary control experiment for these simulations is the uncoupled ($\alpha = 0$) case. In the uncoupled case, units receiving sufficient input exhibit stable limit-cycle oscillations, while units receiving insufficient input quickly approach a low equilibrium value (Figures 7 and 8). Since the initial values were chosen randomly, the units oscillated in random phase with respect to one another. If all ON inputs were the same, these phase relationships did not change over the time course of the simulation, since their frequencies were the same.

Using the Cooperative Bipole Coupling Architecture in (9)–(11), coherent oscillations emerged rapidly (approximately one cycle or less) for both the one bar (Figure 9) and two bar (Figure 10) stimuli. In the two bar case, oscillations were induced in the slit between the two bars and these oscillators could be almost instantly synchronized with the others. In both stimulus cases, the bipole architectures did not induce a spreading of oscillatory activity to the outer regions beyond the stimuli. Inward boundary completion without outward spreading of oscillatory activity was found to be a robust property of bipole coupling.

Adaptive Filter coupling also yielded rapid synchronization for single bar stimuli (Figure 11). As shown in Figure 12, the disconnected regions that were activated by a double bar stimulus were synchronized with respect to each other. This is a robust property of adaptive filter coupling. If the fan-in and fan-out are chosen broadly enough to include only one bar, then the cellular units corresponding to that bar become synchronized. If the fan-in and fan-out are chosen broadly enough to also include the region spanned by both bars, then units corresponding to both bars are synchronized. Depending upon cell parameters, the intervening units may undergo synchronized subthreshold or suprathreshold oscillations. There is also a strong tendency for cells flanking the exteriors of the bars to undergo analogous oscillations.

Nearest neighbor coupling and random coupling could also cause synchronized oscillations to emerge, although this synchrony did not occur as rapidly, nor for as robust a set of initial conditions, as it did for the bipole and adaptive filter architectures. See Grossberg and Somers (1991) for examples of these oscillations.

Figures 8 and 10 together with Figures 13 and 14 present two different perspectives of the synchronized behavior for double bar stimulus, using the bipole architecture. Figures 8 and

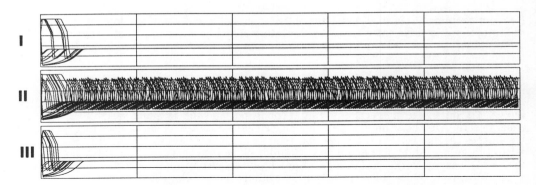

Figure 7: Uncoupled ($\alpha = 0$) case for single bar input. Three windows are displayed in which plots of X_i versus time are overlayed. In window I, positions $i = 1$ thru $i = 18$ are overlayed. In window II, positions $i = 19$ thru $i = 46$ are plotted. Positions $i = 47$ thru $i = 64$ are shown in window III. The positions displayed in windows I and III correspond to the image background, while window II displays activity of X_i along the bar. In this uncoupled case, the activities at positions corresponding to the background quickly approach the same steady-state value, while positions along the bar oscillate in random phase. This uncoupled case represents the control simulation for single bar input.

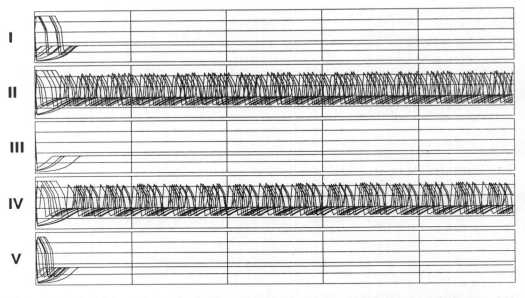

Figure 8: Uncoupled ($\alpha = 0$) case for double bar input. Five windows are displayed in which plots of X_i versus time are overlayed. In window I, positions $i = 1$ thru $i = 18$ are overlayed, while in window II, positions $i = 19$ thru $i = 30$ are plotted. Windows III and IV display positions $i = 31$ thru $i = 34$ and $i = 35$ thru $i = 46$, respectively, and positions $i = 47$ thru $i = 64$ are shown in window V. The positions displayed in windows I and V correspond to the image background, while window III displays activity in the slit between the two bars. Windows II and IV display activity of X_i along the left and right bars respectively. In this uncoupled case, the activities at positions corresponding to the background and the slit quickly approach the same steady-state value, while the activities at positions along both bars oscillate in random phase. This uncoupled case represents the control simulation for double bar input.

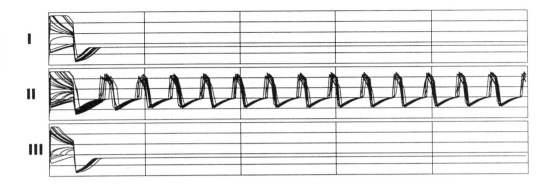

Figure 9: Bipole coupling for single bar input. Using the same inputs and initial conditions which were used to generate Figure 7, bipole coupling with $\alpha = 0.25$ yielded rapid and sustained synchronization of oscillatory activity at positions along the bar without inducing oscillatory activity at positions corresponding to the background. Each bipole flank received input from six neighboring positions (width = 6).

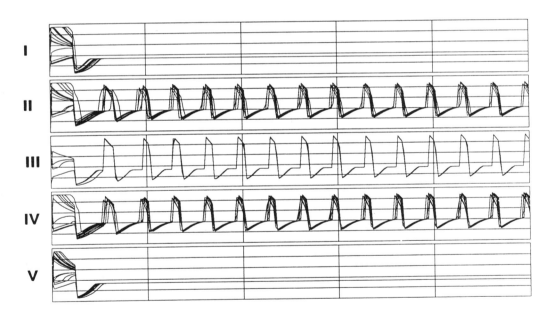

Figure 10: Bipole coupling for double bar input. Using the same inputs and initial conditions which were used to generate Figure 8, bipole coupling with $\alpha = 0.25$ yielded rapid and sustained synchronization of oscillatory activity at positions along both bars and induced synchronous oscillatory activity at the slit positions $i = 31$ thru $i = 34$ but did not induce oscillatory activity at positions corresponding to the outer background regions. This may be interpretted as the completion of a disconnected image boundary, resulting in the perception of a single long bar. Each bipole flank received input from six neighboring positions (width=6).

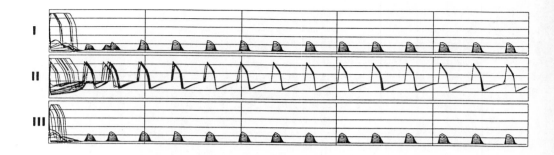

Figure 11: Adaptive filter coupling for single bar input. Shown here with $\alpha = 0.10$ and fan-in and fan-out widths of 9, adaptive filter coupling yielded rapid synchronization of oscillatory activity along the bar. While some small oscillatory activity was induced at background positions, this activity was well below the firing threshold of 0.4. The inputs and initial conditions were identical to those used in generating Figure 7.

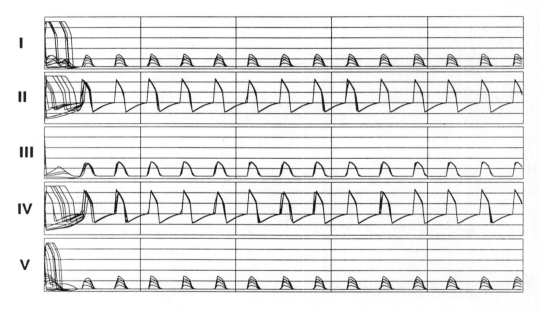

Figure 12: Adaptive filter coupling for double bar input. Shown here with $\alpha = 0.10$ and fan-in and fan-out widths of 9, adaptive filter coupling yielded rapid synchronization of oscillatory activity along each bar and synchronized the bars with respect to each other. While some small oscillatory activity was induced at background positions and in the slit, this activity was subthreshold. The inputs and initial conditions were identical to those used in generating Figure 8.

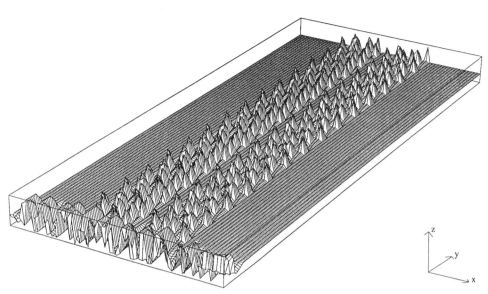

Figure 13: 3-D perspective of uncoupled case for double bar input. The data of Figure 8 is replotted in three dimensions in order to demonstrate the positional structure of the activity. The X-axis represents the position, i, of the $X - Y$ unit; the Y-axis represents time; and the Z-axis represents the activity, X_i. This perspective more clearly displays two regions of incoherent oscillatory activity separated by a slit of region that quickly approaches equilibrium. The larger background regions to the outsides also quickly approach equilibrium from their random initial conditions.

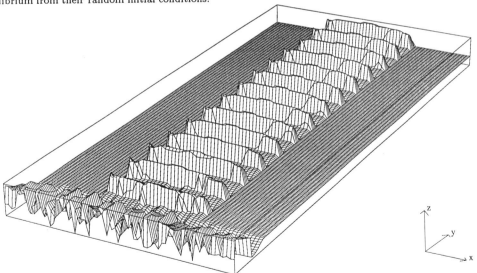

Figure 14: 3-D perspective of bipole coupling for double bar input. By implementing bipole coupling for the inputs and initial conditions shown in Figure 13, two nearby regions of incoherent oscillatory activity are very rapidly synchronized and oscillations are induced in the slit region and these oscillations very rapidly synchronize with the bar regions. This represents a successful boundary completion between the bars which results in a single coherent contour of oscillatory activity that does not spread to the outer background positions. The data plotted is the same that is shown in the overlay plots of Figure 10.

13 display the uncoupled ($\alpha = 0$) case and Figures 10 and 14 display the tight synchrony ($\alpha = 0.25$) of the bipole coupling. Not only are the two bars synchronized, but also synchronized oscillations were induced in the slit region between the two bars, indicating that a subjective contour was generated across the slit and that the features were linked across the "occluding" region, as was also ascribed to bipole cells in the equilibrium point version of the BCS (Grossberg and Mingolla, 1985a, 1985b). A quantitative measure of oscillator coherence as a function of the coupling parameter α in (7) is described in Grossberg and Somers (1991).

8. Discussion

The present results indicate that a wide variety of nonlinear cooperative feedback networks, whose cell units obey shunting or additive equations, can undergo synchronous oscillations if their coupling strength is sufficiently high, and if at least one slow variable, here a slow inhibitory interneuron, exists. These synchronous oscillations can, for example, support a preattentive boundary completion process, as occurs during visual boundary segmentation; an attentive resonant state, as occurs during visual object recognition; and either preattentive or attentive adaptive filtering operations during more general processes of cortical feature detection and short term memory representation.

These results illustrate how the brain may overcome the temporal "jitter" inherent in multi-level processing of spatially distributed data. Such compensatory processing enables mechanisms of short term memory and long term memory to process the coherent spatial patterns of spatially distributed features that are used to represent the events with which we deal in the external world.

REFERENCES

Atiya, A. and Baldi, P. (1989). Oscillations and synchronizations in neural networks: An exploration of the labeling hypothesis. *International Journal of Neural Systems*, 1, 103–124.

Baldi, P. and Meir, R. (1990). *Neural Computation*, in press.

Bregman, A.S. (1990). **Auditory Scene Analysis**. Cambridge, MA: MIT Press.

Carpenter, G.A. and Grossberg, S. (1987a). A massively parallel architecture for a self-organizing neural pattern recognition machine. *Computer Vision, Graphics, and Image Processing*, 37, 54–115.

Carpenter, G.A. and Grossberg, S. (1987b). ART 2: Stable self-organization of pattern recognition codes for analog input patterns. *Applied Optics*, 26, 4919–4930.

Carpenter, G.A. and Grossberg, S. (1990). ART 3: Hierarchical search using chemical transmitters in self-organizing pattern recognition architectures. *Neural Networks*, 3, 129–152.

Carpenter, G.A. and Grossberg, S. (1991). **Pattern Recognition by self-organizing neural networks**. Cambridge, MA: MIT Press.

Carpenter, G.A., Grossberg, S., and Reynolds, J.H. (1991). ARTMAP: Supervised real-time learning and classification of nonstationary data by a self-organizing neural network. *Neural Networks*, 4, 565-588.

Carpenter, G.A., Grossberg, S., Markuzon, N., Reynolds, J.H., and Rosen, D. (1991). Fuzzy ARTMAP: A neural Network Architecture for incremental supervised learning of analog multidimensional maps. *Technical Report Series*, **CAS/CNS-91-016**. Boston, MA: Boston University.

Cohen, M.A. and Grossberg, S. (1984). Neural dynamics of brightness perception: Features, boundaries, diffusion, and resonance. *Perception and Psychophysics*, 36, 428–456.

Eckhorn, R., Bauer, R., Jordan, W., Brosch, M., Kruse, W., Munk, M., and Reitbock, H.J. (1988). Coherent oscillations: A mechanism of feature linking in the visual cortex? *Biological Cybernetics*, 60, 121–130.

Ellias, S.A. and Grossberg, S. (1975). Pattern formation, contrast control, and oscillations in the short term memory of shunting on-center off-surround networks. *Biological Cybernetics*, 20, 69–98.

Ermentrout, G.B. and Kopell, N. (1990). Oscillator death in systems of coupled neural oscillators. *SIAM Journal of Applied Mathematics*, 50, 125–146.

Gilbert, C.D., and Wiesel, T.N. (1989). Columnar specificity of intrinsic horizontal and corticocortical connection in cat visual cortex. *Journal of Neuroscience*, 9, 2432–2442.

Gray, C.M., and Singer, W. (1989). Stimulus-specific neuronal oscillations in orientation columns of cat visual cortex. *Proceedings National Academy of Sciences (USA)*, 86, 1698–1702.

Gray, C.M., Konig, P., Engel, A., and Singer, W. (1989). Oscillatory responses in cat visual cortex exhibit inter-columnar synchronization which reflects global stimulus properties. *Nature*, 338, 334–337.

Grossberg, S. (1968). Some nonlinear networks capable of learning a spatial pattern of arbitrary complexity. *Proceedings of the National Academy of Sciences*, 59, 368–372.

Grossberg, S. (1969). On learning and energy-entropy dependence in recurrent and nonrecurrent signed networks. *Journal of Statistical Physics*, 1, 319–350.

Grossberg, S. (1971). Pavlovian pattern learning by nonlinear neural networks. *Proceedings of the National Academy of Sciences*, 68, 828–831.

Grossberg, S. (1973). Contour enhancement, short-term memory, and constancies in reverberating neural networks. *Studies in Applied Mathematics*, 52, 217–257.

Grossberg, S. (1976a). Adaptive pattern classification and universal recoding, I: Parallel development and coding of neural feature detectors. *Biological Cybernetics*, **23**, 121–134.

Grossberg, S. (1976b). Adaptive pattern classification and universal recoding, II: Feedback, expectation, olfaction, and illusions. *Biological Cybernetics*, **23**, 187–202.

Grossberg, S. (1978a). Decisions, patterns, and oscillations in the dynamics of competitive systems with applications to Volterra-Lotka systems. *Journal of Theoretical Biology*, **73**, 101–130.

Grossberg, S. (1978b). Competition, decision, and consensus. *Journal of Mathematical Analysis and Applications*, **66**, 470–493.

Grossberg, S. (1978c). A theory of visual coding, memory, and development. In **Formal theories of visual perception**, E. Leeuwenberg and H. Buffart (Eds.). New York: Wiley.

Grossberg, S. (1980). Biological competition: Decision rules, pattern formation, and oscillations. *Proceedings of the National Academy of Sciences*, **77**, 2338–2342.

Grossberg, S. (1982a). Associative and competitive principles of learning and development: The temporal unfolding and stability of STM and LTM patterns. In **Competition and cooperation in neural networks**, S.I. Amari and M. Arbib (Eds.). New York: Springer-Verlag.

Grossberg, S. (1982b). **Studies of mind and brain: Neural principles of learning, perception, development, cognition, and motor control**. Boston: Reidel Press.

Grossberg, S. (1984). Outline of a theory of brightness, color, and form perception. In **Trends in mathematical psychology**, E. Degreef and J. van Buggenhaut (Eds.). Amsterdam: Elsevier/North-Holland, pp. 59–86.

Grossberg, S. (1987a). Cortical dynamics of three-dimensional form, color, and brightness perception, I: Monocular theory. *Perception and Psychophysics*, **41**, 87–116.

Grossberg, S. (1987b). Cortical dynamics of three-dimensional form, color, and brightness perception, II: Binocular theory. *Perception and Psychophysics*, **41**, 117–158.

Grossberg, S. (Ed.), (1987c). **The adaptive brain, Vol. II: Vision, speech, language, and motor control**. Amsterdam: Elsevier/North-Holland.

Grossberg, S. (1988). Nonlinear neural networks: Principles, mechanisms, and architectures. *Neural Networks*, **1**, 17–61.

Grossberg, S. and Marshall, J. (1989). Stereo boundary fusion by cortical complex cells: A system of maps, filters, and feedback networks for multiplexing distributed data. *Neural Networks*, **2**, 29–51.

Grossberg, S. and Mingolla, E. (1985a). Neural dynamics of perceptual grouping: Textures, boundaries, and emergent segmentations. *Perception and Psychophysics*, **38**, 141–171.

Grossberg, S. and Mingolla, E. (1985b). Neural dynamics of form perception: Boundary completion, illusory figures, and neon color spreading. *Psychological Review*, **92**, 173–211.

Grossberg, S. and Mingolla, E. (1987). Neural dynamics of surface perception: Boundary webs, illuminants, and shape-from-shading. *Computer Vision, Graphics, and Image Processing*, **37**, 116–165.

Grossberg, S. and Somers, D. (1991). Synchronized oscillations during cooperative feature linking in a cortical model of visual perception. *Neural Networks*, **4**, 453–466.

Grossberg, S. and Stone, G.O. (1986). Neural dynamics of word recognition and recall: Attentional priming, learning, and resonance. *Psychological Review*, **93**, 46–74.

Hopfield, J.J. (1984). Neurons with graded response have collective computational properties like those of two-state neurons. *Proceedings of the National Academy of Sciences*, **81**, 3088–3092.

Kammen, D.M., Holmes, P.J. and Koch, C. (1989). Cortical architecture and oscillations in neuronal networks: Feedback versus local coupling. In **Models of brain function**, R.M.J. Cotterill (Ed.). Cambridge: Cambridge University Press.

Rall, W. (1955a). A statistical theory of monosynaptic input-output relations. *Journal of Cellular and Comparative Physiology*, **46**, 373-411.

Rall, W. (1955b). Experimental monosynaptic input-output relations in the mammalian spinal cord. *Journal of Cellular and Comparative Physiology*, **46**, 413–437.

Rall, W. (1956). Analysis of reflex variability in terms of partially correlated excitability fluctuation in a population of motoneurons. *Journal of General Physiology*, **39**, 397–422.

Somers, D. and Kopell, N. (1991). Properties of oscillators affect the transient time to synchrony in large coupled systems. In preparation.

Sperling, G. and Sondhi, M.M. (1968). Model for visual luminance discrimination and flicker detection. *Journal of the Optical Society of America*, **58**, 1133–1145.

Treisman, A., and Gelade, G. (1980). A feature-integration theory of attention. **Cognitive Psychology**, **12**, 97–136.

von der Heydt, R., Peterhans, E. and Baumgartner, G. (1984). Illusory contours and cortical neuron responses. *Science*, **224**, 1260–1262.

A Quotient Space Hough Transform for Space-Variant Visual Attention

Alan S. Rojer ‡
Eric L. Schwartz ‡ † •

‡ Computational Neuroscience Laboratory
Department of Psychiatry
New York University Medical Center
550 First Avenue
NY, NY 10016

† Courant Institute of Mathematical Sciences
Department of Computer Science
New York University
251 Mercer Street
NY, NY 10012

• Vision Applications, Inc.
611 Broadway
New York, N.Y. 10012

---

supported by AFOSR 88-0273 to NYUMC and DARPA/ONR N00014-90-C-0049 to Vision Applications

ABSTRACT

We consider the problem of "visual attention" in the context of space-variant machine vision: Is there a general theoretical and practical formulation for the "next-look" problem to guide a space-variant sensor to a rapid choice for its next fixation point? This topic is developed in the context of Hough transform methods, by the addition of a third space to the usual feature and object spaces considered in traditional Hough methods. This third space is a "behavioural," or "motor" space, which is typically low-dimensional with respect to the feature and object spaces. For example, the motor space of particular interest for us is the two-dimensional manifold of monocular eye positions. By "collapsing" the generalized Hough transform into a low dimensional motor space, we show that it is possible to avoid a practical difficulty in applying Hough transform methods, which is the exponential dependence of the accumulator array on the dimensionality of the object space. Beginning with a simple and very general Bayesian scheme, we derive in stages the generalized Hough transform as a special case. Since "attentional" applications, by their nature, require only partial knowledge about objects, computation of all the parameters characterizing a scene object is superfluous and wasteful of computational resources. This suggests that for an attentional application, collapsing the (large) object space onto the (small) motor space provides a computationally grounded definition of the term "visual attention". We illustrate these ideas with an example of choosing "fixation" points for a space-variant sensor in a machine vision application for real-time reading of license plates of moving vehicles.

1. Introduction

The space-variant, or foveal nature of human vision has been known since (at least) the time of Helmholtz, and a simple characterization of the primate visual system in terms of the complex logarithmic mapping has provided a specific model for this sensor architecture for more than a decade(Schwartz 1977; Schwartz 1980; Schwartz 1985; Schwartz *et al.* 1989). During the past several years, this vision architecture has begun to attract considerable interest in the machine vision community. One reason is that several groups are in the process of building space-variant CCD (Spiegel *et al.* 1989) or MOS [1]

---

[1] A research group at Vision Applications (R. Wallace, B. Bederson, P. Ong, and E.

sensor chips which are modeled after and inspired by the architecture of the human retino-striate visual system. Another reason is the increasing understanding that the combination of active-vision systems with foveating sensors can provide many orders of magnitude of "speed-up" of machine vision algorithms. In several recent papers, we have provided analyses of the advantages of space-variant vision sensors (Rojer and Schwartz 1990; Yeshurun and Schwartz 1989). Just as the primate brain has evolved a sensor architecture which can cover a wide solid angle, and at the same time provide highly detailed pattern recognition capabilities, certain machine vision applications (e.g. target acquisition, unconstrained robotic vision) can strongly benefit from the same architecture.

Given the likely near-term availability of space-variant VLSI sensors chips, there are several major areas of space-variant image processing which must be developed to allow an exploitation of this type of sensor. Early vision and shape description present a new set of problems and opportunities in a space-variant context. One area which must be approached in a new way is the design of motor control algorithms, which become prominent due to the need for "pointing" and "tracking" targets with this type of sensor. But perhaps the chief problem area is the nature of "visual attention". The advantage of space-variant vision is the concentration of the "spatial-band-width" of the sensor in a small area of interest (the "fovea"), while at the same time maintaining a very wide solid angle under increasingly lower resolution sensing. Such systems must have the ability to rapidly choose a "next-look" position which has a high likelihood of satisfying the organisms (or machines) goal-directed needs. This is a problem which has traditionally come under the area of "visual attention", and it represents a key problem area in the development of space-variant vision systems.

Although a large psychological literature exists in the area of "attention", we have found that the non-computational motivation of this literature limits its utility in the present context. To start with, we have not found a convenient (algorithmic) definition of the term "attention" in the psychology literature. As a working definition of the term "visual attention", we suggest that attention is an attempt to allocate scarce computational resources in order to facilitate a subsequent pattern recognition or classification stage. In the example of space-variant vision, the oculo-motor system is the "consumer" of the attentional computation, using this information to point the scarce resource

---

Schwartz) is currently collaborating with a group from Synaptics, Inc. (T. Allen, F. Faggin, A. Greenblatt and C. Mead), under DARPA sponsorship, to fabricate a space-variant CMOS sensor, with expected first test chips in December, 1990.

of high resolution (foveal) pixels at important regions of the scene. We have implemented this idea by amending the classical object and feature spaces of the Hough transform with a third "motor" space, and then using a "collapsed", or "quotient space" Hough transform which relates the feature space directly to the motor space, thus by-passing the object space. In this way, we provide both a computational definition of the notion "visual attention", and a pragmatic realization of this definition. And, in the important case where the manifold of visual directions is two-dimensional, we achieve an image-like representation of the Hough/Quotient space accumulator array. This allows the use of both image processing algorithms and image-processing hardware to play a role in this method (e.g. smoothing across the accumulator array via convolution, peak detection, thresholding, convenient display for debugging, etc.). Finally, the notion of an image-like two-dimensional Hough-Quotient space accumulator array enforces an additional analogy with and insight into the nature of biological visual processing. Those structures of the brain which are thought to play a role in the control of eye-movement (e.g. superior-colliculus, medial temporal cortex, etc.) generally provide an anatomical version of a two-dimensional, spatial representation of eye-coordinates and/or visual angle.

Figure 1 shows a schematic version of this idea, illustrates our notation, and also makes explicit the various mappings which are involved in both the standard, and in our "collapsed" Hough transform. The principal advantage of this approach is that it avoids a fundamental practical difficulty with the application of Hough space methods, which is the exponential dependence of accumulator bins on the object space parameterization.[2] In fact, we will demonstrate below that the intermediate, high dimensional object space may be replaced by a precomputed composite function, allowing a direct mapping of the feature space onto the motor space. Then the large object space is summarized in a set of look-up tables which map features directly to distributions over the motor fixation points. Thus, the analysis of this paper provides a theoretical insight into the nature of "visual attention" as well as a pragmatic means of computing visual scan paths via a quotient space Hough transform. These ideas will now be developed by demonstrating a Bayesian derivation of the generalized Hough transform, by a demonstration of the collapsed Hough transform, and by the presentation of several examples of this methodology

---

[2]Other methods that also avoid this problem are based on the use of k-D trees to represent the "voting" structure (R. Wallace, PhD Dissertation, Carnegie Mellon University (1989)).

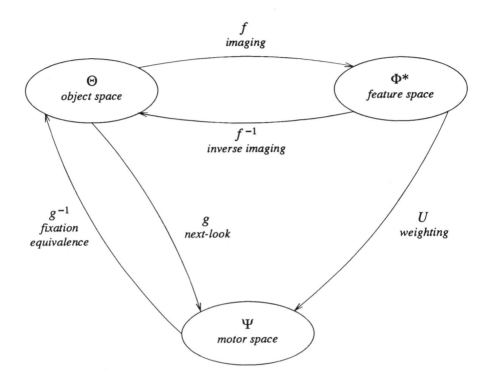

Figure 1 Schema for model-based attention. Θ: space of possible objects. Φ^*: space of possible features. Ψ: space of possible fixation points. $f:\Theta\to 2^{\Phi^*}$: imaging function; maps an object to its set of consistent features. $f^{-1}:\Phi^*\to 2^{\Theta}$: inverse imaging function; maps a feature to the set of objects with which it is consistent. $g:\Theta\to\Psi$: next-look function; maps an object to the preferred next fixation for that object. $g^{-1}:\Psi\to 2^{\Theta}$: inverse next-look or fixation equivalence; maps a fixation point to the set of objects which share the fixation point as their next-look. $U:\Phi\to\mathbf{R}^{\Psi}$: maps a feature to a distribution of interest over the space of possible fixation. See text for further discussion.

applied to choosing fixation points in a space-variant vision system.

2. Hough Transform Methods

The Hough transform has existed for more than two decades in a border territory at the intersection of computer vision, image processing and pattern recognition. Numerous extensions and generalizations to the original scheme have been proposed and utilized and a wide variety of applications have been described. Yet the fundamental characterization of the Hough transform as an information process has been vague. For example, a recent comprehensive survey of the Hough transform (Illingworth and Kittler 1988) characterized the Hough transform as "an evidence gathering procedure" and noted its close relationship to correlation, and the Hough transform for lines has also been shown to be a special case of the Radon transform. However, a fundamental characterization of the Hough transform in information theoretic or statistical terms has not been provided. In the present paper, we will provide a derivation of several versions of the Hough transform, starting from Bayes theorem.

For our experiments in model-based visual attention (Rojer 1990). we wished to utilize a variant of the generalized Hough transform to identify candidate positions for parametric objects in scenes after detection of simple image features like lines or corners. Because we were attempting to identify candidate positions (not to fully characterize the parametric objects) we simplified the problem by lumping all the possible objects at a particular position into a single (equivalence) class, alleviating the necessity for managing a huge accumulator. Because the lumped objects were equivalent with respect to position, and an equivalence relation induces a quotient space, we called this approach the **quotient space method**. Our success with this technique led to consideration of the underlying probabilistic processes which characterize the Hough transform and related algorithms.

In this paper, we introduce a simple and very general Bayesian algorithm (Hough^1) for recognition of parametric objects (i.e. models) from measurements of parametric features. As is usual with Bayesian formulations, the generality of the algorithm underlies its intractability. To construct more practical versions, we introduce a series of restricting assumptions on the nature of the probability distributions of feature and object space, and demonstrate algorithms (Hough^2) and (Hough^3). The result (Hough^3) can then be shown to be equivalent to the generalized Hough transform! We then argue why the extra effort required to use Hough^2 is well justified; actually the points we make have been noted repeatedly over the years by users of the Hough transform and its generalizations, although the link of these algorithms to a Bayesian formulation has not, to our knowledge, been discussed.

Having arrived at the generalized Hough transform, we note that it has generally been dismissed as a practical algorithm due to the exponential dependence of storage and processing time on the dimensionality of the object models. Various palliatives have been introduced over the years to address this problem. Many of them can be unified under the concept of "quotient space" methods. Under this concept, groups of objects which are equivalent under some relation are collapsed, and information about the entire group is accumulated without reference to the individual elements of the group. Such techniques permit successive processing of a sequence of low-dimensional accumulators, in contrast to the one-step processing of a high-dimensional accumulator. For many applications (we consider, for example, attentive vision with a space-variant sensor) it may be sufficient to use the output of one low-dimensional quotient space, depending on the output specifications of the recognition stage.

To summarize, we consider first a simple Bayesian algorithm for recognition of parametric objects from measured features. We show how further restrictions of the algorithm yields the computational equivalent of the generalized Hough transform. The consequences of removing some of the specific restrictions associated with the Hough transform are discussed. We next consider a useful technique for reduction of the accumulator space complexity by quotient space methods. These techniques can be used directly in some problem domains (e.g. space-variant visual attention) and indirectly in others (successive recognition of objects).

3. Notation for Multi-Feature Bayesian Object Recognition

With reference to Figure 1, let Θ^* be a "space" of possible objects, denoted the **object space**. Denote a particular object by $\theta \in \Theta^*$. Θ^* may be a finite set or a continuously parameterized set, characterized by a sequence of parameters. When Θ^* is continuous, we assume that each object θ is characterized by a unique parameter sequence; i.e. there is a one-to-one relationship between objects and their parameterization. Otherwise there is no restriction on the parameters. The elements of the sequence need not be drawn from the same base set (e.g. the first element may be a positive real number, the second element a binary number, etc). Since an object and its parameterization are uniquely associated, we will also refer to the associated parameter sequence as θ. Finally, we will refer also the the probabilistic *event* θ, which may be interpreted as the statement "object θ occurs in the scene". In this context, we may also refer the event θ^c, which is equivalent to the statement "object θ does *not* occur in the scene". The exact meaning of θ should be clear from context.

Let Φ^* be a set of features, uniquely characterized by some other parameter space. As above, $\phi \in \Phi^*$ will refer to a particular feature, the (unique) parametric representation for that feature, and the event that the feature is detected in the image. For example, let Θ^* be the five-dimensional space of planar ellipses, which may be parameterized by center location $(x_c, y_c) \in \mathbf{R}^2$, radii $R_1 > 0$, $R_2 > 0$, and orientation of long axis $\pi > \alpha \geq 0$. A typical feature space Φ^* might be edge elements, parameterized by center location (x, y), length l, and orientation β.

Let the **scene** consist of a possibly empty set of objects from the object space, plus other unspecified elements. The scene induces, via a simple Bayesian imaging, the **image** $\Phi \subset \Phi^*$ consisting of a set of features which is to be analyzed for some potential output behavior. The process of feature detection (identifying occurrences of $\phi \in \Phi^*$) is implicitly included in the imaging model. In general zero, one or many objects from Θ^* may be present. Let $P(\theta)$ denote the a priori probability that an object $\theta \in \Theta^*$ is present in the scene. Similarly, $P(\phi)$ is the a priori probability that a feature $\phi \in \Phi^*$ is present in the image. Note that a feature may be present due to an object or due to some other scene element.

In the event that an object is present in the scene, we have a conditional probability $P(\phi | \theta)$ reflecting the a posteriori probability that a particular feature will be found in the image, given the presence of a particular object in the scene. This conditional probability embodies the imaging model; in general we may determine this probability empirically (counting occurrences) or analytically (by geometric considerations with assumed noise, etc), or some combination of analysis and measurement. In practice, the determination of $P(\phi | \theta)$ will depend on the particular problem and the availability of data. For example, in the example outlined above, with planar ellipses for objects and line elements for features, we may determine analytically (in principle) whether an object should induce a feature in the absence of noise and occlusion. We may then apply a heuristic or empirical correction to account for noise and occlusion.

We are interested determining the presence of objects given a set of detected features (an image) $\Phi \subset \Phi^*$. For any particular object θ, we may ask for the probability, conditioned on Φ, that the object is present in the scene. This is just the Bayes a posteriori probability $P(\theta | \Phi)$. From Bayes law, we write

$$P(\theta | \Phi) = \frac{P(\Phi | \theta) P(\theta)}{P(\Phi)}. \qquad [1]$$

Here, $P(\Phi | \theta)$ is the probability of obtaining a particular image (set of features) Φ given the presence of object θ in the scene, and $P(\Phi)$ is the a priori probability of obtaining a particular image.

The observation of an image $\Phi \subset \Phi^*$ actually comprises two sets of events, the occurrence of a detected set of features Φ, as well as the less obvious non-occurrence of the balance of the features, $\Phi^c = \Phi^* - \Phi$. For any feature $\phi \in \Phi^*$, we denote its occurrence by ϕ and its absence ϕ^c. Concerning the *event* Φ of observing an image $\Phi \subset \Phi^*$, we strictly take account of the occurrence and non-occurrence over the entire feature set

$$\Phi = \bigcup_{\phi \in \Phi} \phi \bigcup_{\phi \in \Phi^c} \phi^c.$$

We can now present algorithm Hough[1]. To estimate the probability of occurrence of an object θ given an image Φ, we simply use [1] to estimate $P(\theta|\Phi)$. Of course, the difficulty lies with the computation of $P(\Phi|\theta)$ and $P(\Phi)$. We are asking for the probability of obtaining a particular image given an object as well as the a priori probability of a particular image. Without some assumptions on the separability of features, it's hard to see how to proceed.

4. Abstract Hough Transform (Hough[2])

We assume that the occurence of features in a scene is independent,

$$P(\bigcup_{\phi \in \Phi} \phi) = \prod_{\phi \in \Phi} P(\phi), \qquad [2]$$

and that the occurence of features is conditionally independent with respect to object occurence,

$$P(\bigcup_{\phi \in \Phi} \phi | \theta) = \prod_{\phi \in \Phi} P(\phi | \theta). \qquad [3]$$

Then recalling that all complementary combinations of independent events are independent, we obtain

$$P(\Phi) = \prod_{\phi \in \Phi} P(\phi) \prod_{\phi \in \Phi^c} P(\phi^c). \qquad [4]$$

Similarly, using conditional independence of occurences [3], obtain

$$P(\Phi|\theta) = \prod_{\phi \in \Phi} P(\phi|\theta) \prod_{\phi \in \Phi^c} P(\phi^c | \theta). \qquad [5]$$

In many cases, we wish to process only features which occur in the image (note that the [4] and [4] involve products over the entire feature space). We can rearrange the preceding expressions with this goal in mind. Note first that

$$P(\Phi) = \prod_{\phi \in \Phi} P(\phi) \prod_{\phi \in \Phi^c} P(\phi^c)$$

$$= \prod_{\phi \in \Phi} P(\phi) \frac{\prod_{\phi \in \Phi} P(\phi^c)}{\prod_{\phi \in \Phi} P(\phi^c)} \prod_{\phi \in \Phi^c} P(\phi^c)$$

$$= \prod_{\phi\in\Phi} \frac{P(\phi)}{P(\phi^c)} \prod_{\phi\in\Phi\cup\Phi^c} P(\phi^c)$$

$$= \prod_{\phi\in\Phi} P'(\phi) \prod_{\phi\in\Phi^*} P(\phi^c), \qquad [6]$$

where we have introduced the composite term $P'(\phi) = P(\phi)/P(\phi^c)$. The latter product term in [6] is over the entire feature space, and hence is independent of any particular image (and therefore can be precomputed). We apply a similar analysis to obtain

$$P(\Phi|\theta) = \prod_{\phi\in\Phi} P'(\phi|\theta) \prod_{\phi\in\Phi^*} P(\phi^c|\theta), \qquad [7]$$

with $P'(\phi|\theta) = P(\phi|\theta)/P(\phi^c|\theta)$. Finally, we can express the original desired posterior probability [1] using [6] and [7] via

$$P(\theta|\Phi) = P'(\theta) \prod_{\phi\in\Phi} H(\phi, \theta), \qquad [8]$$

with

$$P'(\theta) = P(\theta) \prod_{\phi\in\Phi^*} \frac{P(\phi^c|\theta)}{P(\phi^c)}, \qquad [8a]$$

a term independent of the image $\Phi\subset\Phi^*$, and

$$H(\phi, \theta) = \frac{P'(\phi|\theta)}{P'(\phi)}, \qquad [8b]$$

a term which depends jointly on the object θ and the feature ϕ. Note that the product in [8] is restricted to the features present in the image. When $H(\phi, \theta)$ is very large or very small with respect to unity, the feature ϕ contributes strongly to the probability of occurence for object θ. Conversely, when $H(\phi, \theta) \approx 1$, the occurrences of feature ϕ and object θ are nearly independent. We can use this observation to simplify the calculation.

We introduce an **imaging function** $f: \Theta^* \to 2^{\Phi^*}$, which maps an object θ to the set of features $f(\theta)$ which are consistent with the object, where a feature ϕ is **consistent** with an object θ unless $P(\phi|\theta) \approx P(\phi)$; i.e. unless the liklihood of occurrence of ϕ is nearly independent of θ. Thus, a feature is consistent with an object if the presence of the object in the scene significantly modifies the liklihood of observation of the feature in the image from its a priori probability. As stated, the definition includes the case where $P(\phi|\theta) > P(\phi)$ (i.e. the feature is more likely to be observed given θ in the scene), and $P(\phi|\theta) < P(\phi)$, (the less intuitive case where the feature is *less* likely to be observed given the presence of θ).

We can restrict the product in [8] to exclude unity elements; formally, we have

$$\phi \notin f(\theta) \rightarrow P(\phi|\theta) = P(\phi), P(\phi^c|\theta) = P(\phi^c).$$

In this case, $H(\phi, \theta) = 1$, and

$$P(\theta|\Phi) = P'(\theta) \prod_{\phi \in \Phi \cap f(\theta)} H(\phi, \theta); \qquad [9]$$

i.e. we can restrict concern to the intersection between the image feature set and the consistency set for the object of interest.

We can now present an algorithm which uses the foregoing analysis to estimate probabilities of occurences of objects $\theta \in \Theta^*$ given an image $\Phi \in \Phi^*$ of observed features. The similarity to typical Hough transform methods can be seen in the use of an accumulator which is incremented whenever an object/feature consistency pair is found. For consistency with the usual Hough transform, we make use of the inverse consistency function,

$$f^{-1}(\phi) = \{\theta \in \Theta^* \mid \phi \in f(\theta)\},$$

which of course is the set of objects with which feature ϕ is consistent.

Algorithm Hough²(Φ)
  for $\theta \in \Theta^*$ *{A[θ] = log(P(θ)) + $\sum_{\phi \in \Phi^*} log(\frac{P(\phi^c|\theta)}{P(\phi^c)})$}*
  for $\phi \in \Phi$ {
    for $\theta \in f^{-1}(\phi)$ *{A[θ] += log(H(ϕ, θ))}*
  }
  for $\theta \in \Theta^*$ *{P($\theta|\Phi$) = exp(A[θ])}*
  return(P($\theta|\Phi$))

In Hough², the accumulator array is initialized to contain the feature-independent component of $P(\theta|\Phi)$, which we have denoted $P'(\theta)$. The increment to the accumulator A[θ] for feature $\phi \in f(\theta)$ is $\log H(\phi, \theta)$. It is an easy matter to show that $\exp A[\theta] = P(\theta|\Phi)$; the Hough peak detection is replaced by computation of the actual posterior probability. The user can apply an absolute threshold to the computed probabilities. The abstract Hough transform requires additional work in the initialization, but that process is independent of the feature set, and hence may be computed once and stored for all images to be processed.

5. The Generalized Hough Transform (Hough³)

The original Hough transform (Hough 1962) and its early popularizations (Duda and Hart 1972; Rosenfeld 1969) considered the problem of finding linear structure in images consisting of individual pixels. In our notation, the scene consists of lines plus noise, the object space Θ is the two-dimensional space of lines in the plane (e.g. parameterized by slope $m \in \mathbf{R}$ and intercept $b \in \mathbf{R}$, or by orientation $\alpha \in [-\pi, \pi]$ and closest distance to the origin $\rho \in [0, R]$), and the feature space Φ is a set of "edge" pixels $(x, y) \in [-R, R] \times [-R, R]$. The consistency set $f^{-1}(\phi) = f^{-1}(x, y)$ is determined by the set of lines which pass through the pixel $\phi = (x, y)$. The Hough transform is implemented by quantizing the object space, and assigning an accumulator to each of the resulting quanta. An image Φ, consisting of a set of edge pixels, is processed by determining $f^{-1}(\phi)$ for each edge pixel, and incrementing each $\theta \in f^{-1}(\phi)$. After the edge pixels have been processed, the object space is scanned for accumulator entries with large values. The parametric objects associated with heavily assigned accumulators are considered to be present in the scene with a high confidence.

Formally, if $A[\Theta^*]$ is the accumulator array for the object space Θ, and we assume that $\theta \in \Theta^*$ correspond to the quantized elements of the object space, we have the algorithm Hough³:

Algorithm Hough³*(Φ)*
    for $\theta \in \Theta^*$ *{A[θ] = 0}*
    for $\phi \in \Phi$ *{*
        for $\theta \in f^{-1}(\phi)$ *{A[θ]++}*
    }
    Θ = *empty set*
    for $\theta \in \Theta^*$ *{*
        if (θ is a "peak") add θ to Θ
    }
    return(Θ)

The obvious difficulties with this approach include the errors of quantization in the representation of the object space and the ill-defined problem of peak detection in the accumulator array. A less obvious problem is that the size of the consistency set $\|f(\theta)\|$ (the number of features with which an object is consistent, or the volume of the consistency set in the continuous case) varies from object to object. This problem was originally described (Cohen and Toussaint 1977) as the uneven distribution in the accumulator array of random features (i.e. features that are not induced by an object in the scene). The usual fix has been to empirically determine a background subtraction, or to use a nonlinear quantization which normalizes the expected distribution of random features in the accumulator array. A main result of the analysis to follow is to make clear how this effect arises and how it can be neutralized.

It was clear to early workers that the Hough algorithm readily generalized to other simple curves; e.g. circles (Duda and Hart 1972; Kimme et al. 1975). In the formulation above, it is necessary only to substitute the appropriate $f^{-1}(\phi)$ for a new shape. What these early workers did not emphasize was that there was no need to restrict features to edge-detected pixels. The later development of the generalized Hough transform incorporated the direction of the gradient of brightness (a readily available output of many edge-detection operators) to augment pixel location, obtaining a three-dimensional feature vector (Ballard 1981). Further generalization observed that more complex features were permissible as well (Davis 1982).

Early workers considered object spaces for which the parameterization was a simple closed-form function (e.g. lines, circles, ellipses, etc). The generalized Hough transform (Ballard 1981) permitted application of the Hough transform to objects that were characterized only by a table of boundary positions. The object space was considered to represent not the space of possible objects, but rather the space of possible *transformations* from a standard template to observable objects in the scene. The generalization to tabular models for objects is important, but the alternative view of object space does not seem useful. There is a one-to-one correspondence between the space of objects which can be obtained by a class of parameterizable transformations from a base object (the old way of viewing the object space) and the space of permissible transformations of the base object (the new way of viewing the object space). Therefore, the information content of the two representations is identical, so the new representation doesn't offer any quantitative gains. Moreover, the emphasis on computing a transformation detracts attention from the problem at hand, which is to determine which, if any, objects from the object space are present in the scene.

We note that, if we retain the old way of looking at the object space, the algorithm Hough[3] is completely capable of representing the generalized Hough transform of (Ballard 1981). All that is required is to acknowledge that Φ is three-dimensional (incorporating a direction in addition to position), and to adapt $f^{-1}(\phi)$ to represent the set of objects with which a given feature is consistent (all details carry over unchanged from the original work of Ballard). The important point is that all the interesting information concerning the parametric model of the objects and the relationship between objects and features is embodied in $f^{-1}(\phi)$. We now turn our attention to $f^{-1}(\phi)$ to make the link between the Hough transform and the Bayesian model presented above.

In our Bayesian model, we defined $f^{-1}(\phi)$ as the set of objects with which a feature ϕ is consistent. Consistency, in turn, depended on the conditional probability of observing a feature in the image given the presence of object in the scene. If the conditional probability $P(\phi|\theta)$ differed significantly from the a priori probability $P(\phi)$, we assigned θ to $f^{-1}(\phi)$. The Hough formulation is slightly different; the forward definition (from objects to features) of consistency is implicit in the parametric characterization of the objects and the features. The backward definition is usually explicit; the approach is to define $f^{-1}(\phi)$ as the set of objects satisfying a system of parametric equations which depend on the parameterization of the object and feature spaces. In practice, the Bayesian model would probably be implemented the same way; $P(\phi|\theta)$ could be obtained from the parametric relationships of θ and ϕ and an imaging model which would incorporate the possibilities of occlusion, noise, etc. Of course, we leave open the possibility that $P(\phi|\theta)$ might be determined empirically, by measurement of features in a supervised learning setup, or by a combination of analytic and empirical techniques.

We now will show that Hough³ is a special case of Hough². First, we observe that after all the features $\phi \in \Phi$ have been processed in Hough³, the accumulator value A[θ] contains the number of image features which are consistent with the object θ.[3] This is equivalent to noting

$$A[\theta] = \| \Phi \cap f(\theta) \|. \qquad [10]$$

We will now obtain [10] by restriction of the Bayesian model.

Starting from [9], we first take the log to produce a sum.

$$\log P(\theta|\Phi) = \log P'(\theta) + \sum_{\phi \in \Phi \cap f(\theta)} \log H(\phi, \theta). \qquad [11]$$

We introduce two assumptions; let

$$P'(\theta) = k_1, \forall \theta \in \Theta, \text{ and} \qquad [12]$$

$$H(\phi, \theta) = k_2 > 0, \forall (\phi, \theta) \in \Phi \times \Theta. \qquad [13]$$

---

[3]This may be obvious; if not, observe that for each $\phi \in \Phi \cap f(\theta)$, A[$\theta$] is incremented once (since $\theta \in f^{-1}(\phi)$), thus A[θ] $\geq \| \Phi \cap f(\theta) \|$. On the other hand, to increment A[θ], there must exist some $\phi \in \Phi$ with $\theta \in f^{-1}(\phi)$. Obviously, in that case, $\phi \in \Phi \cap f(\theta)$. Then A[$\theta$] $\leq \| \Phi \cap f(\theta) \|$. Conclude A[$\theta$] = $\| \Phi \cap f(\theta) \|$. Q.E.D.

Under these conditions, the log posterior probability becomes

$$\log P(\theta|\Phi) = k_1 + \sum_{\phi \in \Phi \cap f(\theta)} k_2 = k_1 + k_2 \|\Phi \cap f(\theta)\|. \qquad [14]$$

But now the log posterior probability induces the same ranking on the object set as does the generalized Hough transform;[4] formally,

$$P(\theta_1|\Phi) > P(\theta_2|\Phi) \text{ if and only if } A[\theta_1] > A[\theta_2].$$

We have shown that the ranking of candidate objects $\theta \in \Theta^*$ under the generalized Hough transform Hough[3] is the same as that obtained under the Bayesian model Hough[2] subject to two additional assumptions ([13]: $P'(\theta)$ constant, independent of θ; [14]: $H(\phi, \theta)$ constant, independent of ϕ, θ). Hence, the generalized Hough transform emerges as a special case of the Bayesian model. Next we examine the consequences of the additional assumptions which are necessary to restrict the Bayesian model to obtain the Hough transform.

Consider first [13], which requires that $P'(\theta) = k_1$, independent of θ. Returning to [8a], we have

$$P'(\theta) = P(\theta) \prod_{\phi \in \Phi^*} \frac{P(\phi^c|\theta)}{P(\phi^c)} = P(\theta) \prod_{\phi \in f(\theta)} \frac{P(\phi^c|\theta)}{P(\phi^c)}. \qquad [8a]$$

We have used the assumption that $P(\phi^c|\theta) = P(\phi^c)$ when ϕ and θ are inconsistent (i.e. ϕ and θ are independent when $\phi \notin f(\theta)$). When we set this term to a constant independent of θ, we lose two kinds of information. First, any relative a priori probabilities are discarded. In effect, we assume that all objects are equally likely. Second, we ignore the product term in [8a]. To understand the effect of this term, consider a simple situation in which $P(\phi^c|\theta)$ and $P(\phi^c)$ are constant for all features; let $k_3 = P(\phi^c|\theta)/P(\phi^c)$. Then we obtain

$$\prod_{\phi \in f(\theta)} \frac{P(\phi^c|\theta)}{P(\phi^c)} = \prod_{\phi \in f(\theta)} k_3 = k_3^{\|f(\theta)\|}.$$

Taking logs in [8a], obtain

$$\log P'(\theta) = \log P(\theta) + \|f(\theta)\| \log k_3. \qquad [15]$$

The latter term scales with the number of features in the consistency set of the object.

---

[4]Obvious, but formally, let $\alpha > \beta$; then $k_2\alpha > k_2\beta$, since $k_2 > 0$, and $k_1 + k_2\alpha > k_1 + k_2\beta$. Q.E.D.

This term is directly related to the early work (Cohen and Toussaint 1977) which showed that the distribution of uniform random features was non-uniform in the accumulator array. In particular, objects with large $\|f(\theta)\|$ will receive proportionally large accumulation in response to uniform random features. Note that the Bayesian model predicts this effect. In summary, when we ignore $\log P'(\theta)$, we throw away object-specific information about the a priori probabilities and about the relative number of features in the consistency set.

Consider the second Hough assumption, [13]. We have, using [8b],

$$H(\phi, \theta) = k_2 = \frac{P'(\phi|\theta)}{P'(\theta)}. \qquad [13]$$

In this case, we are assuming that the gain in information for a feature $\phi \in f(\theta)$ is the same for all ϕ and θ. This is probably not a bad assumption under many situations, but it can be dangerous if the features vary widely; in particular, if there is more than one class of features (e.g. edges and corners) or if features have a complex representation. Also it does not allow for gradations of probability in the conditional observation of a feature.

6. Quotient Space Methods

In the paper so far, we have introduced the very general but intractable Bayesian model for object recognition (Hough[1]). We then presented the "abstract Hough transform" (Hough[2]), which incorporates featural independence assumptions for tractability provides quantitative values for initialization and incrementation of the Hough accumulators, and allows interpretation of accumulator values as conditional probabilies. We have also shown that the usual generalized Hough transform (Hough[3]) is a special case (with additional, implicit assumptions) of object recognition under the Bayesian model.

Although these alternatives provide theoretical and practical improvements for the generalized Hough transform, we have not addressed the most important limitation in the use of the Hough transform, the exponential dependence of the accumulator array on the dimensionality of the object space. In particular, if D is the dimensionality of the object space, and we require a quantization into n bins for each dimension, the total number of required bins in n^D, which is prohibitive in both space and time when D is large. Much of the research in Hough transform applications has involved attempts to circumvent the exponential dependence of the Hough accumulator array.

Some researchers have used non-uniform quantization of the accumulator with such techniques as iterative focussing (Adiv 1983; Illingworth and Kittler 1987; Li *et al.* 1986; Silberberg *et al.* 1984). Others have utilized sophisticated data structures for efficient accumulation (Li and Lavin 1986; O'Rourke and Sloan 1984). More germane to this paper is the use of sequential accumulation of object spaces (Tsuji and Matsumoto 1978; Tsukune and Goto 1983), and projections of the Hough transform (Davis and Yam 1980; Taylor 1990). We follow this vein of dimensionality reduction in the object space.

When the class of outputs which a system must execute has lower dimensionality than the underlying object space, an immediate simplification results. More precisely, we consider the case when many objects induce the same output. In these cases, there may be no need to process these objects separately; efficiency is enhanced by treating the class of objects as a unit. Formally, let Ψ^* be a space of possible outputs; in general, these may be finite or infinite, discrete or continuous, etc. We assume only that there is a function $g: \Theta^* \to \Psi^*$, which associates with each object $\theta \in \Theta^*$ an appropriate output $\psi = g(\theta) = \in \Psi^*$. In many cases of interest, g is many-to-one; then there is a set of objects which have the same output. Formally, this set is characterized by

$$g^{-1}(\psi) = \{\theta \in \Theta^* \mid g(\theta) = \psi\}. \qquad [16]$$

As far as the output of the process is concerned, all the objects $\theta \in g^{-1}(\psi)$ are equivalent; i.e. they are indistinguishable with respect to the output. We can introduce an equivalence relation $\sim$ such that

$$\theta_1 \sim \theta_2 \text{ if and only if } g(\theta_1) = g(\theta_2). \qquad [17]$$

It is easy to show that this definition satisfies the usual definition of an equivalence relation (Lipson 1981). Clearly $\theta \sim \theta$ (reflexivity) since g is a function hence single-valued; when $\theta_1 \sim \theta_2$ and $\theta_2 \sim \theta_3$, $\theta_1 \sim \theta_3$, since $g(\theta_1) = g(\theta_2) = g(\theta_3)$ (transitivity). Finally the relationship is trivially seen to be commutative, since $\theta_1 \sim \theta_2$ implies $g(\theta_1) = g(\theta_2)$ implies $\theta_2 \sim \theta_1$. The equivalence relation partitions the object space into equivalence classes. We may denote $[\theta]$ as the set of all objects equivalent under $\sim$; note $[\theta] = g^{-1}(\psi)$ for some ψ. Formally, we have $\Theta^*/\sim$ as the set of equivalence classes (or **quotient space**), which is in one-to-one correspondence with the output values.

The key point of this algebra is to emphasize that, from the point of view of the output, all the elements $\theta \in g^{-1}(\psi)$ are identical; therefore it is potentially efficient to ignore the individual elements and process the equivalence class. Moreover, the set of equivalence classes $\Theta^*/\sim$ is in one-to-one correspondence with Ψ^*; thus, in the event that Ψ^* is "simpler" than Θ^*, e.g. of lower dimensionality, the prospect of working with the equivalence classes (and hence structures of complexity $O(\|\Psi^*\|)$ instead of $O(\|\Theta^*\|)$) is very attractive.

When we consider the overall process, another perspective may be introduced. In general, we are processing some set of features Φ to obtain an output ψ. In the case of the Hough transform, the output is a conditional probability for each element of the object space, and hence is no simpler than the object space. In other situations, however, the output might be much simpler than the object space. For example, the problem of "model-based visual attention with a space-variant sensor" under the Bayesian model may be characterized as the selection of a "next-fixation" point, given the current image and possibly some historic information. Under the Bayesian model, we have parametric models of the objects in the scene, parametric models of image features, and the relevant probability relations $(P(\phi), P(\theta), P(\phi|\theta))$. The output Ψ^* is a two-dimensional space of directions (e.g. azimuth and elevation).

In general, the parametric models of the feature space and the output space are fixed, but we might want to consider a variety of object spaces with variable parametric structure. In any case, however, information may be conceptualized to flow from the feature space to the object space to the output space. In effect, we use the feature set to compute an output. The intervening object space represents our model of the objects of interest, but the net result is a function from features to output, via the objects. We may analogize the process to a composite function from $\Phi^* \to \Psi^*$. The range of the function is the output space, regardless of the intermediate complexity of the object space. We can use this concept to greatly enhance the computational efficiency of the overall process. In particular, if we add an assumption of conditional independence of occurence of objects in the scene, we can construct an algorithm which has "on-line" complexity independent of the dimensionality of the object space (a single "off-line" preprocessing of the object space is required, however).

Starting from the Bayesian model [9], we observe the probability of an object's presence in the scene conditioned by an observed feature set is given by

$$P(\theta|\Phi) = P'(\theta) \prod_{\phi \in \Phi \cap f(\theta)} H(\phi, \theta); \qquad [9]$$

Now consider a particular output $\psi \in \Psi^*$. We ask for the conditional probability that *any* object θ in the equivalence class $g^{-1}(\psi)$ is present in the scene. Formally, this is equivalent to the complement of the presence of *none* of the objects $\theta \in g^{-1}(\psi)$ in the scene.

$$P(\bigcup_{\theta \in g^{-1}(\psi)} \theta | \Phi) = 1 - P(\bigcap_{\theta \in g^{-1}(\psi)} \theta^c | \Phi). \quad [18]$$

Now we introduce the assumption of conditional independence:

$$P(\bigcap_{\theta \in g^{-1}(\psi)} \theta | \Phi) = \prod_{\theta \in g^{-1}(\psi)} P(\theta | \Phi). \quad [19]$$

In case events are independent, their complements are also independent. Then we can write

$$P(\bigcap_{\theta \in g^{-1}(\psi)} \theta^c | \Phi) = \prod_{\theta \in g^{-1}(\psi)} P(\theta^c | \Phi). \quad [20]$$

Thus, the probability that any object θ from the equivalence class $g^{-1}(\psi)$ is present in the scene may be represented as

$$P(\psi | \Phi) = 1 - \prod_{\theta \in g^{-1}(\psi)} P(\theta^c | \Phi), \quad [21]$$

where ψ has the event interpretation "the probability that an object $\theta \in g^{-1}(\psi)$ is present in the scene." It is an easy matter to re-derive the Bayesian model in terms of complementary objects, as required by [21]. We obtain

$$P(\psi | \Phi) = 1 - P(\psi) \prod_{\phi \in \Phi} U(\psi, \phi), \quad [22]$$

with

$$P(\psi) = \prod_{\theta \in g^{-1}(\psi)} \left[P(\theta^c) \prod_{\phi \in \Phi*} \frac{P(\phi^c | \theta^c)}{P(\phi^c)} \right] = \prod_{\theta \in g^{-1}(\psi)} P'(\theta^c),$$

and

$$U(\psi, \phi) = \prod_{\theta \in g^{-1}(\psi)} \frac{P'(\phi | \theta^c)}{P'(\phi)}.$$

Here, $P'(\phi | \theta^c) = \frac{P(\phi | \theta^c)}{P(\phi^c | \theta^c)}$ (by analogy to $P'(\phi | \theta)$). The crucial observation is that, in [22], $P(\psi)$ and $U(\psi, \phi)$ may both be precomputed; then the execution time using precomputed $P(\psi)$ and $U(\psi, \phi)$ is independent of the size of the object space.

We can use results from sequential accumluation (Tsuji and Matsumoto 1978; Tsukune and Goto 1983) to permit application of quotient space techniques to typical recognition problems. In this case, we have provided a Bayesian framework for sequential accumulation.

7. Model-Based Attention

Historically, our interest in the Hough transform grew out of research into models of visual attention for a space-variant sensor (Rojer 1990; Yeshurun and Schwartz 1989). We have considered the use of the complex-logarithmic sensor geometry for machine vision. Such a sensor is characterized by a small region of high resolution, surrounded by a concentric regions of decreasing resolution. The visual attention problem for such a sensor reduces to the problem of where to point the high-resolution central region.

In our early work, we attempted to apply the generalized Hough transform to this problem, but we were stymied by the large space and time requirements of the accumulator for arbitrary objects. We noted, however, that the output space in our problem was two dimensional (azimuth and elevation of the high resolution central region of the sensor). This realization led to the concept of a composite function from the feature space to the output space, bypassing the high-dimensional accumulation in object space. Later, we realized that we had formed equivalence classes of objects with common preferred fixation points; this was the basis for the quotient-space techniques.

When we are looking for specific objects in a scene, we can bring to bear whatever parametric information about the objects is available. An object can be characterized by a set of parameters, e.g. its location in the scene, orientation and size. For any particular object, there exists a set of scene elements (features) which are consistent with that object. If the object is not occluded and the observation of the scene elements is reliable, we expect that a substantial subset of the consistent elements will be present when a particular object is present in the scene.

To join the goal of model-based fixation point selection to alleviation of the intractability of Hough accumulator storage, note that we are not interested in the full identification of elements from the object space in this stage of processing. We are only trying to find a "good" selection for the next fixation point. Thus, the space of possible outcomes from the processing of features is really the motor space of possible fixation points, which is two dimensional, not the full space of possible objects, e.g. in the case of ellipses described above, which is four dimensional. Thus, we can dramatically simplify by considering only equivalence classes of objects (which share a common "best-look position"). The space of equivalence classes is the same dimension as the motor space. The relevant function we need to compute is from the feature space through the object space to the motor space. Although the object space may be of high dimension, the *composite* function which links the feature space to the motor space has a two-dimensional range (ref. Figure 1).

From the perspective of the motor space Ψ, any particular point $\psi \in \Psi$ (potential fixation) is likely to be the next-look position for a number of objects; i.e. the next-look function is many-to-one in general. The key point in this analysis is to ignore the diverse set of possible objects and simply accumulate all next-look information at a given point, regardless of the intermediate objects.

To summarize, we have a motor space Ψ of possible fixations, an object space Θ of possible objects, and a feature space Φ^* of possible features. The feature space is linked to the object space by the imaging function f and its inverse (which is the usual Hough consistency function). The object space is linked to the motor space by the next-look function g and its inverse. For a particular set of features Φ, we are concerned with the probability function $P(\psi|\Phi)$ over the motor space, which provides an estimate of the probability that an object from the equivalence class represented by ψ is present in the scene. Maxima of $P(\psi|\Phi)$ are selected for the next fixation in the scene.

The most expensive stage of processing is the computation of the weighting function, $U(\psi, \phi)$ in [22]. If the number of features is not too large or if $U(\psi, \phi)$ transforms in a simple way for various ϕ, ψ combinations, we could precompute $U(\psi, \phi)$ for each ϕ and use a lookup table for a fast implementation. We have concentrated on analytic and heuristic methods for determining $U(\psi, \phi)$, but estimation techniques can also be brought to bear. We could use adaptive techniques (i.e. supervised learning) to estimate $U(\psi, \phi)$ from a set of examples, each comprising an object, its consistency set $f(\theta)$, and its best fixation point $g(\theta)$. Another possible approach is to assume a functional form for $U(\psi, \phi)$ and use statistical techniques to estimate the parameters of the functional form. For example, we might assume the $U(\psi, \phi)$ is bivariate Gaussian over ψ; then we would need to estimate $\mu(\phi)$, $C(\phi)$, with μ the mean of the Gaussian, and C the (2×2) covariance matrix.

8. An example of the Quotient-Space Hough Transform

As an illustration of the quotient space Hough transform approach to visual attention, we consider the problem of pointing a sensor at a license plates (rectangles) in a complex-log mapped scene. The three dimensional feature space was based on an edge detection scheme which assigns position, length, and orientation to edge elements in the scene. The three dimensional object space was based on rectangular shapes(aspect ratio and (x,y)position), consistent with license plate shape. The two dimensional motor space consisted of pixel coordinates, (i.e. eye fixation directions), in the domain image.

For the $U(\phi, \psi)$ function, we have simply assumed a bivariate Gaussian distribution over the motor space, with the orientation, position, and variances of the distribution determined by the orientation, position, and length of the detected edge elements. In a genuine application, the function relating the parameters of the distribution to the $U(\phi, \psi)$ function could be determined by statistical analysis (eg maximum likelihood estimation).

Figure 2 shows several frames from a 35mm movie film of traffic on 2nd Avenue in New York City. On the left is shown the full resolution (512x512) frames, shot with a wide-angle (17mm fisheye) lens. On the right side of figure 2 are shown the corresponding complex logarithmic mappings of these frames,[5] using a simulation of an MOS sensor chip which we have designed in collaboration with Synaptics, Inc. The frames on the left subtend about 150 degrees of visual angle, and contain .25 megabytes of data. The complex log scenes on the right subtend the same visual angle, have the same resolution in the center (i.e. the "fovea"), but contain only 2000 pixels. For a real-time wide-angle target acquisition problem, the complex log geometry is highly favorable, since it provides coverage of a large solid angle, retains the ability to perform fine-grained pattern recognition in the center of the field, yet has a very low pixel-bandwidth burden. But to make effective use of this type of sensor, the "attentional" problem becomes of paramount importance.

The task is to locate the license plates in the complex logarithmic frames, using a Quotient-Space Hough Transform technique in which the feature space consists of edges in the images, the object space consists of rectangular (i.e. license plate "like") regions, and the motor space consists of a (two dimensional) manifold of visual directions for pointing the sensor. Figure 2 shows one of the (complex log mapped) frames. Figure 3

---

[5]The complex logarithmic mapping (also called log polar mapping) uses the complex logarithm function to map image coordinates to new image coordinates. This function is an approximation to the mapping which occurs in the retina->striate cortex mapping of primates (Schwartz 1980), and is frequently used as a specific model for space-variant image processing in the machine vision literature.

shows an example of this two-dimensional quotient space, in which each pixel represents an equivalence class of "interesting" object/feature associations. This approach produced likely candidates for license plates, as shown by the image density in Figure 3, which is an indication of the "best" places in the frame for the sensor to "look".

This application provides an example of a pragmatic, real-time attentional application of the Hough-Quotient space approach, and also indicates the utility of having a two-dimensional, image-like representation of the Hough accumulator array. The ability to interactively visualize the accumulator array, as the algorithm is run (Figure 3) is an important aid of system development. Finally, we hope to gain important insights into the nature of certain areas of the brain, such as the superior colliculus, which are known to have two-dimensional representations of the "interest" aspect of the visual scene, in a way which is at least superficially similar to that suggested by figure 3.

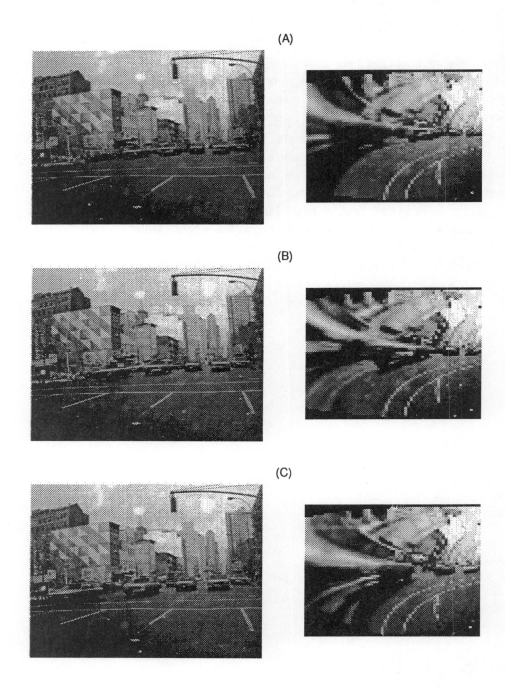

Figure 2 MOVIE FRAMES OF TRAFFIC AND LOGMAPS (**A**) Frame 1, t = 0 seconds (**B**) Frame 5, t = 0.91 seconds (**C**) Frame 8, t = 1.82 seconds

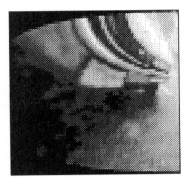

Figure 3 Two-dimensional quotient space representation of areas of interest for "license plates", as represented in a two-dimensional "motor" space of visual directions. Darker peaks represent maximum attentional value.

---

9. Discussion

We have presented a hierarchy of techniques for object recognition which, starting from a simple Bayesian model (Hough[1]), can be specialized by successive addition of assumptions to the abstract Hough transform (Hough[2]) and finally the generalized Hough transform (Hough[3]). We also showed how the Bayesian model, combined with some algebraic insight provides techniques for alleviation of the exponential space complexity of the Hough transform accumulator. In this section, we consider the role of the various assumptions in the successive algorithms.

The Bayesian model for object recognition at its most general (Hough[1]) provides an (asymptotic) upper bound for performance(Duda and Hart 1973). However, it is rarely attainable, since conditional and a priori distributions are typically unknown and must be estimated, calculated, or assumed. The intrinsic intractability of Hough[1] due to lack of this extensive knowledge leads to assumption of independent feature occurrences and conditionally independent feature occurrences.

It is probably safe to assume that features chosen on an a priori basis have occurrences (conditional or otherwise) which are *rarely* independent. On the other hand, if feature occurrences are nearly independent, then the assumption may not lead to incorrect results. The application of the assumption really needs specific analysis on a case-by-case basis; this is an area which could be illuminated by further analytic and experimental studies. In the absence of such data, we can make a few observations.

Conditional and nonconditional independence are separate issues. Conditional independence depends mainly on the imaging model; we consider that an object is present in the scene; we ask whether two features which are consistent with the object have independent conditional occurrences. If we ignore occlusion and consider only noise, the assumption is reasonable. Occlusion, however, will have the effect of making nearby features less independent, since, in general, an occlusion will be more likely to jointly involve neighboring features than distant ones. Thus spatial separation of features is a likely necessary condition for conditional independence.

Unconditional independence is a different phenomenon. Two features which are jointly consistent with a large subset of objects from the object space will probably not be unconditionally independent. The probability of the occurrence of one can be seen to be increased by the observation of the other. Similarly, mutually exclusive features (eg, two lines of different orientations at the same position) can hardly be unconditionally independent. Once again, chances for unconditional independence of features are probably enhanced when features are spatially separated.

Various well-known techniques (e.g. covariance diagonalization) can be applied to reduce correlations between otherwise dependent features. And even if independence of features is unobtainable, at least we have a criterion for what constitutes useful features, and for what the limitations these techniques incur. Obviously, the selection of features can be guided by independence considerations. Thus, just as the Bayes formulation provides an asymptotically optimal pattern classification (under restrictive assumptions about knowledge of the underlying probability distributions), unconditional independence of features provides an optimal, but difficult to obtain, upper bound on the feature selection process.

The generalized Hough transform (Hough³) incorporates not only assumptions of featural independence, and conditional independence, but also further assumptions which were elucidated above. Qualitatively, these additional assumptions may be to considered to embody indifference to a priori probabilities and the size of consistency sets ([32]), as well as indifference to variable information content of features ([33]). Several important practical benefits derive from the reformulation of the Hough transform along the lines of Hough²:

- The problem of nonuniform featural noise in the accumulator array is clarified and mitigated.
- Quantitative values for initializing and incrementing the accumulator array are specified.
- Interpretation of the value of the entries in the accumulator array as conditional (a posteriori) probabilities is supported, in contrast to the usual vague interpretation as a "confidence level."

And, although the requirement to estimate or measure a priori probabilities may seem impractical, it should be recalled that the default usage implies an assumption of prior probabilities that is rarely examined and perhaps could often be improved.

Quotient-space methods have appeared in various specialized guises over the years. The interpretation we have offered has several consequences. Because it is derived in the context of a general Bayesian model, the assumptions which it incorporates are explicit and may be readily identified and examined. The main assumption was the conditional independence of object occurrence, [19]. This assumption is immediately violated if objects are mutually exclusive. Otherwise, it may be considered for each specific task. The main advantage of using this assumption was that it permitted features to be "decoupled" in their influence on a particular output. This is embodied in the representation of $1-P(\psi|\Phi)$ as a product over the feature space, with the terms of the product depending on the object space, but fixed with respect to any feature set. Hence the complexity of the object space does not bear on the complexity of "on-line" calculation of $P(\psi|\Phi)$.

Summary

This paper provides a new way of viewing the Hough transform approach, and its generalizations, in terms of Bayesian pattern classification, by means of identifying the a usual a priori and posterior probability densities of Bayes method with a feature and object space. By adding a third, low dimensional space, which serves the role of a "behavioral" or "motor" space, and by projecting the high-dimensional object space onto it, we achieve the following useful theoretical and practical insights:

- Practical difficulties associated with the exponential dependence of the Hough transform accumulator bins are avoided.

- Theoretical insight into the nature of "visual attention" is provided. In particular, a computationally grounded approach to the general area of attentional control (of space-variant sensors) is achieved.

- In the common case in which the final "motor" space is two-dimensional, we achieve an image-like representation or projection of the Hough accumulator array, which may be visualized, and which is amenable to standard image processing hardware and algorithmic manipulation.

- Hough methods are derived from basic principles of statistical estimation (Bayes rule), and the nature of statistical independence of features, for optimal classification, is clarified.

- Practical examples of this methodology to several space-variant imaging problems is illustrated.

References

Adiv, G. (1983), Recovering motion parameters in scenes contatinng multiple moving objects, *IEEE Conf. Computer Vision and Pattern Recognition*, pp. 399-400.

Ballard, D.H. (1981), Generalizing the Hough transform to detect arbitrary shapes, *Pattern Recognition* **13**: 111-122.

Cohen, M., and Toussaint, G.T. (1977), On the detection of structures in noisy pictures, *Pattern Recognition* **9**: 95-98.

Davis, L.S., and Yam, S. (1980), A generalized Hough-like transformation for shape recognition, Tech. Rep.-134, University of Texas Computer Sciences.

Davis, L.S. (1982), Hierarchical generalized Hough transforms and line-segment based generalized Hough transforms, *Pattern Recognition* **15**: 277-285.

Duda, R.D., and Hart, P.E. (1972), Use of the Hough transform to detect lines and curves in pictures, *Comm. ACM* **15**: 11-15.

Duda, R.O., and Hart, P.E. (1973), *Pattern Classification and Scene Analysis*, NY: Wiley.

Hough, P.V.C. (1962), A method and means for recognizing complex patterns, U. S. Patent 3,069,654, U. S. Patent Office.

Illingworth, J., and Kittler, J. (1987), The adaptive Hough transform, *IEEE Trans. Pattern Analysis and Machine Intelligence* **9**: 690-697.

Illingworth, J., and Kittler, J. (1988), A Survey of the Hough Transform, *Computer Vision, Graphics, and Image Processing* **44**: 87-116.

Kimme, C.D., Ballard, D., and Sklansky, J. (1975), Finding circles by an array of accumulators, *Comm. ACM* **18**: 120-122.

Li, H., Lavin, M.A., and LeMaster, R.J. (1986), Fast Hough transform: A hierarchical approach, *CVGIP* **36**: 139-161.

Li, H., and Lavin, M.A. (1986), Fast Hough transform based on the bintree data structure, *IEEE Conf. Computer Vision and Pattern Recognition*, pp. 640-642.

Lipson, J.D. (1981), *Elements of Algebra and Algebraic Computing*, Menlo Park, CA: Benjamin/Cummings.

O'Rourke, J., and Sloan, K.R. (1984), Dynamic quantization: Two adaptive data structures for multidimensional spaces, *IEEE Trans. Pattern Analysis and Machine Intelligence* **6**: 266-279.

Rojer, A.S. (1990), *Space-Variant Computer Vision with a Complex-Logarithmic Sensor Geometry*, Ph.D. Thesis, New York University.

Rojer, A.S., and Schwartz, E.L. (1990), Design considerations for a space-variant visual sensor with complex-logarithmic geometry, *10th International Conference on Pattern Recognition, Vol. 2*, Atlantic City, NJ, pp. 278-285.

Rosenfeld, A. (1969), *Picture Processing by Computer*, NY: Academic Press.

Schwartz, E.L. (1977), Spatial mapping in primate sensory projection: analytic structure and relevance to perception, *Biological Cybernetics* **25**: 181-194.

Schwartz, E.L. (1980), Computational anatomy and functional architecture of striate cortex: a spatial mapping approach to perceptual coding, *Vision Research* **20**: 645-669.

Schwartz, E.L. (1985), On the mathematical structure of the retinotopic mapping of primate striate cortex, *Science* **227**: 1066.

Schwartz, E.L., Munsif, A., and Albright, T.D. (1989), The topographic map of macaque V1 measured via 3D computer reconstruction of serial sections, numerical flattening of cortex, and conformal image modeling, *Investigative Opthalmol. Supplement*, pp. 298.

Silberberg, T.M., Davis, L., and Harwood, D. (1984), An interative Hough procedure for 3D object recognition, *Pattern Recognition* **17**: 621-629.

Spiegel, J., Kreider, F., Claiys, C., Debusschere, I., Sandini, G., Dario, P., Fantini, F., Belluti, P., and Soncini, G., (1989), A foveated retina-like sensor using CCD technology, in: *Analog VLSI Implementations of Neural Networks*, C. Mead, and M. Ismail, eds. Boston: Kluwer.

Taylor, R.W. (1990), An efficient implementation of decomposable parameter spaces, *10th International Conference on Pattern Recognition, Vol. 1*, Atlantic City, NJ, pp. 613-619.

Tsuji, S., and Matsumoto, F. (1978), Detection of ellipses by modified Hough transformation, *IEEE Trans. Computing* **27**: 777-781.

Tsukune, H., and Goto, K. (1983), Extracting elliptical figures from an edge vector field, *IEEE Conf. Computer Vision and Pattern Recognition*, pp. 138-141.

Yeshurun, Y., and Schwartz, E.L. (1989), Shape description with a space-variant sensor: algorithms for scan-path, fusion and convergence over multiple scans, *IEEE Trans. Pattern Analysis and Machine Intelligence* **11**: 1217-1222.

OPTICS AND NEURAL NETS

David Casasent

Abstract

One version of the role for optics in neural nets is noted with attention to: new associative processor, optimization, adaptive learning, symbolic correlator, and production system neural nets. New algorithms, optical architectures and optical hardware are noted with emphasis being the role for optics. Digital and analog VLSI neural nets could benefit from the new algorithms described.

1. Introduction

Optical processors have many unique advantages that make them suitable for use in neural nets (NNs). We review some of these properties in Section 1.1. Sections 2-5 then detail specific cases and uses of optics in NNs. These include: associative processors (APs), new and efficient AP algorithms, and the use of optical feature space input neuron representations (Section 2); optimization NNs (Section 3) including new formulations without all constraints included within the neuron weights; adaptive learning NNs (Section 4) that combine the best properties of standard pattern recognition and NN techniques; and symbolic correlator NNs and associated production system NNs (Section 5) to accommodate multiple objects in the field of view (FOV).

1.1 Optical Processing

Optical processing offers many attractive properties. Those that we find attractive for NNs are now noted with examples of their use provided in subsequent sections.

Optical systems allow the use of massive numbers of interconnections (Goodman, Leonberger, Kung and Athale, 1984) with no crosstalk and with many-to-one and one-to-many interconnections equally possible. A matrix-vector architecture (Figure 1) is the most straightforward realization of this concept. The output at P_3 is the product vector of the input vector at P_1 and the matrix at P_2. If different elements of the P_2 matrix are activated or "on", the P_3 output can have one P_1 input element distributed to many P_3 elements or many P_1 elements combined into one P_3 element and this one-to-many and many-to-one distribution of data can occur in parallel for all P_1 inputs and P_3 outputs.

A second major operation easily achieved optically is the generation of various feature spaces. These feature spaces are attractive descriptions of an input object. Many optically generated feature spaces are possible (Casasent, 1985). Table 1 provides a subset of these feature spaces. All can be optically produced on a correlator as

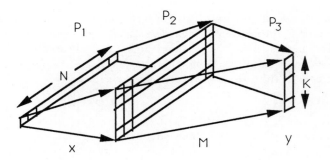

Figure 1. Optical matrix-vector processor architecture.

detailed elsewhere (Casasent, 1985). Figure 2 shows perhaps the most used optically generated feature space (Lendaris and Stanley, 1979), the wedge ring detected (WRD) magnitude Fourier transform (FT) of the input data. Figure 3a shows the standard optical FT system and Figure 3b shows the wedge/ring FT detector concept.

Fourier Transform (FT)
Wedge Ring Detected (WRD) FT
Hough Transform (HT)
Moments
Chord Distributions
Mellin/Polar-log FT Samples

Table 1. Optically generated feature spaces (Casasent, 1985).

Probably the most powerful optical processing architecture is the correlator. This system is a shift-invariant NN in which the input iconic neurons at P_1 are operated upon by a set of weights (filter functions at P_2) with their correlation function appearing at P_3. In the system shown in Figure 3, multiple laser diodes (LDs) at P_0 can be activated to access different filters at P_2. Different spatially-multiplexed P_2 filters are shown and at each P_2 spatial location multiple frequency-multiplexed filters are present. Thus, for each P_2 location selected, we can provide the correlation of the P_1 input with 4 etc. filters (with the 4 etc. correlations occurring in different quadrants of P_3). The most attractive aspect of an optical correlator is that the set of weights (at P_2) is applied in parallel to all regions of the input in plane P_1. When the proper

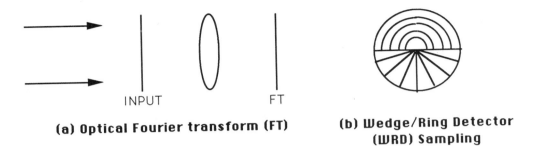

(a) Optical Fourier transform (FT)

(b) Wedge/Ring Detector (WRD) Sampling

Figure 2. Optically generated wedge ring detected (WRD) feature space (b) magnitude Fourier transform (FT) (a).

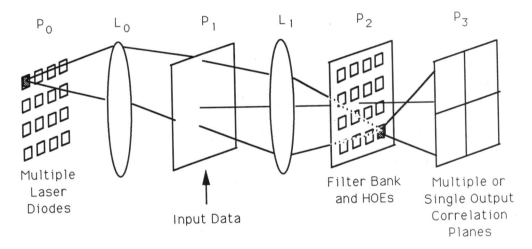

Figure 3. Space and frequency-multiplexed optical correlator.

computer generated hologram (CGH) is present in P_2, then the P_3 output can be any feature space (Table 1). Thus, the optical correlator architecture of Figure 3 is most useful, powerful and general purpose.

2. Optical Associative Processor (AP) Neural Nets (NNs)

We now describe specific optical AP NNs using new algorithms and their realization on the basic optical matrix-vector processor of Figure 1. The P_1 inputs $\underline{x}$ are key vectors and the P_3 outputs $\underline{y}$ are recollection vectors where $\underline{M}\,\underline{x} = \underline{y}$ and $\underline{M}$ is the AP matrix. The keys are of dimension N, the recollections of size K and the number of key/recollection vector pairs stored is M. The basic issues we note concerning any AP are:

1. We require M > N for reasonable storage. The Ho-Kashyap (HK) AP provides the largest M > N.
2. We use storage density M/NK rather than just M as a performance measure.
3. APs must handle dependent keys. The HK AP does.
4. APs must handle input noise and limited input accuracy (inherently, not ad hoc). Robust HK APs do this optimally.
5. Encoded outputs (content addressable APs, CAAPs with non unit-vector output encoding) are required to achieve the best storage density.
6. We use the percentage P'_C of output vectors completely recalled (rather than the fraction of recollection vector elements recalled correctly) as our performance measure.

Table 2 gives the steps in the HK AP synthesis algorithm (Telfer and Casasent, March 1990). The algorithm iteratively refines the initial pseudoinverse estimate of the AP matrix $\underline{M}$ by adjusting the recollection vector elements and the matrix. It handles linearly dependent key vectors and yields the largest storage capacity M = 2N of any general memory 1:1 AP. Convergence has been proven. Iterations are used only in synthesis. In recall (classification), the system requires only one-pass. The Robust HK-2 version of this algorithm (Casasent and Telfer, 1990) includes noise (described by σ^2) and the number of stored vector pairs M, by using the new starting matrix

$$\underline{M} = \underline{Y}\,\underline{X}^T(\underline{X}\,\underline{X}^T + M\sigma^2\underline{I})^{-1} \tag{1}$$

in step (1) of the HK algorithm in Table 2. This yields the optimum (not ad hoc) performance in noise and equivalently optimum performance with analog input accuracy (related to σ^2).

STEP	OPERATION
1	$\underline{M}_n = \underline{Y}_n X^+$
2	$\underline{E}_n = \underline{M}_n X - \underline{Y}_n$
3	$\underline{E}'_n = 0.5[\underline{E}_n + \underline{S} \oplus |\underline{E}_n|]$
4	$\underline{Y}_{n+1} = \underline{Y}_n + 2\rho \underline{E}'_n,\ 0 < \rho < 1$
5	If $\underline{E}'_n \neq \underline{0}$ go to 1

Table 2. Ho-Kashyap associative processor algorithm

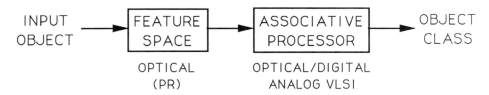

Figure 4. Optical feature space input neurons to an associative processor.

Figure 4 shows the use of feature space input neurons (optically generated) to an AP. This is a very attractive approach for pattern recognition APs since:

1. The number of input neurons is much less than if iconic neurons are used.
2. The feature space (e.g. wedge ring FT samples) is shift, rotation and scale-invariant.
3. Property (2) greatly reduces the required training and learning.

We find such free-space optically generated feature space neurons to be far preferable to higher-order NNs and other techniques to achieve shift-invariance (these require N^4 hardwired interconnections for N input neurons).

We have shown (Casasent and Telfer, 1990) that CAAP HK-2 APs are necessary to achieve storage density better than that of the direct storage nearest neighbor matrix-vector processor. We found L-max encoded output recollection vectors to perform better in APs than other standard error correcting encodings. We have achieved $M > 1.5N$ storage with $P'_C > 95\%$ performance in $\sigma_n = 0.05$ noise with 6-8 bit accurate AP elements (using random analog keys and binary recollection vector elements in a general 1:1 AP).

For many:1 APs (in which many input keys (for different distorted objects in one class) are associated with the same recollection vector), we have achieved (Telfer and Casasent, July 1990) excellent storage density and recall accuracy for multiclass distortion invariant recognition applications. We use an optically generated feature space for our input neurons (Figure 4), specifically the WRD FT feature space of Figure 2. This provides low dimensionality input neurons (we use 32 wedge $|FT|$ samples and $N = 33$ neurons, with the additional neuron used to shift the thresholds and allow the hyperplane discriminant surfaces to not pass through the origin). For two aircraft (Test Sets 1 and 2) with large pitch and roll variations as in Table 3, we used a training set of $N_T = 1250$ vectors (corresponding to 5° increments in pitch and roll). This corresponds to a large storage of $M = 1250$ vectors (Test Set 2) or $M/N > 25$. A Test Set of intermediate 2.5° distorted images not present in the training set were used. Table 4 shows the test results in noise for the standard and the Robust HK-2 (with $\sigma_{syn} = 0.1$) APs. As seen, results were excellent with over 95% correct recognition (P'_C) for the lower angular distortions in Test Set 1 and $P'_C > 90\%$ for the larger distortions in Test Set 2. These APs achieved these results using only 6 bits of input and memory matrix accuracy ($\approx 1\%$). Thus, this shows that a $33 \times 2 = 66$ element AP memory can store over 1000 vector pairs ($M > 25N$) or over $1250(33) = 36{,}250$ vector elements with $P'_C > 95\%$ in $\sigma = 0.1$ of noise ($SNR_I = 5$) with only 6-bit accuracy required. We find these results most impressive.

Test Set	Aircraft	Pitch and Roll Distortions	$M = N_T$ (Train Set) (5° increments)	Test Set (2.5° intermediate)
1	Phantom, DC-10	±50°	$2 \times 441 = 882$	$2 \times 400 = 800$
2	Phantom, DC-10	±60°	$2 \times 625 = 1250$	$2 \times 576 = 1152$
3	Phantom, DC-10, F104	±80°	$3 \times 1089 = 3267$	$3 \times 1024 = 3072$

Table 3. Test sets for multiclass distortion-invariance.

HAP	Test Set 1		Test Set 2	
	Training Set	Test Set	Training Set	Test Set
HK	93.2	92.3	82.3	88.5
HK-2	95.1	95.4	90.6	91.2

Table 4. Test results (P'_C) for associative processors in noise ($\sigma_n = 0.1$).

3. Optimization NNs

We formulate our optimization NNs as iterative algorithms (Yee and Casasent, 1991)

$$\underline{x}_{n+1} = f_Q[\underline{M}\,\underline{x}_n + \underline{v}] = (\text{Matrix}) \bullet (\text{Vector}) + (\text{vector}) \qquad (2)$$

where f_Q is a sigmoid nonlinearity function. This formulation as iterative matrix-vector products plus an added vector is very attractive from many standpoints. We easily modify the optical system of Figure 1 as in Figure 5 to implement the algorithm in (2). This is also very attractive from an optical implementation standpoint, since the matrix $\underline{M}$ is fixed (independent of the input data) for a given optimization problem of given maximum size. Thus, $\underline{M}$ can be realized on film (which greatly simplifies the optical system). Only the external vector $\underline{v}$ varies with the data. The matrix $\underline{M}$ also has a block Toeplitz structure that can simplify realization by partitioning for large problems. This formulation is free from the problems of the standard Hopfield optimization NN (Bruck, 1990). Specifically, our use of a sigmoid nonlinearity at P_3 reduces oscillations and we only enforce certain constraints (e.g. 1:1 correspondence) in the weight matrix and enforce other constraints (e.g. distance) by the added bias vector $\underline{v}$ (this avoids convergence to false states and such problems when the number of neurons used is large). The neuron bias values in the standard optimization NN carry no information or constraints while in our algorithm they include the distance constraints (and they change with the input data).

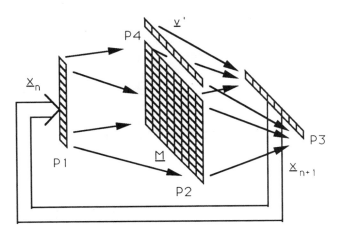

Figure 5. Optical optimization neural net.

We have extensively tested this optimization NN for the data association problem in multitarget tracking. In this case, we consider N_m measurements in frame n and N_m in frame n+1 and use N_m^2 input neurons. We consider the neuron states as an $N_m \times N_m$ matrix to best convey results (Figure 6). The axes of this neuron matrix are measurements in two adjacent frames. We require one "on" neuron per row and per column (the 1:1 constraint) and we require associations that are closest in distance (the added bias vector $\underline{v}$ contains these data). Figure 6 shows the neuron states at several different iterations with convergence in Figure 6d (for the case of 6 targets or measurements). Figure 7 shows the measurements for 6 targets in 5 time frames (Figure 7a) and their associated tracks (Figure 7b) calculated by the data association NN. The data in Figure 6 are for frames 3 and 4. These data contain jitter in each target location plus noise points (SNR = 1) not shown. As seen and detailed elsewhere (Yee and Casasent, 1991), this NN performs excellently.

4. Adaptive Learning NNs

This type of NN has received the most attention since it adaptively determines nonlinear piecewise discriminant surfaces for multiclass pattern recognition applications. We consider the basic three-layer NN of Figure 8. We first advance remarks on the role

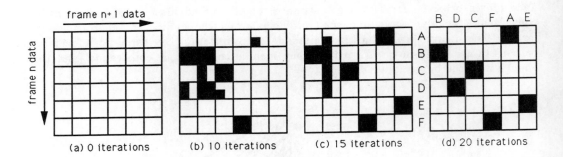

Figure 6. Neuron states in the optimization NN at different iterations.

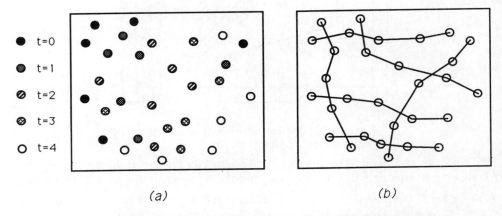

Figure 7. Measurements (a) for 6 targets in 5 frames (with jitter and noise not shown) and their associated tracks (b) calculated by the optimization NN.

for optics in such a NN (Section 4.1), the key issues in our adaptive clustering neural net (ACNN) (Casasent and Barnard, 1990) are then described (Section 4.2). This approach combines pattern recognition and neural net techniques. Initial multiclass distortion invariant object recognition results are then presented.

4.1 Role for Optics

Optical systems are ideal for interconnections since photons cross and do not interact. Thus, free-space interconnections are preferable to hardwired ones. The interconnections can be made with a matrix-vector architecture (Figure 1) or using a volume hologram (this allows operation on a 2-D input neuron array). Fiber optics and integrated optics are also possible but are more limited. The interconnections can be

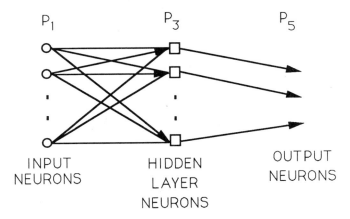

Figure 8. Standard three-layer neural net.

formed optically or using computer generated holograms (CGHs). We prefer CGHs for their better accuracy. To best utilize the potential of optics, we feel that analog neurons and weights are needed.

We consider image processing applications. When there is one object in the field of view (FOV) and its location is now known, we require shift invariance. Optically generated feature spaces can achieve this and thus the input neurons should be a feature space (Figure 9). This reduces the number of input neurons, the number of interconnections, achieves in-plane distortion-invariance and thus reduces training, plus the input neurons are now analog. When multiple objects are present, an optical correlator provides a most useful input neuron space as in Figure 10 (Section 5 describes such a system). Recall that an optical correlator is a most powerful neural net. We feel that neural net learning algorithms are best realized digitally with the calculated weights downloaded to an optical NN (gated learning) for classification. We also consider only supervised learning NNs.

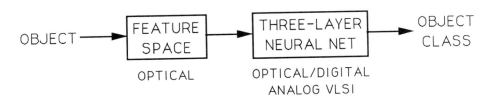

Figure 9. Optical feature space neuron inputs for a shift and distortion-invariant neural net.

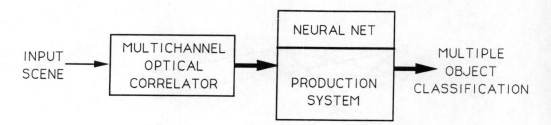

Figure 10. Optical correlation neuron inputs for a neural net to handle multiple objects.

4.2 The ACNN (Adaptive Clustering NN)

Our ACNN adaptive learning NN for pattern recognition is based upon reducing the number of ad hoc parameters in standard NNs and in marrying pattern recognition and NN techniques.

We use clustering techniques to select prototypes or exemplars of each object class. These are each associated with one hidden layer neuron and thus provide a definite algorithm (using pattern recognition techniques) to select the number of hidden layer neurons N_3. At P_3, we employ a winner take all (WTA) maximum selection of the most active P_3 neuron. Binary weights then map the P_3 neurons to the output P_5 neurons. We use N_4 output neurons, where $N_5 = C$ is the number of object classes.

Our initial P_1-to-P_3 weights are not random but are the feature vectors associated with each P_3 prototype neuron. Thus, the initial analog weights are linear discriminant functions (LDFs) chosen by pattern recognition techniques. We then present the training data to the NN, measure the most active true and false class P_3 neuron value and use a conjugate gradient algorithm to adapt the weights. This uses NN techniques to combine LDFs into the nonlinear decision surfaces needed and hence combines NN and pattern recognition techniques.

4.3 Multiclass distortion-invariant object recognition results

As an example of the performance of our ACNN in a difficult pattern recognition problem, we consider the three class (C=3) multiple degree of freedom distortion problem in Test Set 3 in Table 3. Our AP performed very poorly ($P'_C < 60\%$) on this data set. We thus formed an ACNN for it with $N_1 = 33$ input neurons (wedge $|FT|$

samples and an extra neuron), $N_3 = 9$ hidden layer neurons (3 per class), and $N_5 = C = 3$ output neurons (one per class). The NN was trained on $3 \times 630 = 1890$ distorted aircraft (630 per class) and tested on 578 images per class at intermediate distortions not previously seen. Test results were excellent with $P_C' = 98.6\%$. Learning required only 180 iterations. The standard backpropagation NN required a factor of over 100 more iterations. The prototypes and initial weights need not be selected carefully. We showed this by using only 45 random samples (15 per class) to select the initial weights. A negligible difference in performance occurred. We have also showed that the computational load for N_3 prototype and initial weight selection is $< 0.1\%$ of the full computational load. Thus, our algorithm is most efficient. This multiclass distortion-invariant object recognition work represents one of the most complex pattern recognition NN applications with excellent performance obtained.

5. Symbolic Correlator and Production System NNs

When multiple objects are present in the field of view, we require a fully connected shift-invariant NN. This is best achieved in with optical correlator input fact neurons (Figure 10). In one realization of this concept that we have demonstrated in the optical laboratory (Casasent and Botha, 1991), the correlation filters were of generic object parts (posts, boxes, circles, etc.) and distortion-invariant filters were used that could recognize the object parts independent of 3-D distortions. The outputs from a multichannel optical filter indicated the presence and location of each generic object part in an input scene. We used these as an input fact neuron space. We then described each object as various sets of generic parts as a set of if-then statements. To determine the class of each input object, we used a production system NN (Figure 1 with P_3 to P_1 feedback). We allocate the $N_1 = N_3$ input and output neurons to be objects and object parts. The set of weights at P_2 describes the rules. This production system NN (Casasent and Botha, 1991; Botha, Casasent and Barnard, 1988) is thus an iterative matrix-vector multiplier. All possible rules need not be encoded as the system can learn new rules. It iterates until one of the output object class neurons is activated, thus indicating the class of the input object.

References

Goodman, J.W., Leonberger, F.J., Kung, S-Y., and Athale, R.A. (July 1984). Optical Interconnections for VLSI Systems. *Proc. IEEE*, Special Issue on Optical Computing, **72**, 850-866.

Casasent, D. (January 1985). Coherent Optical Pattern Recognition: A Review. *Optical Engineering*, **24**, 26-32.

Lendaris, G.G. and Stanley, G.L. (February 1979). Diffraction-Pattern Sampling for Automatic Pattern Recognition. *Proc. IEEE*, **58**, No. 2, 198-216.

Telfer, B. and Casasent, D. (10 March 1990). Ho-Kashyap Optical Associative Processors. *Applied Optics*, **29**, 1191-1202.

Casasent, D. and Telfer, B. (submitted November 1990). High Capacity Pattern Recognition Associative Processors. *Neural Networks*.

Telfer, B. and Casasent, D. (July 1990). Ho-Kashyap Advanced Pattern Recognition HeteroAssociative Processors. *Proc. SPIE*, **1347**, 16-32.

Yee, M. and Casasent, D. (December 1991). Multitarget Data Association Using an Optical Neural Network. *Applied Optics*, **30**.

Bruck, J. (1990). On the Convergence Properties of the Hopfield Model. *Proc. IEEE*, **78**, 1579-1585.

Casasent, D. and Barnard, E. (10 June 1990). Adaptive Clustering Optical Neural Net. *Applied Optics*, **29**, 2603-2615.

Casasent, D. and Botha, E. (December 1991). Optical Correlator Production System Neural Net. *Applied Optics*, **30**.

Botha, E., Casasent, D., and Barnard, E. (15 December 1988). Optical Production Systems Using Neural Networks and Symbolic Substitution. *Applied Optics*, **27**, 5185-5193.

Neural Networks for Image Analysis

Robert Hecht-Nielsen

HNC, Inc.
San Diego, California
and
Department of Electrical and Computer Engineering
University of California, San Diego
La Jolla, California

Abstract

Neural networks have now been used for a variety of image analysis applications, some with considerable success. This paper reviews some methodologies for using neural networks for image analysis in the context of three case studies (apple sorting, automatic target recognition, and medical image analysis).

1 Introduction

In writing a paper about the capabilities of neural network-based image analysis systems there is a danger that the capabilities demonstrated will be mistakenly viewed as impressive on some absolute scale.

In reality, current-day machine vision systems are vastly different from, and in most ways inferior to, biological vision systems. Machine vision systems build the image of a scene using frame gathering operations with constant-resolution sensors that have the same solid angular density of pixels in all areas of the frame. No vertebrate animal employs such constant resolution, single snapshot imaging. Instead, vertebrates build their mental image of a scene sequentially through hundreds of twitches and tens of slow drifts of their foveated eyeballs at each of tens of saccade fixation points in the scene. As vertebrates, we are not aware of this sequential nature of our visual process. We don't notice our eyes moving to additional fixation points to enhance and expand our awareness of details as we ponder a scene. Somehow, the entire herky-jerky scene building and scene enhancement process is invisible to us. All we see is the illusion of a constant, steady visual reality.

Thus, in discussing machine vision applications we must constantly remind ourselves to stick to the facts and avoid the almost always erroneous and misleading anthropomorphisms that so readily spring to mind. If we are to understand machine vision and its potential we must view it with the knowledge that it is really totally different from vision as we personally know it[1]. This paper provides some new data regarding the effectiveness of particular technological neural network machine vision methodologies applied to three specific problems. In discussing these applications, the "case study" approach is used. Perhaps this information can contribute to the ongoing process of understanding machine vision on its own terms.

---

[1]Note: Machine vision may be changing. Practical eyeball-based machine vision systems could emerge in the future (see [13,14,15,16,17] for some representative ideas in this direction).

2 Apple Sorting

A classic machine vision problem is real-time, on-line product inspection. For products such as apples this has never been practical because the operations that must be carried out involve the measurement of color, texture, and three-dimensional shape. In this subsection a real-time on-line inspection system built by HNC, Inc. and Sumitomo Heavy Industries is described. This system, which is installed in Japan, can inspect three apples per second on a sustained basis.

Apple inspection involves measurement of four properties of each apple: color grade (overall color), discoloration, the presence of defects, and size. These attributes are then combined to determine an overall grade for the apple.

The measurement of coloration involves determining whether the apple's color is *striped* or *uniform*. Striped apples have their color pattern arranged in radial stripes that emanate from the stem and antistem. Uniform apples have a uniform color (with small dots) over the whole apple.

Finally, defects must be discovered. These come in a variety of forms, but the most important are discolorations (e.g., sunburns), surface roughness (russeting), and bruises.

Once the grade, coloration, and defect content of the apple are determined a *sorting function* is applied to determine which of several boxes the apple should be put in. This sorting function is determined mostly by marketing considerations. For example, at a time of year when apples are scarce an apple with a small yellow streak may be placed in a box with apples of the same grade and coloration but without streaks (and therefore sold at the same price). When apples are plentiful the apple with the yellow streak may not be allowed in that same box. The process of determining the sorting function is an art that is well mastered by apple packaging and marketing experts. This process will not be discussed here.

The apple sorting system described here operates on apples as they ride by single-file on a conveyor belt. The apples are placed stem-up in small cups on the belt. Because of this method of transport, the very bottom of the apple cannot be inspected (this is not a significant consideration for the Japanese market).

As the apples move along on the conveyor belt they enter a small tunnel that is brightly illuminated. In the middle of this tunnel are three color CCD television cameras (each having a resolution of approximately 512 × 512 color pixels). One camera looks at the apple from above and the other two look at it from the sides. The belt runs at a rate of 2 to 3 apples per second, so a total of from 6 to 9 color images are taken per second. Each of the color images is digitized by a color frame grabber board and these images are then transferred via a direct Data Translation Corporation DT-Connect™ high-speed digital interface to an HNC, Inc. Balboa™ 860 neurocomputing coprocessor board. The frame grabber board and the Balboa board are plugged into the ISA bus of a PC-compatible computer.

The processing of each of the three images of each apple is essentially the same, so we will concentrate on the image of the top of the apple. This image is a bit more difficult to process than the side images because the stem must be ignored. Stems can come in many sizes and shapes and can cause many defect false alarms unless they are explicitly dealt with.

Typical top camera images are shown in Figure 1 (along with the color-coded outcomes of

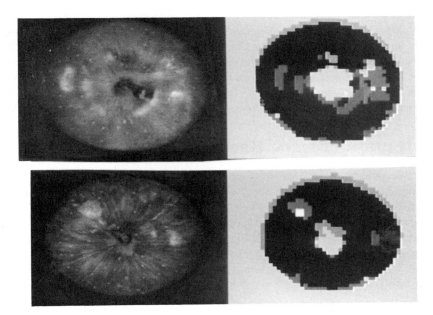

Figure 1: Apple defect detection and classification. The apple in the top left image (originally in color) has a small hole in it just above the stem. The neural network based system has found and correctly identified this defect. The apple in the lower left video image has a discoloration defect (a yellow streak) on the left side of the stem. Again, this has been correctly classified. The stem regions and illumination light glare regions have also been correctly identified on both images. These results are typical.

processing). As can be seen in the figure, glare from shiny spots on an apple cause bright artifacts that look very much like local discoloration defects.

Determination of the grade of the apple is done by simple mensuration of the edge profile of the apple in the three color views.

Coloration (uniform or striped) classification is carried out using color spatial frequency filter features and a simple rule-based system.

The technical approach used for defect classification is to employ a coalition of specialized neural networks. One set of networks carries out local analysis on small (10×10 or smaller) square tiles of color pixel values within the image. These mapping networks (e.g., back-propagation [9,10] or counterpropagation [10] networks) are trained to identify the presence or absence of each specific type of defect. The (essentially binary) outputs of these classifiers are then used as the inputs to other neural networks which carry out regional defect classifications within larger tiles (up to 30×30). The tile classification network outputs, the regional classification network outputs, and other image features (including the size and shape of the apple boundary in the image) are then submitted to a rule-based system which makes the final defect detection and classification decisions. The use of a rule-based back end allows easy insertion of modifications to the overall decisionmaking process as knowledge is gained about the on-line performance of the system.

In early experimental testing of the completed system its performance was compared with that of human apple inspectors in Japan. The conclusion was that the performance of the machine is roughly comparable to that of a human apple inspector in terms of the technical grade, coloration, and defect decisions made on a large varied lot of apples. However, the machine is only able to carry out these technical evaluations. It is not able to adaptively develop the market-driven sorting function that determines which apples should go into which boxes. To compensate for this, the system allows manual adjustment of the decision boundaries, so that the sorting process can be configured to market conditions. This on-line apple inspection system appears to be the first successful system of its kind. It has sufficient speed to operate in real-time and sufficient accuracy to replace a human inspector.

3 Automatic Target Recognition

The neural network machine vision problem of detecting and classifying targets of interest in military images has been studied by many groups (see [12] for a recent survey). This section discusses three components of an end-to-end automatic target recognition (ATR) system recently constructed and tested at HNC sponsored by DARPA's *Comparative Measurements: Neural Networks for Target Recognition* Program (via a subcontract from Martin Marietta, prime contractor to the U.S. Army Center for Night Vision and Electro-Optics – the DARPA management agents for the ATR portion of the Program). The HNC ATR system carries out target detection and classification for infrared images (see [18] for additional technical information).

The imagery used during this project was simulated IR imagery produced on a terrain board by the U.S. Army Center for Night Vision and Electro-Optics (see Figure 2). This method of producing experimental imagery has the advantages that the images are both realistic and relatively inexpensive to produce. Thus, statistically significant numbers of training and testing examples can be produced by this method. The ATR system developed using this simulated imagery can then be applied to real imagery with the reasonable expectation that it will work roughly as well as on the simulated imagery. Although the results will not be reported here, testing of the HNC ATR system on real IR imagery did indeed produce essentially the same results as testing on simulated imagery.

The HNC ATR system consists of three functional modules that are sequentially applied to each image:

1. Potential Target Locator: the first stage in a two-stage detection process. This component identifies regions in the image which are good candidates for target locations.

2. False Alarm Filter: the second stage in a two-stage detection process. This component screens the target locations identified by the Potential Target Locator so that false alarms can be reduced.

3. Target Classifier: analyzes the target locations which survive the False Alarm Filter in order to classify the target type.

All three of these ATR modules operate on a Gabor profile representation of an image region [10]. These profiles consist of a number of Gabor jets which have been computed at selected points in the region [4,5,6,7,8]. The jets in turn consist of the coefficients generated by correlating different Gabor wavelets with the image at the selected points. Circular

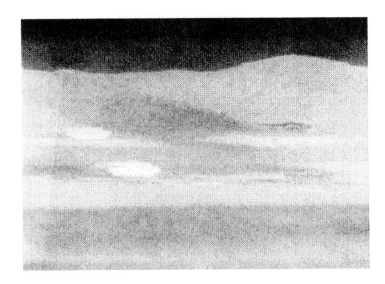

Figure 2: A simulated infrared battlefield image. This image contains four tanks. These are oriented as if they were moving left in the image. The tanks have simulated reflectivities representing different illumination situations (positive contrast, negative contrast, and neutral contrast). Finding and accurately identifying targets of these types is a very difficult problem. Only highly trained and skilled experts operating at peak performance can equal the target detection and classification performance of the neural network-based ATR system described here. No human can equal its speed; nor its relentless, tireless consistency. The nature of warfare may be permanently altered by the emergence of such capabilities.

Figure 3: Location of jets in a Gabor profile.

Gaussian Gabor wavelets of three to six wavelengths (half-octave spacing) and eight orientations (between 0 and 180 degrees) are used to generate the jet coefficients. Each wavelet has approximately 1.5 cycles of sine grating within its 1σ Gaussian circle. The profiles we use consist of five jets computed at the center point of the image region being considered and at four surrounding points. The geometric relationship of these five points is illustrated in Figure 3.

The images used in this study were such that each target subtended an image area of approximately 1000 pixels. Human studies have concluded that this is approximately the smallest target size that can be reliably used for target identification by expert photoanalysts. Thus, if we can use human performance as a guide, performance of an ATR system should drop off rather rapidly when target size decreases below 1000 pixels. Performance with larger target sizes should be about the same as for a 1000 pixel target system.

In the first ATR module, the Potential Target Locator, various regions in the image are identified as Potential Target Locations (PTLs) after comparing their Gabor profiles to the profiles of one or more filled polygons (the centroid of the polygon serves as the center point of the profile). The polygons are intended to be roughly similar in appearance to human-made objects. The entire image is checked by examining the region centered on each point in a grid of equally spaced points. For each such region, a similarity measure is computed between the regions profile and that of each polygon. If the similarity measure for any of the polygons exceeds a pre-selected threshold, the center of the image region is deemed to be a PTL. Multiple PTLs in a small area of the image are eliminated by choosing the one with the highest similarity measure.

The first step in computing the similarity measure is to calculate an individual similarity measure for each pair of corresponding jets. Given the profile pictured in Figure 3, there would be five such pairs. The product of the individual similarity measures serves as the similarity measure for the region. The similarity measure used is:

$$S(i,j;k,l) =$$
$$F(G(i,j) \cdot G(k,l), \|G(i,j)\| \, \|G(k,l)\|) \min[F(\|G(i,j)\|, \|G(k,l)\|), F(\|G(k,l)\|, \|G(i,j)\|)]$$

where F is a real function of two real variables (see [3,18] for insights into the construction of F), $G(i,j)$ is the jet centered at point (i,j) in the polygon, $G(k,l)$ is the jet centered at point (k,l) in the image, and $S(i,j;k,l)$ is the similarity measure between those two jets. Each jet can be thought of as a vector of coefficients, and the similarity measure takes on its highest values when the two jet vectors are similar in both direction and magnitude. This similarity measure is based on one used by Buhmann, Lange, and von der Malsburg [2,3,11,1].

In the second and third ATR modules, the Gabor profiles are used in a different way. Instead of comparing profiles and computing a similarity measure, the profiles are used as input to a feed-forward neural network trained with backpropagation. In the case of the False Alarm Filter, a Gabor profile representing the PTL is used one jet at a time as network input. Thus, the network contains one input processing element (PE) for each coefficient in a jet. The network has a number of hidden layers and hidden PEs, and a single output PE. After a Gabor jet is input to the network, the state of the output PE is a measure of the networks confidence that the jet is centered on a target. This confidence measure is referred to as the figure of merit (FOM). The mean FOM is computed over all jets in the Gabor profile. By comparing the mean FOM to a pre-selected threshold, a decision is made as to whether the profile represents a target.

In the Target Classifier, the input to the network consists of an entire Gabor profile. Thus, the network contains one input PE for each coefficient in the profile. The network has a number of hidden layers and hidden PEs, and one output PE for each possible target type. When a profile is input to the network, the Target Classifier's prediction is taken to be the target type corresponding to the output PE with the highest state.

The overall performance of the HNC ATR system was tested using images that it was not trained on. Testing of the Potential Target Locator portion of the system demonstrated performance in excess of 95% probability of detection with a false alarm rate of 5 false alarms for every real target. The False Alarm Filter was able to reduce the false alarm rate from 5:1 to 1:9 while degrading the probability of detection by about 10%. However, for some practical applications a false alarm rate of 5:1 is not at all a problem. Thus, in development of an operational system the False Alarm Filter might or might not be included (in order to keep the probability of detection as high as possible).

Testing of the Target Classifier portion of the system was carried out on image segments known to contain a target of a given class (or, in some cases, no target – i.e., "clutter"). The results are shown in the form of scatter matrices in Figures 4 and 5. The six-class problem (plus clutter) yielded an overall classification accuracy of 97.5%. A state-of-the-art statistical linear discriminant classifier achieved slightly less than 70% accuracy when trained and tested using the same overall image set. In this comparative test the neural network-based ATR system demonstrated an error rate that was 83% lower than that of a currently used approach. Note that, because of the excellent clutter rejection performance of the system, operation with the False Alarm Filter may not be required, or desirable.

When the perplexity of the problem is increased by increasing the number of target classes from six to 10, performance drops to an average of 91% classification accuracy. Very few unconstrained image object classification systems for any purpose have been built that deal effectively with 10 classes. And none that we are aware of perform at this level of accuracy.

In conclusion, the use of neural networks in automatic target recognition systems seems to hold great promise.

	ATR DECISIONS								
GROUND TRUTH	M1A1	M60	M113	M35	HUMMV	M163	CLUTTER	Totals	%Correct
M1A1	25	0	0	0	0	0	0	25	100.0
M60	0	35	0	0	0	0	0	35	100.0
M113	0	0	33	0	1	1	0	35	94.3
M35	0	0	1	35	0	0	0	36	97.2
HUMMV	0	0	0	0	37	0	0	37	100.0
M163	0	0	1	1	0	28	0	30	93.3
CLUTTER	0	0	0	0	1	0	39	40	97.5
Totals	25	35	35	36	39	29	39	238	97.5

Figure 4: Confusion matrix for the results of testing of the six-target-class plus clutter neural network ATR classification system on 238 images not used for training. This performance is approximately the same as that of the best human experts on their best days.

	ATR DECISIONS										
GROUND TRUTH	M1A1	M60	M2	M113	M35	HUMMV	M163	M730	A10	F15	Totals	%Correct
M1A1	30	1	0	0	0	0	0	0	1	1	33	90.9
M60	0	27	0	0	0	1	0	0	0	0	28	96.4
M2	0	0	31	1	0	0	0	0	0	0	32	96.9
M113	0	0	1	16	0	0	1	0	0	0	18	88.9
M35	2	0	0	1	27	0	0	0	0	0	30	90.0
HUMMV	0	0	0	5	0	39	0	0	0	0	44	88.6
M163	0	1	0	2	0	0	29	0	0	0	32	90.6
M730	0	0	0	2	0	0	0	38	0	0	40	95.0
A10	1	0	0	0	1	3	0	0	36	2	43	83.7
F15	1	0	0	0	0	0	0	1	1	32	35	91.4
Totals	34	29	32	27	28	43	30	39	38	35	335	91.0

Figure 5: Confusion matrix for the results of testing of the 10-target-class neural network ATR classification system on 335 images not used for training. This performance is approximately the same as that of the best human experts on their best days.

4 Blood Testing

This section describes a neural network-based system developed at HNC, Inc. for the pathological analysis of images of chemically prepared blood. The images used in the development of this system were provided by a major medical diagnostic products company.

The overall problem was to classify two types of prepared blood images. Each type was produced by a different chemistry. The first type of image is a texture that is to be categorized into either the "positive" or "negative" class. This first type of image is usually easy to visually categorize correctly. The second type of image was also a texture, but these must be identified as belonging to one of five diagnostic classes. These classes correspond to five different densities of a particular protein. The second type of images are much harder for humans to categorize correctly. This discussion shall concentrate on the diagnostic classification of the second type of prepared blood texture images. A similar system was developed for the first type of images and it performed nearly perfectly. Examples of each of the five diagnostic classes of the second type of image are shown in Figure 6.

The common thread of texture differences in both of the image classification tasks led HNC to investigate and select Gabor logon decomposition as a first step in generating an image feature vector that could be used with a backpropagation neural network classifier. Gabor logon decomposition is similar to a Fourier decomposition, but uses Gabor functions instead of sines and cosines [4,5,6,7,8]. As shown by Daugman [4,5,6,7], Gabor logon image features are excellent for texture classification.

The Gabor decomposition was done by a neural network invented by John Daugman [4,7,10]. Daugman's Gabor decomposition was implemented on the HNC ANZA Plus™ neurocomputing coprocessor. This neural network was then used to generate logon descriptions of each image. These logon coefficients were then submitted to a backpropagation neural network that was trained as a Bayesian classifier. In other words, the network had five output units. During training, the unit corresponding to the correct class was to have an output of 1.0 and the other four output units were expected to have outputs of 0.0. The system was trained on 50 images from each class.

Following training, the system was tested on 247 new images of the second type that were not used for training. These test images were divided approximately evenly between the five diagnostic classes. The results of this testing are shown in Table 1. Note that in each case where the network made an error that the error was no more than one class number (the one exception to this – a class 1 image that was misclassified as being in class 5 – was found to have a large air bubble in it; making it a defective example). An interesting fact is that when we ranked the network outputs by size (with the largest being chosen as the class of the image), in those instances where the network made an error the second choice was always the correct class. Overall, the network demonstrated a classification accuracy of 81%. This is significantly better than human diagnostic technicians, who cannot do better than about 50% on classes 1 and 2 and about 75% on classes 3, 4, and 5.

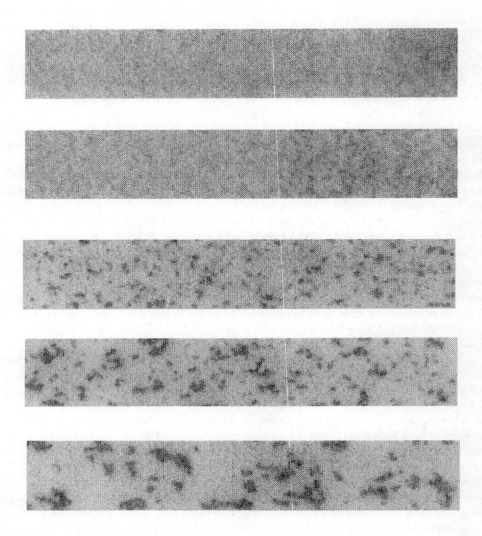

Figure 6: Examples of prepared blood texture images of the second type. From top to bottom, the images belong to classes 1, 2, 3, 4, and 5. Humans have great difficulty correctly distinguishing the first two classes.

Actual Class Number	Network Classification Class Number				
	1	2	3	4	5
1	39	9	0	0	1
2	13	35	2	0	0
3	0	6	40	4	0
4	0	0	2	43	4
5	0	0	0	6	43

Table 1: Performance of the neural network blood image classifier on (type two) test images not used for training. The confusion matrix compares the actual class number of each example with the neural network-determined class number. This performance significantly exceeds human performance on the same task.

5 Conclusions

Three applications of neural network-based image analysis have been presented: apple sorting, automatic target recognition, and medical image classification. In each case, the neural network-based system has demonstrated performance significantly better than than achieved by all previous automated systems. Also, in every case, the performance attained is comparable to, or in excess of, that of the best human experts on their best days. From these facts one must conclude that neural network image analysis is a technical success. What remains now is the translation of this technical success into widespread use. This process is currently underway at companies and government organizations worldwide.

The author thanks Todd Gutschow, Geoffrey Hueter, Robert Means, and Yi-Tong Zhou for help in preparing this description of their work.

References

[1] Bienenstock, E., and von der Malsburg, C., A neural network for invariant pattern recognition, Europhysics Letters, 4, 121-126, 1987.

[2] Buhmann, J., Lades, M., and von der Malsburg, C., Size and distortion invariant object recognition by hierarchical graph matching, Proceedings of the International Joint Conference on Neural Networks, II, 411-416, IEEE Press, NY, 1990.

[3] Buhmann, J., Lange, J., and von der Malsburg, C., Distortion invariant object recognition by matching hierarchically labeled graphs, Proceedings of the International Joint Conference on Neural Networks, I, 155-159, IEEE Press, NY, 1989.

[4] Daugman, J. G., Complete discrete 2-D Gabor transforms by neural networks for image analysis and compression, IEEE Trans. Acoustics, Speech & Signal Processing, 36, 1169-1179, 1988.

[5] Daugman, J. G., Pattern and motion vision without laplacian zero crossings, J. Optical Soc. Am. A, 5, 1142-1148, 1988.

[6] Daugman, J. G., and Kammen, D. M., Image statistics, gases, and visual neural primitives, Proc. of the Int. Conf. on Neural Networks, IV, 163-175, IEEE Press, NY, 1987.

[7] Daugman, J. G., Image analysis and compact coding by oriented 2D Gabor primitives, SPIE Proc., 758, April 1987.

[8] Daugman, J. G., Uncertainty relation for resolution in space, spatial frequency, and orientation optimized by two-dimensional visual cortical filters, J. Optical Soc. Am. A, 2, 1160–1169, 1985.

[9] Hecht-Nielsen, R., "Theory of the backpropagation neural network" in: Wechsler, H. [Ed.], **Neural Networks for Human and Machine Perception**, Volume 2, 65–83, Academic Press, Boston MA, 1992.

[10] Hecht-Nielsen, R., **Neurocomputing**, Addison-Wesley, Reading MA, Corrected Printing, 1991.

[11] von der Malsburg, C., Considerations for a visual architecture, in: Eckmiller, R. [Ed.], Advanced Neural Computers, 323-330, Elsevier North Holland, Amsterdam, 1990.

[12] Roth, M. W., "Survey of Neural Network Technology for Automatic Target Recognition", *IEEE Trans. Neural Networks*, **1**, 28–43, March 1990.

[13] Rybak, I.A., Shevtsova, N.A., and Sandler V.M., Temporal and spatial discrimination of image features in the visual cortex, in: Novak, M., and Pelikan, E. [Eds.] Theoretical Aspects of Neurocomputing, 259-276, World Scientific Publishers, Singapore, 1991.

[14] Rybak, I.A., Shevtsova, N.A., Podlatchikova, L.N., and Golovan, A.V., A visual cortex domain model and its use for visual information processing, Research Report, Institute for Neurocybernetics, Rostov State University, Rostov-on-Don, USSR, 1990. An edited version of this report was recently published in: *Neural Networks*, **4**, 3-13, 1991.

[15] Rybak, I.A., Shevtsova, N.A., and Sandler V.M., The model of the neural network visual preprocessor, Proceedings of the NEURONET-90 Conference, Czechoslovakian Academy of Sciences, Prague, 1990.

[16] Zeevi, Y.Y., and Ginosar, R., Neural computers for foveating vision systems, in: Eckmiller, R. [Ed.], Advanced Neural Computers, 323-330, Elsevier North Holland, Amsterdam, 1990.

[17] Zeevi, Y.Y., Peterfreund, N., and Shlomot, E., Pyramidal image representation in nonuniform systems, *SPIE Proc.*, **1001**, 1988.

[18] Zhou, Y.T., and Crawshaw, R., Contrast, size and orientation invariant target detection in infrared imagery, *SPIE Proc.*, **1471**, 404–411, 1991.

SUBJECT INDEX

2-D surface interpolation, 145
3-D form, 93, 102
3-D shape, 450
3-D structure, 133
3-D structure and motion, 221

ADALINES (adaptive linear
 neurons), 243
adaptation, 1
adaptive clustering, 446
adaptive clustering neural
 net (ACNN), 444
adaptive filter, 369
Adaptive Filter Coupling, 392
adaptive learning, 437, 443
Adaptive Resonance Theory (ART),
 258, 359, 365, 369, 389
affine
 geometry, 96
 representation, 102
 structure, 100
aftereffects, 56, 285
albedo, 115
alertness, 346
alpha cell, 209
amacrine cells, 196
antagonistic rebound, 256, 258
aperture problem, 222, 241, 294
apparent motion, 84, 101, 261
apple sorting, 449, 450
arousal, 346, 371
ART 1, 365, 367, 369
ART 2, 369
ART 3, 369
ARTMAP, 162, 365
artificially thickened edges, 233
association cortex, 346
associative processor (AP), 437, 439
ATR: see automatic target recognition
attached shadow, 128
attention, 40, 47, 84, 343, 346, 349,
 353, 354, 356, 367, 370, 408
 facilitation, 359
 feature integration, 247
 gating, 347, 351

 mechanisms, 365
 recognition learning, 388
 shift, 273
automatic gain control, 221, 223
automatic target recognition, 449,
 452, 455

back propagation, 366, 376
barberpole illusion, 294
baseline vigilance, 376
Bayes law, 414
Bayes theorem, 412
Bayesian classifier, 457
Bayesian object recognition, 413
beta cell, 209
beta motion, 269,
binocular disparity, 94, 98
binocular syncytium, 167
bipolar cell, 195, 196, 204, 221, 293
bipole cell, 304, 307, 389, 392
blind spot, 58
blood testing, 457
boundaries, 161
boundary completion, 190, 312
Boundary Contour System (BCS), 161
boundary detector, 282
boundary segmentation, 162, 165
brightness-reversal, 1

cast shadow, 128
categorical structure, 105
category choice, 368, 373
category learning, 365
category prototype, 369
CC Loop, 249, 260, 300, 303
CCD, 408
cell-gating model, 352, 353
center-surround receptive field, 19, 200,
 221
choice function, 373
choice parameter, 373
clutter rejection, 455
color
 constancy, 349
 discrimination, 346, 355

combinatorial explosion, 347
compactness, 29
competition, 254, 266, 343, 352, 353, 356
complement coding, 374
complex cell, 174, 252, 279, 293
complex off-cell, 263
complex logarithm, 426, 428
computer generated hologram (CGH), 439
computer vision, 222
conditional probability, 414
confidence, 367
conjunctive features, 185
connected scenic figures, 161
contrast, 15
contrast-enhancement, 369
convected activation, 222, 233
Convected Activation Profiles, 227
cooperative feature linking, 256
correlator, 437, 438
cortical maps, 346, 357, 365
CORT-X 2 filter, 161, 162, 174
curvature, 115, 120, 141

decorrelating transform, 145
delta motion, 269
direction-of-contrast, 250
direction-of-motion, 250, 262
directional competition, 312, 326
directional derivatives, 116
directional selectivity, 231
discontinuity, 126, 150
discount the illuminant, 161, 162, 173
disparity, 141
distortion-invariance, 445
distortion-invariant object recognition, 446
DOG (difference-of-Gaussian) filter, 21, 25
double opponent network, 179
dye-injected FBF, 172
dynamic range, 204
dynamic vision, 133

EACH algorithm, 374
early vision, 221, 409
edge
  enhancement, 224, 239
  motion, 222
  velocity extraction, 224
Ehrenstein figure, 280
element motion, 271
emergent boundary segmentation, 247, 385
end cut, 279
end stopping, 253, 282, 310
endstopped simple cells, 314
energy potentials, 156
equiluminance, 1, 50
expectation, 372
expected shading flow field, 122
expert system, 372
extinction, 356
extrastriate visual cortex, 343, 346, 347, 349, 351
Euclidean structure, 97

FACADE theory, 163, 166, 249, 298
face recognition, 2
fading of stabilized images, 214
false alarm filter, 452, 455
fast learning, 368, 374
feature binding, 88, 385
Feature Contour System (FCS), 161
feature integration, 273
feature space, 427
feed-forward shunting network, 200
fibre bundle, 111, 114
field of view (FOV), 437, 445
figure and ground segregation, 77
figure-ground separation, 161, 162, 179
figure of merit (FOM), 455
filling-in, 47, 54, 56, 58, 66, 161, 167, 178
filling-in barrier, 172
first competitive stage, 253
fixation points, 408
FM Symmetry, 247, 254, 277
Fourier transform (FT), 438

fovea, 408, 409
frequency doubling response, 198
full flow recovery, 241
fusion, 171
Fuzzy ART, 365, 367, 373
Fuzzy ARTMAP, 365
fuzzy set theory, 367

Gabor filter, 21, 25
Gabor logon decomposition, 457
Gabor profile, 452
Gabor wavelets, 452
gamma motion, 269
ganglion cells, 195, 197, 204
gated dipole, 247, 283
Gaussian filter, 254
general purpose vision system, 133
generalization, 367
generalized Hough transform, 418
genetic algorithm, 366, 380
geometric discontinuities, 129
global segmentation, 301
go signal, 354
grating, 19
group motion, 271

hierarchical resolution of
    uncertainty, 279
highlights, 12
Ho-Kashyap (HK), 440
hologram, 439
horizontal cells, 196
Hough transform, 408, 411, 415
hypercolumn, 281, 350
hypercomplex cells, 189, 253, 279, 293
hypothesis testing, 369
hue, 19

I-Beam, 19
illumination discontinuities
    (shadows), 128
illusory contours, 47, 67, 69, 306,
    344, 385
illusory figures, 261
image, 414

image analysis, 449
image irradiance partial differential
    equation, 111
image segmentation, 46
image velocity extraction, 221
imaging function, 416
inferior temporal (IT) cortex, 258, 343;
    also see IT
infrared battlefield image, 453
inhibition, 352
inhibitory interneuron, 222, 402
Inner Plexiform Layer (IPL), 204
input gating model, 350, 351, 353
intensity discontinuities (edges), 126
interstimulus interval, 269
invariant filter, 162
invariant pattern recognition, 161
IR imagery, 452
isoluminant pattern, 25
isophotes, 115
IT, 346, 347, 349, 354, 355, 357, 359
iterative focusing, 423

jitter, 385
just noticeable difference (JND), 225

Kalman filter, 134, 135, 137, 141, 148, 159
    range estimates, 154
    with accelerations, 139
    with velocities, 138
Kanizsa square, 70, 252
Karhunen-Loeve transform, 145

Lambert's reflectance function, 112
Lambertian surface, 113
laser radar, 166
lateral geniculate nucleus (LGN), 282
learning, 258
letter recognition, 368
license plate recognition, 427
light adaptation, 221, 222, 224, 239
light source, 111
light source constraint, 118
lightness, 19, 30
linearity, 198

local contrast, 201
local scene element, 112
local scenels, 117
local shading flow field, 124
log-polar map, 428
logarithmic mapping, 408
long-range
    apparent motion, 226
    cooperation, 190
    Gaussian filter, 255
    motion aftereffect (MAE), 285
    spatial filter, 266, 326
long term memory, 386
luminance decreasing off-cell, 308
luminance increasing on-cell, 308

machine vision, 409, 449
Mackay afterimages, 285
map field, 375, 376
Markov process autocorrelation, 149
match function, 374
match tracking, 367, 375, 376
matching, 357
material changes, 111
material discontinuities, 129
matrix-vector architecture, 438, 444
medical image analysis, 449
memory, 344
memory formation, 343
memory matching effects, 357
memory search, 247, 372
Minimax Learning Rule, 366
minimum-energy state, 141
mismatch reset, 374
modal analysis, 143
modal coordinate transform, 143
model-based attention, 426
Mondrian, 170
monocular syncytium, 167
MOC Filter, 250, 260, 263, 293, 309, 317
MOCC Loop, 293
MOS, 408
MOS sensor chip, 428
motion aftereffect, 285

Motion Boundary Contour System (BCS), 247, 248, 260, 293, 297
motion
    capture, 302
    detectors, 255
    direction, 255
    grouping, 294
    parallax, 93
    pathway, 221
    perception, 253
motor control algorithms, 409
motor space, 408, 427
moving edge features, 221
moving form, 247
moving-form-in-depth, 260, 336
moving waves of activity, 227
multidimensional fusion, 171
MT, 222, 231, 241, 247, 248, 260, 336, 346

Nearest Neighbor Coupling, 392
negative afterimage, 256
next-look position, 409
nonlinear compression, 198
nonlinear diffusion, 165
nonspecific arousal, 371
nonstationary environment, 372
normalization, 374
noticeable difference (JND), 225
nuclear layer, 196
null response, 198, 199
null test, 197

object recognition, 343, 344, 346
object space, 413, 427
observer, 135
observer system, 135
observer theory, 135
occipitotemporal cortex, 359
occipitotemporal pathway, 344
optical optimization, 443
optimization, 437, 442
oculomotor control, 354
oculomotor system, 343
off-cells, 161, 196, 263, 275

off-surround shunting network, 200
on-cells, 161, 196, 255, 275
on-off cells, 198, 255
on-off response in X cells, 212
opponent directions, 283
opponent process, 161, 247
opponent rebounds, 283
optical
    blur, 51
    correlator, 439
    hardware, 437
    processing, 437
optics, 437
optimal linear observer, 134
optimal observer, 133, 134, 137
ordinal structure, 102
orientation, 346
orienting, 359
orienting system, 343, 372
orthogonal wavelet bases, 145
orthographic projection, 112
oscillation model, 352

Panum's limiting case, 97
parallel cortical systems, 247
parallel On and Off retinal
    circuits, 203
parallel processing, 185
parietal lobe, 344
passive relaxation, 203
pattern motion cells, 242
pattern recognition, 437, 443
Pdm (pulvinar), 355
pedestal, 19
peripheral discriminability, 38
peripheral viewing, 51
PET, 355, 359
phantom contours, 47, 53
phase discrimination, 50
phase locking, 304
phase velocity, 240
phi motion, 269
photographic negatives, 2
photoreceptors, 195, 196, 198, 211
photoreceptor adaptation, 195

photoreceptor model, 209
plexiform layers, 196
pop-out, 66
posterior parietal cortex, 354
potential target locator, 452
preattentive grouping, 15
preattentive vision, 388
prediction, 133, 365
prefrontal cortex, 354
primate, 343
product inspection, 450
production systems, 372, 447
pulvinar, 354, 355
push-pull
    bipolar cells, 196
    mechanism, 204
    model, 217
    network, 212
    shunting network, 195, 205, 209
pyramidal cell, 151

quotient space, 422, 425
quotient space Hough transform, 407,
    410, 428
quotient space method, 412

Random Connection Coupling, 392
reading of license plates, 408
reality mismatch, 12
receptive field, 147, 151, 196, 242,
    344, 347, 348, 349, 353
reciprocal connections, 344, 365
recognition, 185
recognition learning, 365
recognition memory, 357
rectification, 39
rectified outputs of time-averaged
    signals, 255
recursively using separable filters, 147
regularization, 135, 142, 148, 152, 159
relaxation labelling, 118, 122, 124
reset, 247
resonance, 247
retina, 195, 196, 221, 231, 344, 346
retinal ganglion cells (RGCs), 196

saliency map, 356
scene attributes, 119
scene element ambiguity, 124
scenel bundle approach, 114
scenels, 111, 112
search, 247, 372
second competitive stage, 253
segmentation, 46
segmenting time-varying
    imagery, 221
selective attention, 346, 348, 349
sensor chips, 409
sequential accumulation, 423, 425
serial search, 185
shading flow discontinuities, 126
shading flow field, 111, 115
shadow boundaries, 113
shadows, 1, 12
shape, 95
shape from shading, 1, 71, 111, 112
shift invariance, 438, 445
shift property, 202
shifter-circuit model, 350
shifts of attention, 354
shifts of gaze, 354
short-range space-filter, 254, 320
short term memory, 386
shunt, 200, 223
shunting equations, 195, 393
simple cell, 174, 279, 293
simulation: circle-in-the-square, 376
simulation: letter image
    recognition, 377
sinusoidal gratings, 206
size-disparity correlation, 172
slow variable, 402
smearing, 256
SOC Filter, 249, 260, 299
space-variant machine vision, 408
space-variant sensor, 426
space-variant visual attention, 407
spatial
    attention, 349, 354
    contrast enhancement, 221
    contrast sharpening, 222
    frequency, 147, 346
    frequency analyzers, 15, 19
    frequency channel, 39
    impenetrability, 259
    modulation, 199
    pattern coding, 386
    perception, 344
    recognition, 344
    vision, 146, 195, 200
spatiotemporal
    analysis, 202
    Gabor filtering, 222
    processing, 195, 221, 225
    response characteristics, 196
    stimulation, 195
split motion, 269
spring model, 140, 141, 159
stabilized images, 195, 200, 214
standing wave, 258
static BCS, 247, 260
static form, 247
static nonlinearity, 199
stereo
    algorithms, 140
    disparity, 134
    matches, 140
    networks, 140, 142
stereopsis, 140
striate cortex, 344, 356, 428
superior colliculus, 356, 429
superior temporal cortex, 344
supervised learning, 365, 427
surface interpolation, 152
surface material constraint, 117
surface shape, 111
surface smoothness constraint, 117
sustained cell short-range filter, 264
sustained cells, 247, 255
sustained-transient
    cells, 247
    gating, 266
    simple cells, 319
symbolic correlator, 437, 447
symbolic representations, 46

symbolic substitutions, 372
synchronized oscillations, 385

tangent bundle, 115
target classifier, 452, 455
temporal
    adaptation, 198, 222
    dynamics, 69
    jitter, 385
    lobe, 344
    modulation, 199
    phase discrimination, 50
    response, 200
Ternus display, 270
testing the hypothesis, 369
textural grouping, 259
texture segregation, 15, 39, 40
thalamic brain damage, 355
tomography, 355, 359
topological structure, 104
tracking moving objects, 221, 409
transient
    cell, 247, 255, 318
    cell filter, 265
    detection, 222
    nonlinearity, 199

unexpected, 359

V1, 247, 260, 336
V2, 247, 260, 336, 344, 346, 351, 353
V4, 248, 343, 346, 347, 348, 349, 351, 353, 354, 355, 357
velocity extraction, 221, 222, 225, 241
Velocity Functional Method, 222, 241, 242
velocity space, 294
vigilance, 346, 370, 376
vigilance parameter, 373
virtual point source, 119
visual
    attention, 343, 344, 408, 409
    cortex, 346
    cortical areas, 345
    illusions, 53

    inertia, 275
    module, 285
    motion, 344
visuomotor performance, 344
VLSI, 437
voting strategy, 365

waterfall illusion, 247, 305
wavelet, 145, 147, 149, 151, 454
wavelet-based surface interpolation, 152
wavelet basis, 146
wavelet modes, 148
Weber law, 201
Weber-Fechner law, 225
wedge ring detected (WRD), 438, 439
weighting function, 427
Wiener filter, 135, 142, 159
winner take all, 356, 369, 446
working memory, 357

X cells, 195, 197, 198, 208, 211

Y cells, 195, 197, 198, 208, 211